Page 24 - Existance
Page 18 - Buzz Aldrin
Page 9 - yin and yang
Page 13 - full moon

More Praise for

HOW IT BEGAN

"Here is a universe wrapped in a dream of spacefaring for everyone to share. Author Chris Impey combines the vision of a practicing scientist with the voice of a gifted storyteller. He has crafted some of the finest metaphors imaginable for capturing astronomy's most grandiose concepts. *How It Began* glitters with a sprinkle of Moon dust."

—Dava Sobel, author of *A More Perfect Heaven: How Copernicus Revolutionized the Cosmos*

"Astronomer Chris Impey takes us on a celestial road trip into deep space and time. . . . Each leg of the trip packs in science, history and anecdote, and is topped and tailed with imagined descriptions of each starry port of call."

—*Nature*

"Written with the wit and enthusiasm of a man who truly enjoys his work, this book is an excellent launching pad for readers who want to broaden their understanding of the cosmos."

—Paul Wine, *Tucson Weekly*

"What will especially impress readers is just how entertaining Impey can make science as he regales them with his own piquant experiences as a researcher and translates arcane mathematics into metaphors . . . drawn from everyday life. Readers will never find more intellectual adventure packed into fewer pages."
—Bryce Christensen, *Booklist*, starred review

"In clear, enthusiastic and occasionally lyrical prose, Mr. Impey takes the reader on a mind-blowing tour back through eons, stopping along the way to explain the formation of the solar system, the birth and death of stars, white dwarfs, supernovas, spiral galaxies, cosmic inflation, string theory, black holes and M-theory. . . . *How It Began* deserves to be a well-thumbed guidebook for—and in—this universe."
—Manjit Kumar, *Wall Street Journal*

"An astute tour of the cosmos by a skillful teacher." —*Kirkus Reviews*

"Having dispensed with *How It Ends* in his impressive last book, Impey has now turned his attention to the start of everything. It is no less a treat. . . . Impey is in a class of his own when it comes to guiding the reader through the infinite reaches of the universe. . . . This is not just a book about the history of the cosmos, it is a compelling insight into what it is to be an astronomer grappling with the fundamental questions of existence."
—Michael Brooks, *New Scientist*

"Impey . . . takes readers on a mesmerizing journey through the 'labyrinth' that is our universe. . . . [He] vividly illustrates the most complex topics, like string theory and dark energy, bringing a fresh, original voice to a much-told tale, making cosmology pleasurable to all readers."
—*Publishers Weekly*

"If I had to recommend a single book today to a friend wanting to learn the basics of what's out there, Impey's *How It Began* would be my first choice. . . . The whole volume seems designed to blow you out of the water. And it delivers. . . . An entertaining and illuminating tour de force, which deserves to be read and savored slowly. For it is clearly a work of love."

—Robert Perez-Franco, *The Tech*

ALSO BY CHRIS IMPEY

How It Ends

The Living Cosmos

HOW IT BEGAN

A TIME-TRAVELER'S GUIDE TO THE UNIVERSE

CHRIS IMPEY

W. W. NORTON & COMPANY New York • London

To Dinah: my perfect beginning

Printed in the United States of America
First published as a Norton paperback 2013

For information about permission to reproduce selections from this book, write to Permissions, W. W. Norton & Company, Inc., 500 Fifth Avenue, New York, NY 10110

For information about special discounts for bulk purchases, please contact W. W. Norton Special Sales at specialsales@wwnorton.com or 800-233-4830

Manufacturing by RR Donnelley Harrisonburg
Book design by Dana Sloan
Production manager: Julia Druskin

Library of Congress Cataloging-in-Publication Data

Impey, Chris.
How it began : a time-traveler's guide to the universe /
Chris Impey.1st ed.
p. cm.
Includes bibliographical references and index.
ISBN 978-0-393-08002-5 (hardcover)
1. Cosmology—Popular works. 2. Space and time—Popular works.
I. Title.
QB982.I47 2012
523.1'2—dc23

2011052855

ISBN 978-0-393-34386-1 pbk.

W. W. Norton & Company, Inc.
500 Fifth Avenue, New York, N.Y. 10110
www.wwnorton.com

W. W. Norton & Company Ltd.
Castle House, 75/76 Wells Street, London W1T 3QT

1 2 3 4 5 6 7 8 9 0

CONTENTS

PREFACE

I thought of a labyrinth of labyrinths, of one sinuous spreading labyrinth that would encompass the past and the future and in some way involve stars.

—JORGE LUIS BORGES

WHERE DO WE COME FROM? As busy as we might be in our everyday lives, it's natural to pause sometimes and ask this question. After a quick recognition that we're the genetic product of our parents and forebears, and a trivial response in terms of birthplace, the answers get harder.

We can draw larger and larger boundaries around our situation and ask about the journey that led to "us." The Earth condensed out of a gas cloud that also formed the Sun and the other planets. More than 4.5 billion years old, the story of our origins involves the journey of many of our atoms through the core of multiple generations of stars, where they were fused in fierce nuclear reactions. It's a narrative that involves thousands of long-dead points of light, and many times more atomic journeys, strands that eventually converge and embody us.

The next larger stage is a city of stars called the Milky Way, and the hundred billion similar cities sprinkled through the visible universe. With nothing special about the Milky Way or the processes that led to a fecund planet called Earth, there are likely to be a billion habitable worlds in each of these galaxies, and countless biological experiments. The enormity of space challenges us in trying to discover other worlds and find out whether we are alone.

As galaxies sail silently apart on a backdrop of velvet night, they point to a time when the universe was smaller, hotter, and denser. All the disparate threads of the universe—every person, every planet, every star, and every galaxy—were drawn tightly together in the big bang. All cosmic narratives converge into an iota of space-time. Our personal atoms were conjoined with all other atoms in an unspeakable intimacy. This instant of creation was 13.7 billion years ago.

The frontier of cosmology involves probing as close to the big bang as possible. The two main ingredients of our universe, dark matter and dark energy, are still enigmatic, so cosmologists hope to understand them and fit them into theories of microscopic physics. There's also speculation that the richness of our observable universe is part of a labyrinth of labyrinths called the multiverse.

> *Begin at the beginning and go on until you come to the end: then stop.*
>
> —LEWIS CARROLL

This book seeks to answer the question, How did it begin? Just as the geologically active Earth has obliterated most traces of its formation, the restless universe would seem to have erased all evidence of its formation. How can we tell the story of the universe when there was no one present at the big bang?

Cosmic archaeology is based on the vastness of space and the finite speed of light. To see back in time we simply look out in space. Light spans the Earth in an instant and the Solar System in a couple of hours, but it takes millions of years to reach us from the nearest galaxies and billions of years to reach us from the most distant ones. Astronomers are armchair time travelers, using their telescopes to grasp ancient light, see the formation of stars and galaxies, and peer into the alien conditions of the early universe.

How It Began aims to be a time traveler's guide to the universe. The narrative of this book marches outward in space and backward in time, and it's organized in three equal parts. The first covers the proximate universe, starting with the Moon, the only object in space bearing human footprints other than the Earth. We then move to Jupiter and its moons, in the outer Solar System, exotic and maybe habitable worlds that lie at the frontier of our direct exploration of space. Next is the nearest star system, Proxima Centauri. Astronomers want to know if the Earth is unique so they're looking for remote planets indirectly, by the way they tug their parent stars and by the fleeting shadows they cast on them. Then we venture to the Orion nebula, over a thousand light-years away, where new stars are flickering to life behind veils of gas and dust. The last stop in the Milky Way is the center of our city of stars. Gravity there grips light so tightly that it can't escape. Through the looking glass is a black hole millions of times more massive than the Sun, the edge of the abyss.

The second part of the book explores the remote universe from the nearest galaxy out to the first star. The journey spans 99 percent of the observable volume and 99 percent of the time since the big bang. First is Andromeda, doppelganger to the Milky Way, and then the Coma Cluster, whose galaxy swarm is held together by dark matter. The following two chapters explore the great construction project that assembled galaxies from smaller pieces and grew gigantic black holes in their centers. This part ends with the beginning of the story of light and life—the time when stars first congealed from expanding gas.

We finish our exploration with the alien realm of the infant universe. Evidence for the big bang centers on detailed baby pictures taken in microwaves, beyond which is impenetrable neonatal fog. In a version of the Red Queen's race, light travels as fast as it can, yet is unable to keep up with the galloping expansion. Evidence from earlier epochs is indirect. In our quest for the big bang, we pause at a time when the entire universe was a fusion reactor, a time when matter was created from pure energy, and a time when the forces of nature were melted together. Along the way, and throughout the narrative, we meet the cosmologists who reconstruct cosmic history from slender shards of evidence, some of whom work at the edge of speculation and believe as many as six impossible things before breakfast. In the end we find out just how deep the rabbit hole goes: invisible dimensions of space, timelessness, and multiple universes.

> *Your mind, this globe of awareness, is a starry universe. When you push off with your foot, a thousand new roads become clear.*
>
> —RUMI

If the universe contained nothing more than forces operating on inanimate matter, it would not be very interesting. The presence of sentient life-forms like us (and perhaps unlike us) is the zest, or the special ingredient, that gives cosmic history dramatic tension. We're made of tiny subatomic particles and are part of a vast space-time arena, yet we can hold both extremes in our heads.

Are we nothing more than cosmic flotsam, contingent outcomes of evolution, collections of atoms that got a little lucky? Or are we built deeply into the architecture of the universe?

Scientists can't answer these questions yet and perhaps they never will. Meanwhile, this book humanizes the universe by adding a personal layer of narrative. Vignettes that open and close each chapter place the reader in the scene, with an increasing sense of dislocation from the here and now as the narrative progresses toward the big bang. Analogies of the senses might operate in the early universe when the conditions were extreme. And when sight, sound, smell, and taste fail us, we can still touch the universe with our minds.

I'm grateful to many colleagues, at the University of Arizona and elsewhere, who've shaped my understanding of cosmology over the years. Any errors, omissions, or confusions that remain in this book are my responsibility alone. Portions of the initial manuscript were written at the Aspen Center for Physics, where the atmosphere is always conducive to thinking and writing. It has been a pleasure to work with Angela von der Lippe at Norton on this project and on its predecessor, *How It Ends*. Thanks go to my agent Anna Ghosh, for patiently nurturing my career and my writing. Special acknowledgment goes to Phillip Helbig for an extremely close and careful reading of the hardcoveredition resulting in many significant edits, corrections, and clarifications in this edition.

Chris Impey
Tucson, Arizona
July 2011

PART I
PROXIMATE

1

SEPARATED AT BIRTH

I OPEN MY EYES ONTO A STARK AND BEAUTIFUL LANDSCAPE. *A plain strewn with boulders stretches in front of me. The contours and shadows are etched with perfect precision, as if by a sharp knife. Rolling hills define the horizon, and, above it, a vault of absolutely black sky. The only hint of color is an ochre tinge to the gray rocks and soil.*

The sound of labored breathing echoes inside my space suit. I hear my heart thumping. These rhythms slow and become more regular. Bending my knees, I feel rather than hear the crunch of the lunar soil, its consistency like sugar. Looking down, a fine patina of dust clings to my boots. Minerals in the soil glisten and glint like crushed diamonds in the harsh light.

Something else: footprints. Cleats have sculpted perfect ridges and grooves in the surface. I'm alone. I know that no human has been here for 40 years, but on this airless world these footprints look like they were made yesterday.

Gingerly, I test the gravity. A small jump propels me above waist height. I tilt forward in midair and gently crash onto the surface on hands and knees. Alarmed, I hold my breath and listen for the telltale sound of air rushing out of my suit. Nothing. I get up clumsily. Small steps. Then bigger, but I lose my balance again and tumble to the ground, sending up lazy plumes of dust so fine it looks like talcum powder or smoke.

Then I recall the grainy video beamed back decades ago from the Sea of Tranquility to a keenly waiting world. The trick is not to take giant steps or hops, but to execute a slow loping motion. Galumphing across the Moon with my arms cartwheeling, I whoop with delight. My heart feels light. And so it should—it weighs less than 2 ounces.

—

THE ROCK NEXT DOOR

A quarter of a million miles straight up. That's the distance to the Moon, and it's not just a guess but a very well-determined number, even though the distance varies by 10 percent due to the Moon's elliptical orbit. We know the Moon's distance with a precision of 1 millimeter, far more accurately than you might know the distance to the nearest wall, or the distance to the person across from you at the dinner table.

How? Three of the Apollo landing sites host suitcase-sized reflectors (they're not really mirrors but arrays of reflecting cube corners) with the special property that they send back light in exactly the direction from which it came—think of a racquetball hit into the corner of the court. A 3.5-meter telescope in New Mexico is used as a giant laser pointer. The laser light diverges slightly as it travels to the Moon, so only 1 in 30 million photons hits the small reflector. The spreading also happens on the way back, so (by coincidence) the telescope gathers the same tiny fraction of photons that left the reflector. Just one in a quadrillion photons makes the round-trip but luckily the laser packs a wallop, so a lot of photons are retrieved each second. The telescope "pings" the Moon with ultrashort bursts of laser light and their return trip is timed with an accuracy of a trillionth of a second. The two and a half seconds taken for the trip, combined with knowledge of the speed of light, give the distance.[1]

The Moon is our first stop in a long journey. In a cosmic sense, it's a stone's throw from Earth. Looking up from its surface, the astronauts saw the constellations in their familiar shapes. The stars are so remote there's no discernable change in perspective from hopping to the rock next door; it's like shifting one seat over in the back row of a theater. But for a place so familiar and romance-tinged, the Moon is stark and alien. It's a rock too small to hold even a whiff of atmosphere, with a surface pounded to talc by impacts and exposed to sterilizing radiation from space.

This forbidding place is just ten times the Earth's girth away. A quarter million miles doesn't seem very otherworldly. Many people fly this far in their lives, and road warriors and businessmen fly this far every few years. In 25 years of recreational running, I've been a fifth of the way to the Moon and it makes me tired thinking about it. But normal travel barely deviates from the Earth's surface and these excursions are tucked within the constraints of gravity's implacable grip.

If you jump as high as you can you won't get more than 3 feet off the ground. Children know that the best way to get something beyond reach is to stand on a chair or on someone else's shoulders. Getting enough people to stand on each others' shoulders might do the trick. The population of the United States would be enough, if we ignore the practical problems of how they'd balance and all that weight on poor Yertle at the bottom. We need another idea.

The answer involves a firecracker the size of a skyscraper.

Years of research and billions of dollars in funding led to the vast building that rises out of featureless Florida scrub and surrounding swampland. It would comfortably contain St. Paul's Cathedral. A low rumble announces the opening of the doors and a huge rocket slowly rolls out, tethered up and down its length to a launch tower. The rocket is twice the height of the Statue of Liberty from base to torch. The crawler-transporter weighs 3000 tons and rides on four tracks, each the size of a city bus. It takes five hours for the Saturn V rocket to travel the 3 miles to Launch Complex 39.

As the launch sequence begins, it's eerily quiet. Vapor seeps from the rocket and rises into the humid air. A flock of herons passes overhead. The liquid fuel onboard has the explosive power of half a kiloton of TNT, the equivalent of a modest atomic bomb.

Nine seconds before launch, the ignition sequence starts. The central engine ignites a fraction of a second before the outer four, and the fuse is lit on 2000 tons of kerosene and liquid oxygen. White smoke billows from the engine nozzles as thrust builds to equal the force of all the rivers in North America. A mile from the launch pad it takes five seconds for the din and coruscating heat to arrive. The rocket's force is felt even more than heard, a pounding sensation to crush the chest and penetrate every bone of the body. On the launch pad, four hold-down arms swing to the side, and the rocket creeps agonizingly upward, taking 10 seconds to clear the tower.

The light from the engines is blindingly white. The rocket now seems to leap into the sky, accelerating to 1000 mph by the time it's a mile up. Two minutes later, it's 40 miles high and traveling at 6000 mph. Less than three hours after that, all three stages have been jettisoned and the 40-ton payload of *Apollo 8* is traveling at 25,000 mph through the pure vacuum of space toward the Moon.

The three men in the capsule are all too aware of their mortality as the launch progresses. The Saturn V leapt over previous technologies and was untested.[2] The first launch of the competing Soviet N-1 rocket was a disaster, completely destroying the rocket and the launch tower and shattering windows 30 miles away. There are air pockets in the huge Saturn V fuel tanks so combustion is uneven, with momentary surges of power. Eight million pounds of force create ferocious pressure and vibrations. Bill Anders kept his game face on in the later debriefing, talking about "positive control motion." But in conversation he said it was like being "a rat in the jaw of a giant terrier." He thought they'd been shaken to within an inch of their lives.

This is a very special club.[3] Among the hundred billion people who have ever lived, only 27 have escaped Earth's gravitational umbilical cord and traveled to another world.

When they got there, they were reminded that they weren't in Ohio (home of the largest number of astronauts) by the two-and-a-half-second delay in communications. Light—the fastest thing there is, so fleet its travel on Earth is instantaneous and imperceptible—breaks into a sweat going to the Moon and back. This hiccup gave the dialog between the astronauts and Mission Control a strangely formal and stilted quality. Even though they were still in Earth's hip pocket, the astronauts had a sense of the vastness of space unfolding.

SHEER LUNACY

The Moon looms large in the sky and it also looms large in culture (Figure 1.1). Its prominence and its waxing and waning have made it a symbol of time and change in civilizations around the world. The word shares a root in the oldest Indo-European languages with the verb "to measure."

If the oldest human artifacts are a guide, human culture started with storytelling and timekeeping. The Lebombo bone is the fibula of a baboon carved with 29 notches. It was found in a cave in Swaziland and it's been dated to about 35,000 BC.[4] The number 29 also occurs in the exquisite cave paintings at Lascaux in France, which date to 15,000 BC. There are indisputable examples of astronomical phenomena depicted at Lascaux, but as in all the cases where no written culture

FIGURE 1.1. *The nearest celestial body and the only one humans have ever visited, the Moon is a quarter of a million miles away and has had a profound effect on human culture, as well as a physical influence on our planet.*

survives, we're guessing at the intentions of our distant ancestors.[5] It's certainly plausible that early humans used the Moon to mark time and also created portable calendar sticks.

In the imprint of the cosmos on human affairs, tension has always existed between the sprightly rhythm of the lunar cycle and the slower seasonal variation of the life-giving Sun. Consider the example of the yin and yang, a concept shared by all the major schools of Chinese philosophy. The familiar interlocking tadpoles were inscribed on bones used for divination practice as early as the fourteenth century BC and the symbol has come to represent the coherent fabric of nature and mind, the ebb and flow of the cosmic and human realms, and the dynamic balance of all things.

We associate yang with movement, day, male nature, and the Sun.

We associate yin with rest, night, female nature, and the Moon. While the origin of the popular yin and yang symbol remains elusive we can create the symbol itself with a simple gnomon, or vertical stick, recording the Sun's shadow over the course of a year. The Chinese divided a circle into 24 segments based on the solstices, equinoxes, and positions of the Big Dipper. Recording the length of the shadow every day and connecting the points formed by its changing angle divides the circle in two just as the yin and yang symbol does (Figure 1.2).

The additional Chinese concept of chi, or vital energy, resolves the dichotomy of yin and yang. Yin and yang are both forms of chi, and chi arises from the interplay between yin and yang, forming the basis for a Chinese cosmology.[6] The ancient text *Zhuangzi* by the third-century-BC thinker Guo Xiang explains it this way: "when the two have successful intercourse and achieve harmony, all things will be produced."

The Moon was the basis for the calendars of the earliest civilizations. To match the 29.5-day synodic period, or time for a complete cycle of lunar phases, the Babylonians used alternating months of 29 and 30 days. To the Babylonians, months with 30 days were full, while those with 29 days were defective. In a similar vein, Greeks considered

FIGURE 1.2. *The yin-and-yang symbol is an ancient artifact of Chinese culture, originating 2500 years ago. We can create it from a map of the varying projection of the Sun's shadow cast by a vertical stick over the course of a year. The interplay of dark and light represents Moon and Sun.*

the short months hollow and Celts considered them unlucky. Civilizations using a lunar calendar had to regulate their agriculture so they made occasional adjustments to match the solar year, at first inserting an extra month three times in eight years. Later, Meton, a Greek astronomer of the fifth century BC, calculated that 19 solar or tropical years equal 234.997 synodic months, so solar and lunar calendars could be almost perfectly aligned every 19 years. In ancient Rome, priests observed the night sky and announced a new lunar cycle to the king. The first day of each month was called Calends, from the Latin verb "to proclaim," and the word *calendar* stems from this tradition.[7]

The world's major religions have always set their sails by the Moon. Easter, the most important Christian festival, falls on the first Sunday after the first full moon following spring equinox. The Jewish Passover is on a full moon, and the Chinese New Year occurs on the second new moon after winter solstice. The major Hindu festival of Diwali happens during the new moon when the Sun first enters Libra. Islamic countries regulate their civic and religious affairs by a direct observation of the first sliver of crescent moon visible after sunset.[8]

The Moon has always had ceremonial or religious uses but it was also a catalyst in the birth of science. The Greek philosopher Anaxagoras was the first to argue that the Moon was a mountainous world whose light was reflected from the Sun. He even presaged a modern theory for its origin by imagining that the Moon was a mass of stone torn from the Earth. This kind of thinking got him in trouble; he was forced to flee Athens in 450 BC for daring to suggest the Sun was larger than the Peloponnese peninsula.

Less than two centuries later, the Greek astronomer Aristarchus used ingenious arguments to estimate the size of the Moon and its distance from the Earth. During a lunar eclipse, the Moon passes through the shadow of the Earth (Figure 1.3). The Moon must be smaller than the Earth since it fills less than half the Earth's shadow. Also, the Earth's

FIGURE 1.3. *Time-lapse series of images of the lunar eclipse of February 20, 2008. As the Moon passes through the Earth's shadow, the curved shape of the Earth is obvious and simple geometric arguments estimate the relative sizes of the Earth, Moon, and Sun.*

shadow converges relative to parallel by a half degree since that's the angle the Sun subtends in the sky. Aristarchus made a geometrical construction and placed the Moon at a distance of 60 Earth radii, very close to the correct number.[9] He was also uncomfortably prescient on the peripheral status of the Earth, speculating about a Sun-centered cosmology 1800 years before the world was ready to accept it.

The menstrual cycle is the most noticeable alignment of human affairs with the Moon. The word *menstruation* derives from the Greek for moon and many authors have speculated that women in prehistoric cultures ovulated with the full moon and menstruated with the new moon. However, there's no evidence for synchronization and the 28-day average for the female cycle differs from the 29.5-day cycle of lunar phases. Our closest relatives—chimpanzees—have a cycle of 35 days. Other placental mammal females undergo an estrous cycle, not a menstrual cycle, with a period that ranges from 5 days for rats to 16 weeks for elephants. The similarity of a woman's cycle to the phases of the Moon is likely a coincidence.

The Moon exerts a firm grip on our imagination. It can inspire us, and create fear and awe, during a solar eclipse. The near-perfect blotting out of the Sun by the Moon only takes place because the 400-times-larger Sun just happens to also be 400 times farther away. Creatures in distant solar systems are probably denied this spectacle.

It seems plausible that the Moon's gravity should affect us. Perhaps you can imagine the full moon tugging gently on your head at night, giving you slightly better posture. The Moon heaves the enormous weight of the oceans enough to raise them by 4 to 6 feet, and since we're 75 percent water, surely the Moon should affect us too. It's true that the Moon causes ocean tides and corresponding stresses on the continents, or land tides, but the effects are large because the bodies of water are large. Tidal force is differential, scaling with the size of the object being affected. The Earth is large but a person is small, so a mother holding her infant exerts 10 million times more tidal force on the child than the Moon does. Also, the Moon can only affect unbounded bodies of water and the water in us is tightly contained.

The belief that the Moon can cause mental and behavioral problems is widespread. Some cops and ER nurses claim that "it gets crazy out there" during a full moon. In news reports and in articles by apparently respectable academics you'll read that a full moon triggers elevated rates of murder, suicide, postoperative bleeding, accidental death, heart attack, and traffic accidents. You'll read that the scientific establishment tries to suppress or cover up the shocking truth. You'll also get a whiff of the type of conspiracy theory thinking that extends to UFOs, Elvis, 9/11, and the assassination of JFK.

The trouble is, there's no reliable basis for these claims.

Eric Chudler, Research Professor in the Department of Bioengineering at the University of Washington, compiled 75 different studies of the influence of the full moon on human behavior. They span suicide, violent behavior, depression, ER visits, drug overdoses, even animal bites. In 65 of the investigations, carried out by different researchers, there was no statistically significant result. In the 10 where there was an apparent effect, the sample size was small or the methodology was suspect.[10] In another "meta-analysis" of 37 different studies, the Moon accounted for no more than 0.03 percent of the monthly variation.[11]

This topic is a microcosm of the pitfalls of doing half-assed science. Sometimes the claims are based on paltry samples. Others crumble after a more careful analysis of statistical significance. For example, when 30 unrelated variables are compared, two may be correlated at the 3-sigma level by chance. (Normally, a 3-sigma effect is significant with 99.5 percent confidence, but not if you've been fishing in a large, well-stocked "pond" of variables.) Investigators may fail to notice another cause for the effect. In a study of traffic accidents that showed peaks around a full and new moon, the original researchers missed the fact that a disproportionate number of full and new moons in their sample fell on weekends, when the accident rate is known to be higher. Potential connections sometimes fail on closer examination. Positive ions can have an adverse effect on human behavior and they're more abundant during a full moon, but the lunar variation is dwarfed by the contributions from air-conditioning and air pollution.

Several hundred years ago, before artificial lighting, it was plausible that the Moon could affect behavior. Criminals in England got reduced sentences for being "moonstruck" or "lunatic." The best time to move around outside at night was during a full moon, leading to a greater opportunity for mischief. Now we have no such excuse.

OUR MISBEGOTTEN TWIN

It's a dark family tale. Gaia is born on cue and emerges shrieking into the night. Her twin sister Luna is born slightly later, after having been starved of nutrients inside the womb. Luna is a runt, small and fragile. She's active for a while but then she fades away. Gaia can't bear losing her sister so she keeps her close, even after her body becomes lifeless and her skin pockmarked with time. Determined to avoid the fate of her sibling, Gaia lives life hard. A woman now—sad and sweet—she's always making herself over in a frantic and unbecoming effort to hang on to her youth. Yet perversely, the proximity of her dead twin keeps her stable and stops her from becoming completely unhinged.

The Moon is unusual. It's the fourth largest satellite in the Solar System, by far the largest relative to its parent planet. It also has no magnetic field, and density measurements confirm it has a puny metallic core. When any large Solar System object forms, the heavy elements like iron and nickel sink through the molten magma and fall to the center. For some reason, the Moon is just a big rock.

Scientists explain this puzzle with a story of violence 4.5 billion years ago. We can imagine the scene. In a scant 50 million years—the blink of a cosmic eye—dust in the solar nebula congeals from flakes of rock into boulders, mountains, and then planetary embryos the size of the Moon or Mars. This process accelerates as it progresses, with most of the time taken for the weak gravity of the small pieces to nudge them together. Growing from 100 kilometers to 10,000 kilometers takes just 100,000 years. Each embryo grows by sweeping up material from the vicinity of its orbit. The process is orderly.

The next stage is mayhem. Gravity's long reach deflects embryos from their orbits and soon many roam ominously through the Solar System. Some embryos move in and others move out; most of the interactions are at arm's length. In all the shuffling some protoplanets approach each other gently and stick together while others are ejected into the void of interstellar space. A few impacts are catastrophic.

The scene now shifts to the infant Earth.

The surface of the young planet is tacky and nearly molten and oceans have recently condensed from steam. The young Sun is dim and veiled by placental dust. Looking into the night sky, there are no recognizable bright stars or constellations. Jupiter and Saturn are invisibly faint, not having yet gathered gassy envelopes to surround their rocky cores. There's no Moon.

With no warning a Mars-sized rock approaches the Earth and fills the sky. It strikes a glancing blow but with enough energy to vaporize trillions of tons of rock and send it fizzing into space. Most of the heavy core of the intruder falls into the Earth, while a mix of mantle material from the Earth and the intruder is flung into a molten magma disk surrounding the Earth. In under a thousand years, the Moon has congealed into a viscous mass, although it takes another 100 million years for the lunar magma ocean to cool and crystallize into rocks. Just after it forms the Moon looms 10 times closer and larger in the sky than it does today (Figure 1.4).

Is this nothing more than a "just-so" story—a convenient narrative wrapped around a handful of facts? Probably not.[12] The impact theory not only explains the Moon's relatively large mass and small core, it also accounts for the slightly younger age for the Moon than the Earth, the large angular momentum of the Earth-Moon system, the nearly identical abundance of oxygen isotopes on the Earth and Moon, and the tilt of the Moon's orbital plane relative to the Earth's equator.[13] Yet however many facts can be explained by an impact, nobody was there to witness it and some facts could have other explanations. Historical science is like archeology so we may never be sure.

Billions of years later the stories of the Earth and Moon continue to be intertwined. Our barren twin applies a stretching force that cre-

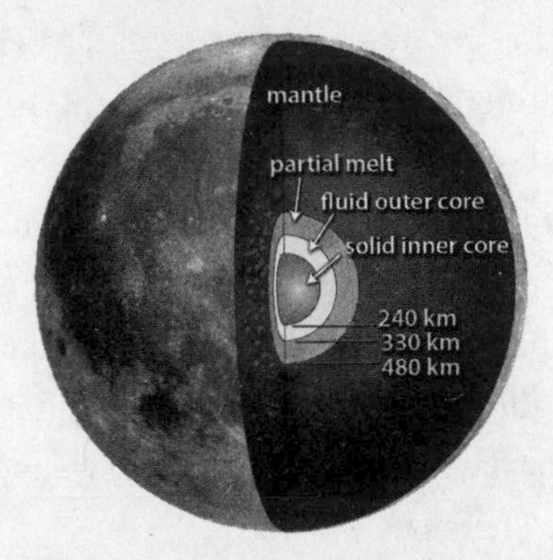

FIGURE 1.4. *The Moon is notably different from the Earth in that it has a weak magnetic field and a small metallic core. The impact hypothesis for the Moon's formation maintains that about 4.4 billion years ago, an impact "splashed off" the Moon and delivered angular momentum to the Earth-Moon system.*

ates tidal bulges on both sides of the planet. (Which is another rebuttal to the notion that the full moon influences human affairs; all the effects would be expected twice a month—not just once.) Since the Earth spins much faster than the Moon moves around it, the watery bulges lead the imaginary line connecting the two bodies. The tug drains angular momentum from the Earth, boosting it into a higher orbit with a slightly lower speed. The Earth, meanwhile, is dragged into spinning slightly slower. So the Moon is slipping from our grasp, receding 1.5 inches every year and making it slightly harder for us to return there. The good news is we can get more done in our busy lives, as the day is 15 microseconds longer every year.[14]

A chance encounter 4.5 billion years ago played a significant role in making the Earth hospitable for advanced life. Tides create a unique transition zone where creatures can experiment with the adaptations needed to move from the sea to the land. The Moon stabilizes Earth's tilt axis. Without a similarly large Moon the axial tilt of Mars wobbles from zero to 60 degrees. The result is that the Earth has less extreme climatic variations. When it was closer to the Earth, the Moon helped generate the tectonic crust that life needed to get established.

It's dangerous to push these arguments too far, but the early impact dusted the Earth's mantle with metals, without which humans might not have progressed beyond flint arrows and stone wheels. We've already seen that lunar cycles spurred thinking about regularities in nature. The Moon was pivotal in the ancient Greeks' awareness of the Earth as a sphere, in Galileo's concept of the "plurality of worlds," and in the first test of Einstein's theory of gravity.

The story of the Moon's creation spurs the question, What if? With a slightly different trajectory, the Mars-like object would have sailed harmlessly by. Subjected to more violent extremes of climate, the planet might never have become hospitable for the hairless ape that can only survive a slender temperature range. And if the collision had been more direct, the Earth would probably have been obliterated rather than augmented. It's sobering to think that our existence hinges on serendipity in the chaos of the early Solar System.

THE GREAT ADVENTURE

The 12 people who have walked on another world were never intended to represent human diversity. All are men. All are white. All but one studied aeronautical engineering at college. Almost all were military test pilots and saw active service during the Cold War or the Korean War. They'd engaged in high-risk professions, where many of their colleagues died, so luck was on their side as they entered the astronaut corps in the early 1960s.

The great adventure that got us to the Moon came down to one man with a cool head waving off the possibility of an aborted landing and bringing the lunar module down with 25 seconds of fuel left. Tension was heightened by the several-second delay for round-trip radio transmissions to the small craft. Heading down, the crew of *Apollo 11* experienced a series of alarms and they chose to ignore data from the onboard computer, which was far less powerful than a modern digital watch. Mission controllers realized with horror the astronauts were closing in on a boulder-strewn moonscape 4 miles downrange from the planned landing site. With the mission in jeopardy, pilot Neil Armstrong seized control from the computer and guided them to a safe landing, while in Houston they tersely read out the fuel countdown. When Armstrong announced, "The Eagle has landed," the mission controller spoke for the millions watching on TV when he said, "Tranquility, we copy you on the ground. You got a bunch of guys about to turn blue. We're breathing again. Thanks a lot."

Walking on the Moon had a profound effect on all 12 men, and on another 15 who made the trip but didn't land, but it's an urban legend that they went crazy or had religious awakenings. Buzz Aldrin suffered from depression and battled alcoholism, and Edgar Mitchell devoted his life to the study of psychic phenomena, while Jim Irwin and Charlie Duke became evangelical Christians. The rest genially accepted their role as grizzled veterans and dimly remembered heroes. It was bound to be life-changing for a set of confident and technical-minded men, not prone to introspection, to see their home from such an extraordinary vantage point (Figure 1.5).[15]

James Irwin, lunar module pilot for *Apollo 15*, described it this way: "The Earth reminded us of a Christmas tree ornament hanging in the blackness of space. As we got farther and farther away, it diminished in size. Finally, it shrank to the size of a marble, the most beautiful you can imagine. That beautiful, warm, living object looked so fragile, so delicate, that if you touched it with a finger it would crumble and fall apart, and seeing this has to change a man."[16]

A persistent 6 percent of the American public believe humans never set foot on the Moon and the entire Apollo program was an elaborate hoax.[17] Presumably these are the same people who hold other unten-

FIGURE 1.5. *The iconic image of Buzz Aldrin's footprint on the Moon is a reminder that we haven't been there for 40 years. Only 12 of the Apollo astronauts set foot on the Moon. For the small minority of the public who don't believe we went there, the Lunar Reconnaissance Orbiter has provided additional evidence of what was left behind on the surface, including the tracks of the lunar rovers.*

able ideas or are vested in conspiracy theories, and perhaps the number is low enough not to be of concern. Astronaut reactions to Moon deniers vary from personal to reflective. Gene Cernan put it this way: "Truth needs no defense. Nobody, nobody can take those footsteps that I made on the Moon away from me." Harrison Schmitt, formerly a U.S. Senator, said, "For most of them, I just feel sorry that we failed in their education." Buzz Aldrin didn't just get mad, he got even. The YouTube video of the *Apollo 11* astronaut losing his patience with skeptic and filmmaker Bart Sibrel and punching him hard in the nose has been viewed more than 2 million times.

Setting aside the lunatic fringe, the passage of time creates a general sense of unreality about Apollo. Less than a third of Americans were alive during the Moon landings so the episode inevitably takes on the flavor of a distant cultural memory.

In one sense, the "great adventure" was an anomaly. President John Kennedy's commitment to reach the Moon within a decade was born of an intense superpower rivalry with the Soviet Union. NASA's budget rose by a factor of 10 from 1960 to 1967 (in fixed-year dollars) from $3 billion to $33 billion and then fell a factor of 3 from that peak.[18] Much of the technical talent that NASA assembled for the immense challenge of Apollo subsequently dispersed and went to work in the new semiconductor and computer industries. And when Gene Cernan stepped up the ladder into the lunar module on December 14, 1972, he never imagined he'd be the last person to set foot on the Moon for what will probably turn out to be more than 50 years. Hopefully, he turned the lights off. NASA's mission is cloudy, so it's no longer clear who'll be the next to explore the rock next door.

Perhaps it won't be humans; the Moon may be inherited by robots. In September 2007, the next race to the Moon was kicked off by the X Prize Foundation, who announced a $20 million prize for the first team to "soft land" an unmanned rover on the Moon, send it 500 meters across the surface, and beam back video, before the end of 2012. The prize drops to $15 million if the landing happens before the end of 2014. Second prize is $5 million and teams can win $5 million in bonuses for finding Apollo-era relics, detecting water, or staying alive through a lunar night.

Experts reckon it will cost at least $50 million to accomplish the goal, so nobody's doing this for the money. Inspired by the previous, highly successful, Ansari X Prize, the foundation hopes to kick-start a wave of entrepreneurial innovation in space travel. Only private companies and individuals and nongovernment entities are eligible; NASA is reduced to a bemused and interested bystander. Moon 2.0 is being sponsored by Google. Twenty-one teams representing 11 countries are feverishly at work on their designs. While the backdrop is competitive, there's enough collegiality and sharing of ideas to make this totally different from the first space race.

The Moon is for tourists. Not yet, but it's coming if the feverish activity in the private space industry is a guide. Virgin and XCOR are engaged in a full-throttle competition to send civilians on suborbital flights, for a cool $200,000 each. Space tourism has attracted over $1 billion from "angel" investors, although the entire enterprise still rests on the work of Space Adventures, who brokered trips on Soyuz to the International Space Station for seven private citizens, who paid $20 to $35 million each. Apart from Charles Simonyi, these very rich people share an aversion to the phrase "space tourist." Rick Tumlinson, cofounder of the Space Frontier Foundation, puts it this way: "Tourist is someone in a flowered shirt with three cameras around their neck." Space Adventures has a plan for flyby trips to the Moon in a modified Soyuz spacecraft. The cost: $100 million, so start saving now.[19]

The Moon is also for lovers. Every time I teach a large class of non-science majors, I fill out the complex paperwork for the loan of a lunar sample package from NASA's Johnson Space Center. Five weeks later a sleek aluminum attaché case arrives by registered mail, containing a Perspex disk and set of mounted slides, along with explanatory audio-visual materials. I joke with my colleagues that the case emblazoned with the NASA logo is cooler than the objects it contains: samples of soil from each Moon landing. They're embedded in an indestructible disk of plastic, so I pass it around the class and watch the teenagers shed their finely honed indifference as they handle dirt from another world.

I'm about to return the sample package to Houston when my eye is caught for the first time by the slides. Each one is a sandwich with a thin slice of lunar rock glued between sheets of glass and set into a stiff metal frame.

I have an idea. After two years of near solitude following separation and divorce, I've met an amazing woman. In my fantasy, I pick a slide containing dark and ancient basalt, primeval material of the Solar System. I borrow a tiny hex key and industrial solvent from work and the slide reveals its precious cargo. I scrape a small amount into a paper funnel and dust it onto the top of a small chocolate dessert. We share this otherworldly treat. It's gritty and granitic, but sublime. Days later, long after it has left me, I still feel altered. All lovers should dine on the Moon.

—

I want to bound over the gray, pockmarked surface forever, but I feel burning in my leg muscles so I stop and lean against a large boulder, heart pounding. The moonscape is utterly barren, and devoid of any landmarks or points of reference. With no air, it's difficult to judge distances. I realize I have no idea where I am.

A ripple of anxiety passes through me. How did I get here? How do I get home? If it's a dream, it's more lucid than any dream I've ever had. If it's a reverie, the fidelity and detail are extraordinary. My imagination is incapable of such hyperrealism, or so I imagine.

Considering my situation I look up and notice the Earth for the first time. It hangs high in the sky, like a bauble of the Gods, lovingly painted a gauzy blue-white and brown. From this great distance there's no hint of human activity, no sign that the hairless apes with the insatiable appetites for land and resources are throwing the whole ecosystem out of balance. However realistic it seems, the Moon must be the illusion. Relaxing and breathing steadily, I close my eyes and try to reel myself in, back into my body on Earth.

Instead I have the strangest sensation. My presence on the Moon is concrete, inviolable. The ebbing burn in my legs, my breath fogging the space suit facemask, the giddiness of my reduced weight—all of these are absolutely real. In my mind's eye I look back toward the Earth and see myself in the backyard of my suburban house. It's a summer day and I'm reclining in a deck chair, iced lemonade in one hand. My chest slowly rises and falls and it dawns on me that it's not synchronized with my breathing here on the Moon. It lags by just over a second.

2

PLANETARY ZOO

THE ICE FIELD STRETCHES INTO THE DISTANCE LIKE A CRUMPLED WHITE BLANKET. *I'm on a low ridge and can see for miles, although there are no familiar objects to give a sense of scale. I turn around slowly and the view is the same in all directions. The horizon is a knife-edge between blinding white and absolute black.*

I look more closely at my immediate surroundings. The ridge I stand on is serrated ice, translucent blue-white in some places and stained brown and red in others. Down the slope is a chaos of ice boulders, angled and tilted, as if an angry giant had smashed them from a smooth block. In the distance, I see a sinuous frozen river, meandering between rugged ice cliffs. The icescape is forbidding, impenetrable. Navigating it would be impossible. This is Europa, Jupiter's frozen moon.

I take a step and spring upward surprisingly. It's a familiar sensation; Europa is similar in size and density to the Moon so I weigh just 30 pounds. But even with gentle gravity the landing sends jarring pain through my legs. This ice is so cold it's like gran-

ite. I'm acutely aware of the thin membrane of space suit fabric that separates my skin from the nearly perfect vacuum beyond, 250 degrees colder.

Above me, Jupiter is a milky orb hanging in the sky, two dozen times larger than the Moon in my more familiar sky. It takes a while, but finally I locate the Earth. The Pale Blue Dot is so small it's barely distinguishable from a star. The hopes and dreams of 7 billion fit on the head of a pin.

Home. I am there. Or am I actually here, a billion miles away in the realm of the gas giants? This reverie started with me in my living room, with the credits of a movie I'd just watched scrolling on the TV, and the remains of my lunch pushed aside. But looking across the gulf of space to my home, I'm just starting my sandwich and the movie is only halfway through. A rift in time has opened up. I'm here now, but there then. Bifurcated across space and time, I'm not really sure where I am.

—

ROBOT MESSENGERS

We were never made to leave the womb. Gaia, our mother, keeps us at the right temperature, soothed by wind and rain, the atmosphere protecting us from UV radiation, cosmic rays, and most meteors. Five miles up, at the summit of Mount Everest, a human wouldn't survive more than a few hours without oxygen. Several brave (and probably crazy) adventurers are preparing to free-fall to the Earth from over 20 miles high. Even with the protection of an advanced pressure suit and life support system, they're taking extreme risks. Unprotected at that altitude, blood boils and death follows swiftly. We aren't built to withstand the rigors of space.[1]

So when we travel beyond the Moon, we send our robotic progeny. It was mostly a cavalcade of failures during the first years of the space program, as the United States and the Soviet Union cut their teeth on the emerging technologies of space flight while engaged in an intense superpower pissing contest.

In 1958, a year after the launch of *Sputnik* kicked off the Space Race, the countries tried to reach the Moon and racked up seven combined failures before the first successful flyby in 1959. It took seven more years, and 22 more failures, before the Soviets managed to soft-land the Luna 9 probe. By then, the United States had raised the stakes with an extraordinary bet on the Apollo program and manned landings. With Mars and Venus, it was the same story. The first attempt was in 1960 and the two superpowers shared seven failures before *Mariner 4* returned the first close-up images of Mars in 1965. There have been successes since then, but also 18 more disappointments. Venus was even more of a graveyard for spacecraft. The 1960s witnessed 17 lost or destroyed probes before *Venera 7* became the first spacecraft to land on another planet. It sent back just 23 minutes of data before giving up the ghost in the blistering heat and the crushing pressure of the Venusian surface.

If you remember the 1970s as the decade of the oil crisis, TV sitcoms with raucous laugh tracks, cheesy stadium rock, and mullets and bell-bottoms, then you'll be pleased to recall it was also the time we first explored the outer Solar System. *Pioneer 10* was the first probe to cross the asteroid belt and in 1973 it made a flyby of Jupiter before heading to Saturn. Mercury got a visit from *Mariner 10* mid-decade. Near the end of the decade *Voyagers 1* and *2* were launched. Each visited Jupiter and Saturn, and the latter whizzed by Uranus and Neptune before both left the Solar System.

During this decade the space agencies went up their learning curve and worked the bugs out of their control systems and other hardware. The result was a dramatic improvement in the success rate of missions to the inner Solar System, up from 35 percent in the 1960s to 73 percent in the 1970s, and 91 percent in the most recent decade.[2] This was accompanied by an increase in the degree of difficulty of the "dives" and "jumps."

We've come a long way in just a few decades. The Space Age kicked off on October 4, 1957, with a Russian satellite no bigger than a beach ball and no heavier than a large man. *Sputnik* functioned as little more than a glorified radio transmitter. The modern era is epitomized by spacecraft like NASA's *Cassini*, launched in 1997 and actively exploring Saturn and its moons since 2004. *Cassini* is as large and heavy as a laden school bus, and it's crammed with sophisticated instruments. Complexity isn't cheap—the price tag since the construction phase is roughly $3.5 billion.

Mission designers learned how to use "gravity-assist" maneuvers to save time, propellant, and money when exploring the Solar System. Also called a gravitational slingshot, the gravity of a planet or moon can be harnessed to alter the path and speed of the spacecraft. The spacecraft can gain (or lose) up to twice the larger object's speed.[3] From a long way off it would look like the spacecraft bounces off the planet, although no contact actually occurs. In physics you can't get something for nothing, so the momentum and energy gained by the spacecraft are matched by momentum and energy lost by the planet (Figure 2.1). Just as an elephant barely feels a mosquito, the effect on the planet's orbit is negligible.

The most spectacular vision of gravity assist was a proposal for The Planetary Grand Tour. Guy Flandro was an undergrad intern working for NASA's Jet Propulsion Laboratory (JPL) in California in the late 1960s when he realized that the outermost planets would be aligned a decade later, an arrangement that wouldn't recur for 176 years. A proposal was developed for four probes to seize on this opportunity. Like stones skipping off a lake, and gaining energy with each hop, they would fly past the giant planets and be ejected from the Solar System.

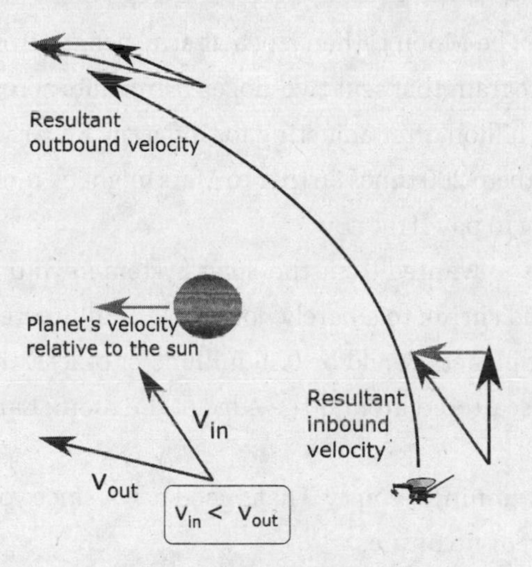

FIGURE 2.1. *In a gravity-assist maneuver, as in this example using Jupiter, the spacecraft approaches a planet and gets a boost from the planet's orbital motion around the Sun. Since energy is conserved, the planet loses energy but it has such a large mass that the change in its speed is negligible.*

NASA budget cuts in the 1970s killed the grand vision and even other proposals for a "mini grand tour." But *Voyager 2* did use gravity assist to visit all four giant planets. Its sister probe, *Voyager 1*, flew by Jupiter and Saturn and then on out of the Solar System, where it's the most distant manmade object, more than 11 billion miles from the Earth.[4] Gravity assist is now part of the standard toolkit of mission designers. Among current NASA missions, *Messenger* used it in early 2011 to slow down enough to go into orbit around Mercury, while *New Horizons* has used it to accelerate toward Pluto and the Kuiper Belt, with a scheduled arrival at Pluto in 2015.

Solar System exploration is challenging because the distance between planets is far larger than their sizes. Imagine the Sun scaled down to a 10-foot-diameter ball of glowing gas. On this scale the Earth is a grape 1000 feet from the Sun, and the edge of the Solar System is

6 miles away. The Moon is then a pea at arm's length from the Earth. Apollo, the program that sent two-dozen astronauts across that small gap, cost $30 billion after adjusting for inflation. So you can imagine why sending them 200 times farther to Mars might be more than we're able or willing to pay right now.

Suppose you wanted to fit the Solar System in your living room. The Sun would shrink to a barely noticeable 1-millimeter dot of light and the giant planets would be 0.10 millimeter or less, the width of a human hair, scattered around the edge of the room. Earth would be invisibly small.

Space is stunningly empty. That's good news, since you'll still have room for all your furniture.

MEET THE FAMILY

Saul shrugs when asked about his family. Who could keep track of so many children? There was the loss of his wife, Nuage, which he still doesn't like to speak about. She was so young; it was tragic. So soon after having the children, she just wasted away. She had always been slight, like a ghost with those diaphanous clothes. But she clung to him loyally and stayed cool when Saul became hot and headstrong. To lose her so swiftly and face child-rearing alone was a hard road.

In no particular order—it's not right to have favorites when you're a parent—Saul talks about his kids. Ares is the difficult one to love, with his volatile temper and his dark moods. The two pit bulls he keeps would give anyone the willies. Fear and Terror; not the most subtle of names. They're brown and misshapen and fiercely loyal. It's easy to tell that Ishtar is Saul's favorite. She is warm and has a beautiful singing voice. She aspires to be like her father although she has a contrarian streak that means she never follows the crowd. Her twin sister Gaia took to science at a young age and is the nerd in the family. First she was downstairs cooking up strange chemicals in bubbling flasks. But what started as scum on the shower curtain got way out of hand and spread everywhere. Saul rolls his eyes and chuckles when he thinks of the trouble it caused.

Hermes is still the closest to Saul, but he never developed his own interests and never became independent. He still dotes, and always looks to Saul for approval, but he's no longer the heir apparent. That role falls to Jove, who has the girth and the ego to think that someday it might all belong to him. He throws his weight around and always has. The others stay well out of his way. Cronus, the pretender. Saul chuckles as he says this. Cronus has the finery but not the substance to be the head of the family. Caelus and Nereus keep to themselves and are no trouble to Saul. He still gets them confused and has to remind himself that Caelus is the one who fell as a small child and has never been the same since (he also has a notorious potty mouth), while Nereus is the compassionate one who has taken in a foster child of her own.

It's a big brood, and Saul has done his best to sustain them over the years, or at least keep them in line.

We've anthropomorphized the planets for thousands of years. The Sun, the Moon, and the five planets visible to the unaided eye—Mercury, Venus, Mars, Jupiter, Saturn—have imprinted deeply in culture, giving us the names of the days and one of our major demarcations of time.[5] The names for the planets originated with the Sumerians and were mapped into the Babylonian culture, followed by translation to Greek and finally Roman, or Latin.

Mercury, who never strays far from the Sun, has nifty winged shoes and a winged hat. He's the fast-moving messenger of the gods and also a god of trade, with a side job of escorting newly dead souls to the afterlife. Venus is the goddess of love and fertility, but the myth has ambiguity that suggests enticement with a downside; the Latin word shares an etymological root with "poison." Maybe the Romans foreshadowed the toxic inferno that lies under Venus's atmospheric veils. Mars is a military god, in honor of the blood-tinged planet that roams the sky. He was the father of Romulus, founder of Rome, and all Romans considered themselves his descendants. With Jupiter, we reach the pinnacle of the pantheon of gods. Son of Saturn, father of Mars and Mercury, Jupiter was the god of sky and thunder. In Imperial Rome, consular oaths were sworn to him, with an offering of a white, castrated ox with gilded horns. Saturn was the god of agriculture and harvest. During the feast held in his honor, the roles of master and slave were reversed, etiquette was ignored, and the normal moral restrictions were tossed out the window.

As new planets were discovered after the invention of the telescope, the tradition continued. Uranus was benign in Greek myth, known as Father Sky, a primordial god. Inevitably, the Romans added a lurid back-story, where Uranus came every night to cover the Earth and then mate with Gaia. But Uranus hated the children she bore and he imprisoned them in the gloomy underworld Tartarus. Gaia persuaded Kronos, the boldest of her children, to castrate Uranus and overthrow him. Neptune is the god of water, the sea, and horses. The Romans honored him after their naval victories, and many modern navies and maritime fleets still carry out initiation rites for any sailors crossing the equator for the first time, under the watchful gaze of King Neptune.[6]

What of poor, misbegotten Pluto? With a stony heart I'll omit Pluto, explaining why as we move to the astronomical view of the planets.

The Greeks paved the way for thinking about the planets as worlds in space, rather than deities that perambulated the ecliptic, taking in the animal zodiac: ram, bull, crab, lion, scorpion, fish, and a strange sea-goat. The modern idea of planets didn't arise until Galileo. We imagine him gazing at the Moon through his spyglass, pausing to add touches to a watercolor painting of lunar features, which he works on by light from an oil lamp. After a break inside for wine and bread, he returns to locate Jupiter in the southeastern sky. The images shimmer slightly in the summer night air, but he can clearly see four outriggers to Jupiter. Their arrangement has changed slightly since his last observation and he notes it in his sketchbook.[7]

Through his telescope, Jupiter is a disk of milky light, a distant world suspended in space, but the moons are just dots engaged in a stately minuet with the parent planet. Galileo's observation of Jupiter's moons provided further support of the Copernican model; the Earth is clearly not the center of their motions. He named them after his patrons, the Medici family, but they were eventually named after Zeus's lovers. It wasn't until a spacecraft named after Galileo arrived at Jupiter late in 1995 that we got to see the moons up close.

The Galileo probe did a series of daring flybys of the major moons. It found exotic worlds of fire and ice. Ganymede is the largest moon in the Solar System, larger even than Mercury. It has a molten iron core, a rock and ice surface, and very probably a salty ocean 200 kilometers below the surface. Callisto is the next largest, and its surface is heavily cratered, with a possible ocean 100 kilometers down. Then comes Io, a bizarre moon with 400 active volcanoes. Io is the most geologically active body in the Solar System, and each year the surface is coated with an inch of fresh sulfur from the volcanoes. Europa is the smallest of the four, slightly smaller than our Moon. It's like an icy billiard ball; several kilometers of ice lie over an ocean 100 kilometers deep. This water world a billion miles from Earth is one of the most compelling targets for future space probes (Figure 2.2).

Our family of planets has moons with individual and very distinctive personalities and some quirks, but there are two main categories and groupings in the Solar System. The terrestrial planets lie close to the Sun, no more than five Earth-Sun distances. These objects are small, dense, and rocky. They have solid surfaces and are mostly made of rocks and metals. These siblings—Mercury, Venus, Earth, and Mars—are unmarried; they've few children of their own and wear no rings. The Jovian planets lie far from the Sun, 5 to 30 times the Earth-

FIGURE 2.2. *This image from the Galileo probe shows the fractured icy surface of Europa. The structures resemble ice floes seen in the polar regions of the Earth, and the cracks and rifts are suggestive of water seeping in and causing movement. The smallest features visible are about 1 kilometer in size.*

Sun distance. They're large, have low density, and are composed of the same chemical ingredients as the Sun: hydrogen, helium, and few heavier elements. It's likely, though not proven, that they each house a small rocky core of 3 to 10 Earth masses. These closely related family members—Jupiter, Saturn, Uranus, and Neptune—have many small progeny (four of whom we've met) and sport fancy rings (Figure 2.3).[8]

The Solar System also contains small rocky objects, most notably between the orbits of Mars and Jupiter in the Asteroid Belt. They're likely to represent the debris left over when a planet failed to form at that location. Mighty Jove was jealous and so he used his influence to thwart a rival, we might guess. Jupiter has doting asteroids that track its orbit, moving at a respectful distance ahead and behind. There are also asteroids traveling through the inner Solar System, some of which are on orbits that cross the Earth's, and rocky objects called Centaurs that live among the giant planets. The Kuiper Belt extends from 30 to 50 Earth-Sun distances and includes misshapen objects that are more ice than rock: about 100,000 larger than 50 kilometers. The limit of the Solar System is marked by the Oort Cloud. This spherical swarm contains a trillion comets but added together they amount to much less than the mass of the Earth.

By now you probably thought I'd completely dispensed with Pluto, the king of the underworld, and the god of the dead and the terminally ill. For 76 years after its discovery in 1903, Pluto had a seat at the table as a planet. But astronomers became increasingly uncomfortable with

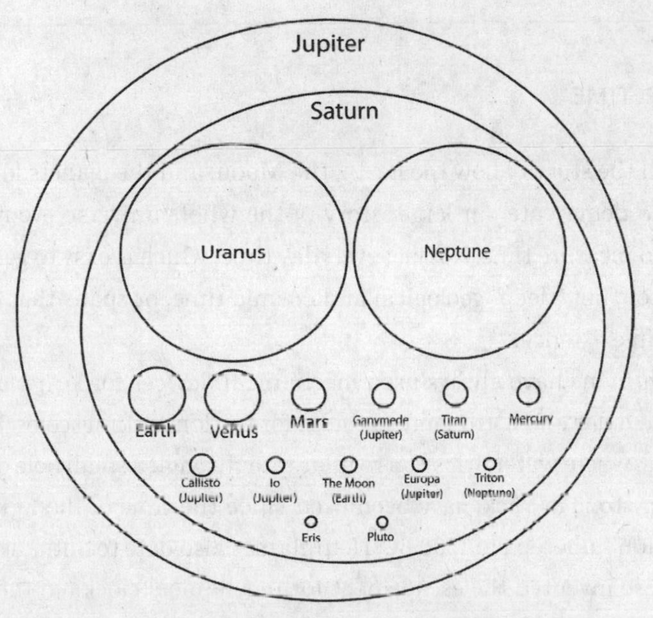

FIGURE 2.3. *Relative sizes of the eight planets and other large objects in the Solar System. The demarcation between the terrestrial planets and the gas giants is obvious, as is the anomalous smallness of Pluto. The seven largest moons in the Solar System are also represented.*

its status because it's such an oddball. Odd in its highly eccentric and inclined orbit that takes it from inside Neptune's orbit to 49 times the Earth-Sun distance. Odd in its tiny size and mass compared to Jovian planets. And odd in having a satellite, Charon, that's over half its size.

The clincher came with the discovery of objects beyond the orbit of Neptune that rival Pluto in size, culminating in the discovery of an even larger object, Eris. So if you want to make Pluto a planet, you have to also make Eris a planet.[9] Since researchers think that the netherworld of the Solar System contains thousands of similar objects, including several bigger than Pluto, the table would get crowded and people would get headaches trying to remember all the new names. So Pluto was unceremoniously demoted to the status of "dwarf planet" and it scurries around under the table getting whatever scraps it can.

DEEP TIME

To tell the story of how the Earth, the Moon, and the planets formed, and to demarcate our larger story of the whole universe, we need a way to measure time. Not just everyday time, which is easy to get from a watch, but "deep" geological and cosmic time, or spans that dwarf human existence.

Humans have always used mechanical devices for keeping time within a day and astronomical cycles for the longer timescales. Water clocks, where water drips at a regular rate through a small hole drilled in gemstone or rock, have been used since the time of the Egyptian pharaoh Imhotep I in 1500 BC. Hourglasses also date to antiquity. The Chinese invented the escapement for mechanical clocks in the eleventh century and some European clocks have worked continuously for centuries. Galileo improved the precision of clocks by understanding and perfecting use of the pendulum.

Lunar and solar calendars have been used for millennia but if we only used human artifacts to track time the trail would then go cold after tens of thousands of years. The geological record is used for looking back millions of years. The Earth is a restless planet and erosion and tectonic activity disrupt the smooth layering of rocks. But seasonal variations of water and airborne dust falling on large continental land masses are often very regular. Layers of ice in Antarctica and dust in Mongolia are two examples where time can be tracked for nearly 10 million years. Long-term cyclic changes in the orbital period and spin axis orientation of the Earth imprint a pattern in the rocks that can be read back 250 million years, though not with good precision.

To measure times of billions of years (and times as short as a billionth of a second), scientists turn to radioactivity. Some naturally occurring heavy elements are inherently unstable. Atoms of these elements will spontaneously decay into a lighter element. Radioactive decay has one very peculiar and one very useful characteristic. The decay of a single radioactive atom is completely unpredictable. But average decay time of a large collection of the atoms is very well-determined. Normally, a lot of "I don't knows" don't combine to make an "I know very well" but quantum mechanics is weird and counterintuitive. The time it takes for half of a collection of radioactive atoms to decay is the half-life. Radioactive decay is contained within the fortress-like atomic nucleus and it's impervious to temperature, pressure, or chemical reactions. This means radiometric dating works perfectly in the geological chaos of an active planet like the Earth[10]—as long as you can find rocks old enough.

Geologists have scoured the Earth to find unaltered rocks from the time of formation. If a rock becomes molten, as happens when it's cycled into the mantle, the endproducts of radioactive decay escape or are redistributed, so geologists look in ancient continental masses like Australia and South Africa and some parts of North America. The oldest unaltered rock is from the Acasta Gneiss in Canada, part of an ancient mountain range exposed by glaciation. It's 4.03 billion years old.[11] The oldest material of any kind yet found on Earth is a zircon crystal from the Jack Hills formation in Western Australia. This tiny, semiprecious stone is 4.40 billion years old.[12] Good.

Since the Moon is geologically quiet, it presents a better prospect for gathering old, unaltered rocks. The biggest haul is the 842 pounds brought back by Apollo astronauts, most from the last three landings. Three Russian Luna spacecraft returned from the Moon with less than a pound of rocks, and 120 lunar meteorites have been gathered since 1980, "free samples" with a total weight of 105 pounds. Even people whose eyes might glaze over at a geology talk sit up and take notice at the value of lunar meteorites. From reputable dealers online the cost can be $10,000 to $20,000 per gram. Compare that to $1000 to $2000 per gram for a high-quality diamond or $50 per gram for mere gold.[13] The oldest rocks on the Moon are anorthsites from the highland regions, pale crystallized rocks that floated like a scum layer on the Moon's magma ocean soon after it formed. Study of samples brought back on *Apollo 16* gives an age of 4.46 billion years.[14] Since the Moon formed very early in the history of the Earth, this gets us close to the age of the Solar System. Better.

The most reliable tracers of the formation of the planets are certain meteorites that are primitive, unaltered relics of the collapse of the solar nebula (Figure 2.4). The Canyon Diablo meteorite, which blasted out the Barringer Crater 49,000 years ago in northern Arizona, was first used for this measurement in the early 1950s. Any sample with uranium includes an automatic cross-check, as uranium-235 decays into lead-207 with a half-life of 700 million years while uranium-238 decays to lead-206 with a half-life of 4.5 billion years. These studies give concordance on an Earth and Solar System age of 4.57 billion years.[15] Best.

The precision of the chronometers has reached the point where the planetary geologists can get into arguments over a million years here and there. (An image of bald men fighting over a comb comes to

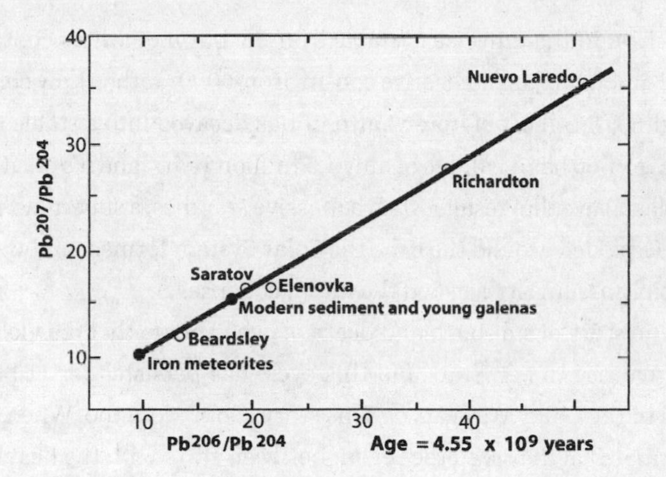

FIGURE 2.4. *The age of the Solar System is determined by the relative abundance of the decay products of different isotopes of lead formed from different isotopes of uranium. For two different decay paths, this graph shows near-perfect agreement on a single age of the Earth from different rock samples and iron meteorites.*

mind, somewhat unfairly.) These intervals are small compared to the overall age but they matter because the Solar System was put together in a hurry.

COSMIC BILLIARDS

Why would a vast diffuse cloud of gas and dust suddenly begin a set of runaway processes that turned it into a blazing young star surrounded by a small set of rocky objects on orderly orbits?

This question puzzled astronomers and planetary scientists for a long time. The answer they came up with: fire from fire. Our Solar System bears the imprint of one or more stars that died violently in the vicinity of the solar nebula 4.5 billion years ago.[16] Meteorites are the primeval material we can use to reconstruct the process of formation.

Most iron in the universe is stable iron-56, but meteorites contain a small amount of its radioactive cousin, iron-60. Or rather, they contain nickel-60, the ghost of iron-60 after it has decayed into a stable form. Since iron-60 has a half-life of only 1.5 million years, and it's created in the blast wave that results when a massive star dies, a supernova must have exploded around the time the Solar System formed, and we can plausibly assume it triggered the initial collapse.[17]

The story of star birth and death is cyclic. Stars that die violently can give rise to new stars and this cycle has persisted for 11 billion years in the Milky Way galaxy. This story is our story too. When stars eject material they seed the region between stars with the heavy elements needed to make planets and people. Every carbon atom in our bodies was once forged in the cauldron of a star somewhere far off in space, long before the Solar System formed.

What happens next is extraordinary: the evolution from motes of dust to mountains of rock in the blink of a cosmic eye.

After its gravitational collapse, the infant Solar System is a dense disk of gas and dust surrounding a young volatile star. Before it settles to fuse helium from hydrogen the Sun is energetic and variable. Within several million years, microscopic dust grains made of graphite and rock and ice coagulate into delicate loose clusters, like dust bunnies. Electrostatic forces initially hold these aggregations together but as they grow to the size of boulders and then mountains, gravity takes over. With gravity firmly in charge the accretion process can accelerate and it takes just a few hundred thousand years to create several hundred planetary embryos. Then things get nasty.

These embryos, which range in size from cities to large countries, are moving in elliptical orbits so they interact and collide. Some collisions are destructive and produce smithereens while others are gentler and lead to a larger object. The construction phase lasts about 100 million years and leaves us with eight planets. Closer to the Sun than a few times the distance to the Earth, any residual gas in the disk is driven

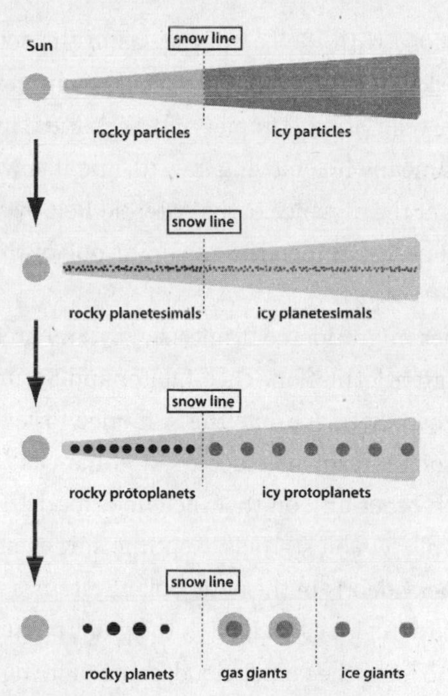

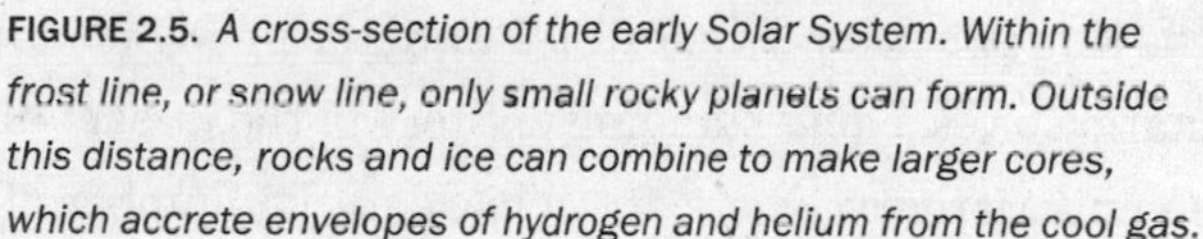

FIGURE 2.5. *A cross-section of the early Solar System. Within the frost line, or snow line, only small rocky planets can form. Outside this distance, rocks and ice can combine to make larger cores, which accrete envelopes of hydrogen and helium from the cool gas.*

away by radiation from the young Sun so planets are rocky. Farther out, vapors like water and ammonia and carbon dioxide are frozen so larger objects made of rock and ice form (Figure 2.5). The rocky cores have enough gravity to gather hydrogen and helium and they grow into gas giant planets.[18] Along the way, the Moon splashes off the molten Earth, Venus is knocked on its ass such that it rotates opposite to all the other planets, and Uranus is kicked onto its ear to make the most dramatic seasonal variations in the Solar System.

Even after this period of mayhem is over, the planets haven't finished their fun and games. None of the planets is at the same distance from the Sun now as it was then. The terrestrial planets move in a bit due to interactions with the last of the gas in the solar nebula, and the gas giants slowly change their orbits due to interaction with the remaining embryos. Saturn, Uranus, and Neptune move steadily outward. Jupiter is meanwhile moving inward, and it thwarts the formation of a planet at the distance of the Asteroid Belt by scattering most of the material either in toward the Sun or out of the Solar System entirely.

Outer planet migration continues slowly and steadily until half a billion years after formation. Then Jupiter and Saturn reach a temporary state where Saturn orbits the Sun once for every two Jupiter orbits. When such a resonance occurs, it enhances the effect of the planets on each other and on their neighborhood.[19] Planets in resonance are like planets on steroids. Neptune surges past Uranus and ploughs into rocky debris in the outer Solar System, sending a lot of it careening inward. As it passes Jupiter, the giant planet flings some of it far outward to form the comet cloud, some of it slightly farther out to form the Kuiper belt, and some of it toward the inner Solar System.[20] This scenario solves a long-standing mystery in the impact history of the Moon and Mercury, where an increase in the number of craters appears 3.8 to 3.9 billion years ago, long after the rate of giant impacts should have died down. "Late heavy bombardment" is now a standard part of the story of the Solar System. It's not clear if early life on Earth could have survived the devastation.

The Solar System is no longer an unruly child, but it hasn't become entirely dull. Calculations and computer simulations show that the orbits of the planets are chaotic and subject to long-term variations. Billions of years from now it's likely that the Earth's axial tilt will grow and the orbits of Mercury and Mars will become unstable enough that they may collide with the Earth or be ejected from the Solar System.

Moons form with scaled-down versions of the same processes. The big moons of Jupiter and Saturn sit close in, orbit in the same direction as their planet, and occupy the equatorial plane. They formed with the same material that left each giant planet with a ring system. The outer moons tend to have eccentric orbits, arbitrary inclinations, and orbit in the opposite direction to the planet rotation—the signature of captured bodies. Phobos and Deimos, for example, are asteroids that were captured by Mars. Neptune's moon Triton was captured from the Kuiper Belt. Pluto's large companion Charon is thought to have arisen from an impact, like the Earth's Moon.

As Galileo tracked the side-to-side motion of four dots of light as they orbited Jupiter, he could scarcely have imagined how interesting these moons would turn out to be. Each is an icy rock, but the composition varies with distance from Jupiter. Moving outward, Io is mostly made of metal and rocks, having been baked dry by Jupiter when it was hot and young. Europa is 90 percent rock and 10 percent ice, Ganymede is 60 percent rock and 40 percent ice, and Callisto has equal proportions of rock and ice. They orbit in almost perfect harmonic resonance, with Io orbiting Jupiter in 42 hours, and each successive moon taking twice as long to orbit as the next. It will be at least a decade before we can send a spacecraft to take a closer look.

In voyaging through the Solar System, we begin to diverge from the here and now. It takes light between 35 and 50 minutes to reach us from Jupiter, depending where the planets are in their orbits. Mission controllers have to wait a lazy hour or so to find out if the command they issued was actually executed. Even to nearby Mars, the round-trip light journey takes between 8 and 40 minutes. We may imagine NASA engineers driving the cute little rovers like crazed gamers, cowboy hats on, whooping and hollering. But in fact they're fastidious and methodical, glued to their checklists. Each command to move forward a foot is followed by a half hour wait to make sure there was no mishap, and then the next command is given. Roll forward a bit. OK. Did you do it? Yes. Are you OK? Yes, I think so. We've got a lot riding on this; I need you to be sure. Really, I'm fine, some red dust in my gearbox, that's all. Alright, move forward 3 more feet, easy now. Back off, I'm going as slow as I can! In a Mars day, the rovers would barely cross your living room.

An hour-long delay feels like instant gratification compared to the time it takes to execute a planetary mission from start to finish. People who are cherubic postdocs in the planning phase become gray-haired, grizzled veterans by the time the data arrive. Careers are torpedoed by any mission that fails, and sending complex hardware a billion miles from home for a close look at a distant world is a risky undertaking.

But why should robots have all the fun? If we're serious about space travel, the time will come when humans will venture beyond the Moon to more exciting Solar System locales. What would it take to realize the situation of the opening vignette of this chapter? There have been plans for manned missions to Mars for decades but they've receded like a mirage on hot blacktop.[21] Current proposals have astronauts on Mars no earlier than 2030, at a cost of over $30 billion. To bring the cost down, there are a growing number of advocates for a one-way mission, including NASA Ames Director Simon "Pete" Worden and astronaut Buzz Aldrin, who says: "Forget the Moon. Let's head to Mars!"

The Jovian system seems beyond the pale; it's 10 times farther and gets 25 times less sunlight than the Earth. Arthur C. Clarke envisaged us getting to Europa with a joint U.S.-Soviet mission in 2040. In the real world, there is an unmanned Europa orbiter on NASA's manifest, due to launch in 2020 and arrive in 2026, at a cost of $4.7 billion. Perhaps for the addition of a few billion dollars, the probe could be modified to include a pressurized "monk's cell" with enough life support for the six-year voyage. As with a one-way trip to Mars, there'd be no shortage of volunteers to go (and die) but see what nobody had ever seen before.

Meanwhile, the more accessible alternative of viewing Europa through a small telescope is appealing. And, sometimes, the first moment of observation is delight enough.

As a postdoc in Hawaii, I did most of my observing on Mauna Kea, the dormant volcano that rises 14,000 feet above the shoreline of the Big Island. Mauna Kea is host to the largest collection of telescopes in the world. Most of them are behemoths of steel and glass controlled by professional telescope operators, and they're deemed too valuable and delicate to be taken on joyrides by the astronomers who use them to gather their data. But the University of Hawaii operates a 24-inch scope that can be used for teaching and outreach.

One evening, I bring a good friend with me to the summit. Mike is a psychologist and the head of a research clinic on campus; he tells me he's never looked through a telescope before. The scene is surreal, with the volcano sloping away in every direction before dipping into the tops of clouds lit by moonlight. Stars blaze above us with Byzantine waste.

Mike is bundled up and he stamps his feet in the thin chill air as he waits for me to settle the telescope on Jupiter. I work as quickly as I can, hands rapidly getting numb. Then I have it and I beckon for Mike to climb the movable metal stairs and look through the eyepiece. Mike is a debonair man, quiet and self-contained, not one to draw attention to himself. But when he climbs the stairs and peers into the eyepiece, and Jupiter and its moons swim into view, four centuries after Galileo first laid eyes on this scene, Mike yelps exuberantly.

Staring into the black sky, I momentarily lose sight of the Earth and a wave of panic passes over me. Despite the unsettling dislocation in time, I'm umbilically connected to the home planet. Compared to Earth, with its soft, billowy clouds and rolling green hills, this place is stark. The jagged ice floes look absolute and unyielding.

I try to set off in the direction of a nearby summit and am surprised to find I can't move. Looking down, my feet are encased in ice up to above my ankles. Another wave of anxiety. What's happening? Then I figure it out: the power plant in my space suit is generating enough heat to melt the ice under my feet and that water is seeping upward and freezing. Slowly and steadily. In the time I've been watching I've settled another few millimeters into the ice. I twist and turn and try to raise one leg with all my strength, but it's useless. I'm held fast by the ice.

Time stretches out as I sink slowly into the blue-white mass around me. I've given up struggling. There's no one to help me and nothing to do. As the ice level creeps up over the visor of my helmet, I console myself with the thought that this is a dream within a dream. Or so I want to believe.

I'm entombed in Europa's icy grip. She's like the Inuit goddess Sedna, so voracious that her father cast her overboard and chopped her fingers off as she tried to climb back into the boat. She's a cold and unforgiving lover. My only hope to avoid a frozen sarcophagus is to keep sinking through ice until I emerge into a subterranean ocean. I try to imagine what I will find there, but my mind is blank. That metallic taste in my mouth is fear. With an act of will I set it aside and concentrate on my slow slide into the abyss.

3

DISTANT WORLDS

THESE SHADOWS WILL TAKE SOME GETTING USED TO. *I hold my hand out in front of me. It casts a strong shadow downward from the red star overhead and two weaker shadows to the side from a pair of pale yellow stars on the opposite horizon. I wave my hand around. The silhouettes dance. My hand is a tentacled sea monster.*

The air is breathable, with a slightly sour tinge. Ozone. I'm standing in a corrugated landscape of ridges and valleys, with no vegetation. The star overhead is dull and red, like a darkroom light, and it removes all sense of perspective or distance. The twin stars low in the sky add a pale overtone to the scene. They're somewhat fainter and dimmer than the red star. Setting behind a jagged mountain range in the distance they send shadows like dark bony fingers in my direction.

I climb up a steady incline toward a promontory. There's a rubble of small stones underfoot, 6 inches deep, and they scatter in all directions as my boots hit them. The grade isn't steep but I'm quickly winded. Gravity is stronger than I'm used to. My legs feel thick and leaden; my heart thumps hard. I reach a point where I can look out over the terrain. With no trees or vegetation, there's nothing to give a sense of scale. I see no water or liquid of any kind. It's barren and forbidding. A shiver passes through me.

Everything's off dead center and not quite right. The light, the gravity, the air. After a lifetime of breathing Earth's air, I take it for granted. But it's sweet and fresh compared to what my lungs are sucking in now. Suddenly I feel quite alone and very far from home.

—

TOUCHING THE STARS

What would it take to touch the stars? For the past 50 years we've been puttering around in our own backyard, sending humans to the nearest rock and robotic probes to most of the significant objects in the Solar System. To reach even the nearest star means crossing a breathtaking gulf of space.

Let's go back to the model where the Solar System fits in your living room. That reduces real distances by a factor of a trillion. The Sun is roughly a millimeter across and the planets are dust motes scattered through the room, unnoticed since you haven't vacuumed for a while. On this scale, the nearest star is 25 miles away. We imagine in a nearby town there's another living room with a tiny dot of light in it and a similar set of orbiting dust motes. Back in the real world, that 25-trillion-mile chasm to the nearest star mocks the primitive chemical rockets of our nascent space technology.

We can also think of the model in terms of light travel time. Shrinking space by a factor of a trillion also shrinks light speed by a factor of a trillion, making its motion visible, or at least imaginable. From the dust-grain-sized Sun, light moves outward at an inch a minute, taking eight minutes to reach the Earth and several hours to reach the most distant planets. Meanwhile, light from the nearest star is approaching at the same glacial pace. It takes four years to creep down the road from the neighboring town, being overtaken along the way by ants and slugs.

We see stars as they were, not as they are.

Creeping light in the scale model is the fastest way information can travel. Light trickles in moving as fast as its little legs can carry it. In the real universe, light moves swiftly but the gulfs of space are so vast that it takes measurable time to reach us. The far reaches of the Solar System are seen in hours-old light, the nearest stars are seen in years-old light, and the faintest stars in the night sky are seen as they were centuries ago. If light traveled with infinite speed, we'd see the entire universe as it is now. Everything everywhere would happen in parallel, a cacophony of events clamoring for our attention. Instead, we view remote space as it was, with concentric spheres of increasing distance representing older and older information. You can ask what the bright star Sirius is doing *right now*, but we can't know—that question can't be answered. We'll have to wait nine years for the light to reach us. Someone right next to Sirius might know, but to tell us they'd have to send a signal and that would take nine years to get to us too. We're just stuck with musty information.

If you were the ruler of a European empire in medieval times, the only way you could learn about your domain would be by messages carried by horsemen. The horses are your photons and their speed is limited (and is affected by how much time the riders spend in taverns along the way). It might take several days to hear about nearby provinces, and weeks to hear about more distant lands. If there were a rebellion at the edge of your far-flung empire, you'd get the information a month late, helpless to act any quicker.

Let's see how astronomers came to realize that the stars are so far away that their light is old. Ancient cultures saw stars wheel overhead as if painted on a dome. They imagined stars couldn't be much farther away than the most distant countries on Earth. To the ancient Greek astronomer Ptolemy, the most extravagant distance imaginable for the stars was about a million miles.[1] Eratosthenes estimated Earth's girth as 25,000 miles, so the crystalline sphere of the stars was 40 times farther, but that still vastly underestimated the actual distance to the nearest star.

The vital conceptual leap was to realize that stars are like the Sun.

Greek thinkers Democritus and Epicurus had broached the idea that the stars were other suns, and the idea was further developed in the twelfth century by the Persian polymath Fakhr al-Din al-Razi.[2] Giordano Bruno argued eloquently for the stars as burning gas balls like the Sun, complete with attendant planets hosting life. In 1600, this was among the heresies that got him burned at the stake by the Catholic Church. Galileo added weight to the argument with the power of his simple telescope. His *Sidereus Nuncius* included engravings showing 30 previously unobserved stars in the Pleiades, and 80 in the Orion region, and he saw that the smooth light of the Milky Way was really composed of the pinpoint light of a myriad of stars. It's a factor of 400 from the brightest star in the sky to the faintest star visible with the naked eye, and Galileo expanded this to a factor of 10,000. If all the stars were intrinsically the same, that range would be a factor of 100 in distance. This sense of a third dimension of "depth" was one of the insights provided by the telescope.

In the mid-seventeenth century, Christian Huygens sharpened the argument by measuring the bright star Sirius to be 600 million times fainter than the Sun. Light dims by the inverse square of the distance traveled, so Sirius is therefore 25,000 times farther from Earth than the Sun. This logic is sound but the assumption that Sirius and the Sun are exactly the same is flawed. The brightest stars in the sky, including Sirius, are more massive than the Sun and intrinsically more luminous so we can see them at larger distances than we can a star like the

Sun. Because of this, Huygens seriously underestimated the distance to Sirius.[3]

It took over 200 years of refinement to the telescope before the distance to a star was measured by a direct geometric method. As we move in our annual orbit of the Sun, our perspective on stars changes. Stars closer to us shift relative to stars that are farther away. This effect is called parallax. If you're in a room with objects at different distances, try covering one eye and then the other. Objects closer to you in the room will shift more than objects farther away. Another familiar example is when you look out the window of a moving car and see nearby objects shift faster than the background scenery. The biggest change in perspective available to us is to make observations six months apart, from opposite ends of the Earth's orbit of the Sun, a separation of 200 million miles.

In the 1830s a number of astronomers were consumed in an intense competition to measure the first distance to a star.[4] Attention focused on the set of stars with "proper motions"; in other words, they moved by a noticeable amount across the sky from year to year. This was a good indication that they had rapid motions and so might be relatively close. Thomas James Henderson, the first Astronomer Royal for Scotland, made the first successful measurement. He saw a parallax of roughly an arc second for Alpha Centauri in 1833 but didn't publish due to concerns over the accuracy of his instrument. (No doubt he was also influenced by the long history of failed measurements and false alarms.) Five years later, Friedrich Wilhelm von Struve measured the parallax of Vega; von Struve was the second in a lineage of illustrious astronomers that spanned five generations. But both were beaten to the punch by Friedrich Bessel, who published his data in 1838. Bessel made major contributions to both astronomy and mathematics despite never having been to university.

Bessel measured a parallax of 0.314 arc seconds for 61 Cygni, which has the largest transverse or proper motion of any star visible to the naked eye. The angular shift is very subtle, and is equal to the angle between the opposite edges of a quarter at a distance of 7.5 miles.[5]

Parallax measurement is an example of triangulation, such as when a surveyor can measure the distance to a distant landmark by making sightings from two different perspectives. Given the huge distance to the stars compared to the Earth-Sun distance, the geometry is a long, very skinny triangle (Figure 3.1). The parallax method also depends on an accurate knowledge of the distance from the Earth to the Sun. The solar parallax—the angle made by lines connecting the center of the Sun to opposite sides of the Earth and used to derive the Earth-Sun distance—is 8.8 arc seconds.[6] Today, the Earth-Sun distance is known with very high precision using radar reflection and timing data to Mars and Venus, and then scaling with Kepler's laws of motion, which give the relationship between distance from the Sun and the time to orbit the Sun.

Astronomers have used satellites to measure the apex angles of those long skinny triangles made by the Earth's orbit out to several hundred thousand stars, and the distances are known to an accuracy

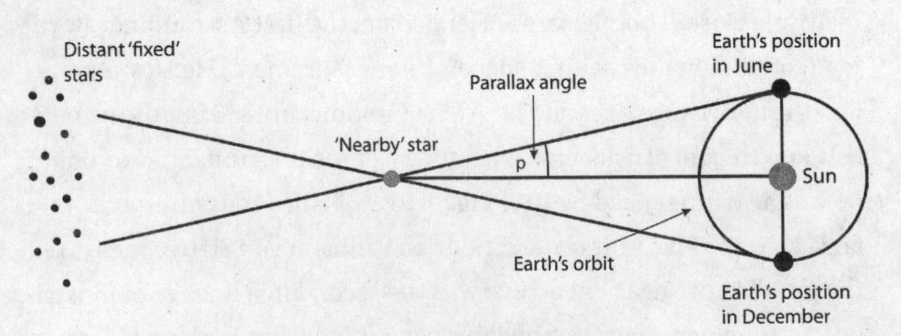

FIGURE 3.1. *Parallax is the angular shift of a nearby object referenced to a distant backdrop when observed from two different locations. In the stellar context, a nearby star will shift with respect to a more distant star when observed from opposite ends of the Earth's orbit of the Sun. In this diagram the parallax angle is exaggerated; in reality the nearby stars are hundreds of thousands of times the Earth-Sun distance.*

of a few percent. Just a handful of these stars are within a few light-years. Most are seen as they were a few centuries ago, and some are seen as they were at the time of the Crusades or Imperial Rome. Don't worry about the old light; stars lead quiet lives and probably aren't doing anything different now than they were then.

DISCOVERING DISTANT WORLDS

The logic sounds compelling. If the stars are fiery balls of gas like the Sun, and the Copernican principle that our situation isn't special holds, they should each be surrounded by a retinue of planets and moons.

The difficulty of testing this proposition was driven home by the great distance to even the nearest stars. Our Solar System has four small, rocky planets within 1.5 times the Earth's distance from the Sun and four gas giant planets ranging from 5 to 40 times the Earth-Sun distance. Neptune is 4 billion miles from the Sun, but Proxima Centauri is 25 trillion miles from the Sun, or 6000 times farther. And there's another obstacle: planets make no light of their own; they just reflect part of the glory of their parent star. Light escapes from a star into space in all directions and only a tiny fraction of that light will be intercepted by a planet billions of miles away. Then, a fraction of that tiny fraction is reflected back toward the remote observer. It doesn't leave us much to work with.

Let's use Jupiter as an example since it's by far the largest planet. If Jupiter can't be detected, the prospects are worse for other planets. Imagine the Sun as a 100-watt lightbulb. In this scale model, Jupiter would be a pale yellow-white marble 150 feet away. Earth in this same model is 5 times closer to the lightbulb but 10 times smaller—like a small bead. It would be hard enough to see the marble by the light it reflects from a lamp half a football field away. But if the lamp is now Proxima Centauri and the marble is now a hypothetical Jupiter in that system, both objects are 1500 miles away! A big telescope might be powerful enough to see a house light half a continent away, but seeing the light reflected by the little marble would be a formidable challenge. It reflects 1 part in 100 million of the light in our direction, so it is like a millionth-of-a-watt lightbulb.

This task seemed impossible, so astronomers used stealth instead.

Newton's law of gravity says that the forces exerted by two objects on each other are equal and opposite. It may seem like your daily battle with the Earth is unequal; the planet grips you relentlessly and some mornings it's difficult to get out of bed. But you pull on the Earth with just the same force as the Earth pulls on you. Any two objects in space orbit a common center of gravity. For objects of equal mass that's the midpoint between them. When the masses are unequal the center of gravity shifts toward the more massive object.

The same is true of solar systems. The planets don't orbit a stationary Sun; they tug it around the center of gravity for the whole system. So the Sun pirouettes about its edge like a corpulent ballet dancer, with almost the entire motion caused by the most massive planet, Jupiter. The wobble isn't detectable—that would be as difficult as measuring a lightbulb pirouetting around its edge 1500 miles away. However, astronomers had hopes of detecting the periodic Doppler shift caused when a star was tugged to and fro by a massive planet. Jupiter makes the Sun pirouette every 12 years (the period of Jupiter's orbit) at 13 meters per second, slower than the speed limit in most urban areas.[7]

Both the wobble and the Doppler shift can be understood by analogy with two objects attached to the ends of a meterstick. If the objects were equal mass, the meterstick could be balanced with a pushpin at the midpoint and spun into an orbit. Make the objects very unequal in mass and the balance point moves much closer to the end where the massive object is attached. Spinning the stick gives a fair facsimile of the situation when a star has a massive planet.

The Doppler method allows planets to be detected indirectly, by

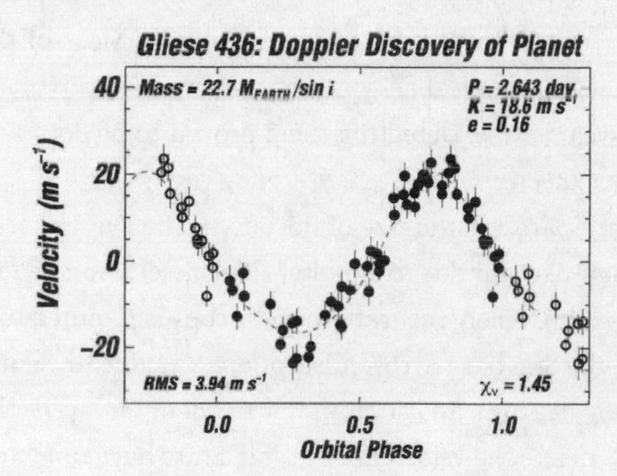

FIGURE 3.2. *The star Gliese 436 has a Neptune-mass planet that orbits in an incredibly rapid 2.6 days, tugging the star in a periodic variation shown in this graph. The Doppler shift has an amplitude of 19 meters per second and the planet's orbit is mildly eccentric. Gliese 436 is a red dwarf 30 light-years from Earth. Its planet was subsequently seen to have eclipses, which allowed astronomers to estimate radius and density. The planet is composed of roughly equal proportions of rock and water, with small amounts of hydrogen and helium.*

their effect on the parent star. The planet causes a sinusoidal variation in the star's velocity, with an amplitude that scales with the planet mass, and with a period equal to the period of the planet's orbit (Figure 3.2). If there are other planets, they add sinusoidal variations of their own, or harmonics, like echoes of Pythagoras. It's a modern version of the harmony of the spheres dreamt of by Kepler.

Look: a large man in a white suit twirls around the darkened dance floor. His partner is slight, and clad in black from head to toe. She's invisible. He leans away from her but the small dark ghost keeps him in perfect balance. We hear the strains of a waltz and see his gloved hands gently cradling and supporting nothing but air.

In 1995, after decades of fruitless searching, scientists detected planets orbiting other stars, or exoplanets, by using the Doppler method. Over 700 exoplanets are now known. Most of these new worlds are in our cosmic backyard, from a few dozen up to a few hundred light-years away. Detecting them proved to be both harder and easier than expected.[8]

The hard part was the size of the effect. The Doppler wavelength shift of a Sun-like star due to a Jupiter-like planet is less than 1 part in 10 million so extremely accurate spectroscopy is required to detect it. A star like the Sun has narrow absorption lines due to elements such as hydrogen, calcium, and sodium in its cool outer layers; these lines are used to define the velocity of the star. Stars have random motions in space that are 1000 times larger than the motion caused by a giant planet. A star's trajectory in space doesn't change, while a planet causes a small periodic variation, and it's this regularity that allows the planet to be detected. By the mid-1990s half a dozen research groups had developed the technical skills to detect Jupiters.[9]

The easy part was the time it took to detect remote planets. Based on our Solar System, astronomers expected to have to gather data for a decade or more—the orbital period—before being able to dig out the Doppler signal of a distant Jupiter. Amazingly, they found the first exoplanet with two weeks of data.

In 1995, Geoff Marcy and Paul Butler had been patiently accumulating data at Lick Observatory in Northern California for several years, using throwaway telescope time when the Moon was up, nights that nobody else wanted. They had painstakingly ferreted out all the imperfections and imprecisions in their spectrograph until it was the perfect tool for the job. They had heard skepticism and worse from their colleagues, and they knew the road to exoplanets was littered with failed claims and tattered reputations. The thrill of the chase sustained them. They were patient because they thought the giant planets they were hunting had leisurely orbits. Few scientists get to answer an age-old question about our specialness in the universe.

Meanwhile, Michel Mayor and Didier Queloz were using a small scope at Geneva Observatory to study binary stars and they extended their methods to detect giant planets. Binary stars orbit each other swiftly so Mayor and Queloz gathered data every night. When they reduced data for the single star 51 Pegasi, an anonymous Sun-like star in the constellation Cygnus, they were stunned to see the Doppler signature of a planet more massive than Jupiter. This super-Jupiter wasn't in a slow orbit remote from its star; it was in a blistering four-day orbit, scorched by the nearby star and at a temperature of 1200°C. Nobody had expected or predicted giant planets so close to their stars. Marcy and Butler had data for 51 Pegasi sitting on their computer but hadn't analyzed it since they assumed it would take years to nail the orbit of a giant planet. They quickly confirmed the discovery, but if a Nobel Prize is awarded it will probably go to the Swiss pair.[10]

The surprises kept coming. Over the next decade, the leading teams trawled the sky for exoplanets and the haul grew to several hundred. They improved the sensitivity of the Doppler method until they could detect planets like Neptune and Uranus. However, gas giants continued to turn up that were on much closer orbits than gas giants in the Solar System. Also, most exoplanets travel on more elliptical orbits than the familiar planets. Their properties are mysterious and puzzling.

The Doppler method is still the workhorse for exoplanet detection, but a second method has grown in importance over the past decade. Distant planetary systems have random orientations, but if they're aligned so that we're in the orbital plane, we can detect an eclipse or a transit as the planet crosses the face of its star. The alignment has to be almost perfect, so a small fraction of all exoplanets will transit, but the dip in light intensity is 1 percent for a Jupiter-mass planet, which is easy to detect with a small telescope. The first transit was detected in 1999 and since then more than 200 exoplanets have had transits observed by ground-based telescopes.[11]

Transits can only be seen for a small subset of exoplanets but they're important because they reveal size. Doppler detection gives the mass and orbital period of a planet and nothing else. In a transit the light of the parent star dips by a fraction given by the ratio of exoplanet area to star area. Patience is required because a transit lasts for a tiny part of the orbit—a few hours every 12 years for Jupiter transiting the Sun. Imagine staring at a lightbulb for a decade and trying to detect the few hours when it dips from 100 to 99 watts. Size plus mass gives density so the Doppler and transit methods together reveal whether a planet is big and gassy, like Jupiter, or small and rocky, like the Earth.

Look again: the man in the white suit rests after his exertions on the dance floor. He slumps in a chair and mops his brow. His partner has gone, but as she left a feather became detached from her boa and it flutters on the evening breeze. It casts a tiny shadow on his white suit. Stirring from his reverie, he smiles at the reminder of her.

What about the most obvious detection method: imaging? After all, seeing is believing.

The difficulty of digging out the feeble reflected light of an exoplanet from the edge of a vastly brighter star image gave rise to many false alarms and failed detections. Imaging did not succeed until 2007 (Figure 3.3), and just a few dozen of the hundreds of known exoplanets have ever been seen directly. Exoplanet imaging will become more routine as large telescopes get fitted with adaptive optics, in which the shape of the secondary mirror is rapidly modified to compensate for the jumbling of incoming light by the atmosphere. This turbulence limits the sharpness of images made with ground-based telescopes and so it bleeds starlight into the nearby and much fainter reflected light from the planet.

Imaging exoplanets is critical to the search for life in the universe. If enough reflected light can be gathered to spread it into a spectrum, astronomers will be able to look for features that indicate the chemical composition of the planet's atmosphere. No surprises are expected for giant planets, which will be dominated by hydrogen and helium.

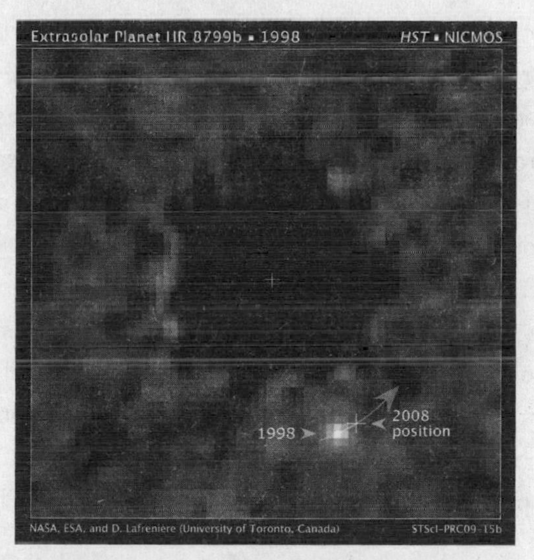

FIGURE 3.3. *The exoplanet HR 8799b was discovered with images from the Keck and Gemini North telescopes taken in 2007 and 2008, but it's also seen in this infrared image taken with the Hubble Space Telescope in 1998. New image processing techniques led to its discovery in the older data. The planet is 100,000 times fainter than the star, which has been removed in this image. It has a period of 400 years and is one of three orbiting the star, which is 130 light-years away.*

Small, rocky planets might show signs of carbon dioxide, nitrogen, and water, but there will be banner newspaper headlines if oxygen or ozone is detected. Those gases are volatile; they react with rocks and metals and quickly disappear. On Earth, they are replenished as a result of biology. This may be the way we learn we're not alone in the universe.

HOW TO MAKE SOLAR SYSTEMS

It seems simple enough. Take a large, diffuse cloud of gas and dust. Give it a nudge. Watch as it collapses by gravity. It has some spin so that spin gets amplified as it shrinks, and the result is a much smaller and denser rotating disk. At the center of the cloud a new star ignites. Grit particles in the disk collide and grow, from dust grains to planets. Radiation from the young star drives out leftover gas close in, leaving rocky planets at small distances and rocky cores that can attract large gassy envelopes farther out. Voilà.

This picture is complicated by collisions, and interactions between the planets that can rearrange their orbits, but it seems to hang together. However, with only one solar system to study it might be a just-so story we've invented to account for the quirks of our neighborhood. With more than 700 exoplanets to learn from, including over 100 in multiple planet systems, the goal is a general theory of how planets form.

The first problem is explaining hot Jupiters. It's almost certain that they didn't form where they are now. Even if a rocky core could form very close to a star it couldn't accrete a big envelope, because there's little gas available, and any gas it did manage to gather would be blasted into space by the heat of the young star. Orbits that deviate by 20 to 60 percent from circles are hard to understand because after the collisions and mayhem of planet-building orbits should become boring and circular (Figure 3.4). Elongated orbits tend to be unstable and so they shouldn't persist.

Theorists were left with egg on their faces by their failure to predict these properties, and they scrambled to come up with explanations. Observers were exhilarated by the pace of discovery but had plenty of questions of their own. Are there any gas giants on more "normal" or familiar orbits? Do all Sun-like stars have planets, or only a fraction? How many exoplanets does a typical system have? And the biggest question of all: are there any Earth clones out there?

A new ingredient in solar system formation is migration. In the early phase of formation, many of the first small planetary embryos interact with gas in the disk and viscosity or drag sends them spiraling

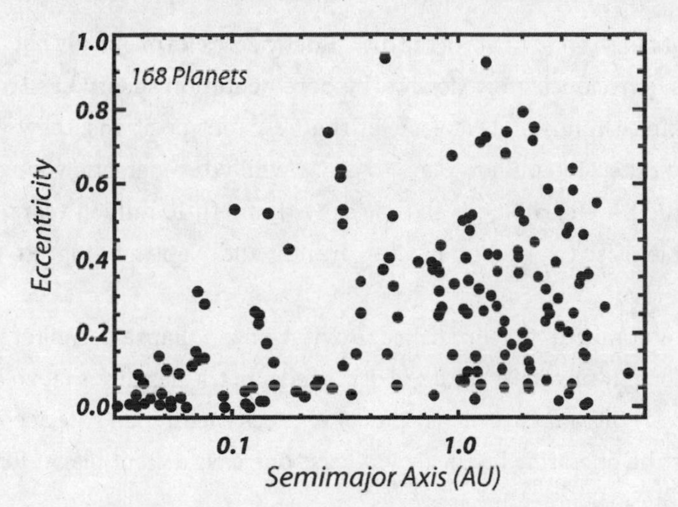

FIGURE 3.4. *Orbital eccentricity, or the fraction by which the orbit deviates from a circle, plotted against half the size of the orbit (in units of Earth-Sun distance) for about a third of the known exoplanets. Planets in our Solar System generally have eccentricities less than 0.1, deviating from a circle by less than 10 percent. For small radii the orbits are tidally locked and so nearly circular. But at large radii the prevalence of elliptical orbits may indicate instability.*

into the star. After the gas has been mopped up or has dissipated, an embryo tends to stay at the distance where it formed. Those few that migrate inward move in a gradual spiral. When a migrating planet gets close to the star, gravitational interaction makes it "park," or stop moving any closer, and "lock," or orbit with the same side facing the star. This is a neat idea, but recent discoveries show that migration can't be the only mechanism involved in the formation of hot Jupiters.[12]

Being wrong-footed by nature was good news for theorists of planet-formation, because it liberated them to try out different ideas.

The standard theory of planet formation is called core accretion. Rocky cores are formed steadily by collisions to mop up most material in the disk and those cores draw in hydrogen and helium to make gas giants in the outer part of the disk. Rocky cores form quickly but large, gassy envelopes grow slowly. The core accretion theory has trouble making Uranus and Neptune at their current positions; they would have taken 100 million years to grow, while the energetic young Sun should have cleared out all the gas within 3 to 10 million years. This problem isn't solved by having Uranus and Neptune migrate from closer in.

A competing theory called gravitational collapse has giant planets form by instabilities or gravity "seeds" in the gas disk that collapse directly into massive planets. A faster mechanism than core accretion must be operating because in at least one case a giant planet formed in only a few million years.[13]

Recent observations have added a third idea to the mix.[14] In 2009, a massive planet, 5 to 10 times the mass of Jupiter, was found orbiting a brown dwarf. A brown dwarf is a substellar object, too puny and cool for fusion to take place. This particular brown dwarf was about a million years old, much too young for such a planet to have formed by core accretion. And the planet was too massive to easily have formed by gravitational collapse. It's likely it formed from its own collapsing cloud of gas and dust, and then fell into a binary system with the nearby brown dwarf. So planets may not always form as contingent outcomes of star formation. They might form as isolated objects with no host star. Confusion and complexity reign in the subject of planet formation. More data aren't always leading to more clarity.

Why is it so hard to say how our Solar System or any of the exoplanet systems formed? Equivalently, why is it so hard to predict properties of planets given that they formed from such simple initial conditions? Nonlinear gravitational dynamics. The phrase is a conversation-killer when people who ply that trade say what they do at a party. But it can be reduced to a less intimidating concept: chaos.

Chaos in science isn't the same as chaos in your kitchen or chaos in a crowd of people. In a dynamical system, chaos doesn't mean disorder.

It's regular behavior mixed with randomness, and extreme sensitivity to initial conditions. The ancient Greek philosophers thought that the universe was in a balance between cosmos and chaos. Both are Greek words; order and harmony were the pleasing aspects of the cosmos but they also were aware of unpredictable behavior. Isaac Newton seemed to resolve the tension between determinism and randomness in favor of determinism. His universal law of gravity led to the metaphor of a "clockwork universe."

But this conclusion was premature. Newton was only able to solve the laws of motion exactly for a system containing two bodies, like the Sun and the Earth, or the Earth and the Moon. With three or more objects, the solutions are just approximations. This left unanswered the question of the stability of the Solar System. Will the planets stay in their current orbits, or will the cumulative effect of small changes build up over time, causing a planet, perhaps the Earth, to crash into the Sun or leave the Solar System entirely? The answer was of more than academic interest, so in the late nineteenth century, King Oscar II of Sweden, who was an amateur mathematician, sponsored a prize of 2500 krona to anyone who could solve the simple situation of three gravitating bodies.

A young French mathematician named Henri Poincaré accepted the challenge, and while he failed to solve the three-body problem, his insights led to a deeper understanding of gravity and he was awarded the prize anyway. Poincaré explored the behavior of planetary orbits and discovered wild variations in three-body systems. Sometimes the orbits were regular and periodic and in other situations the orbits were complex and unrepeatable.[15] He also saw that the outcomes were very sensitive to the starting configuration; if the initial conditions changed just a little bit, the situation after many orbits might be unrecognizably different. One of the triggers of chaos was resonance, when the orbital or spin periods of two bodies were related by a simple numerical ratio.

Chaos entered the public consciousness with meteorologist

Edward Lorentz. In the 1960s he used early computers to solve apparently simple equations for convection, a process that's very important in weather systems. Lorentz saw Byzantine behaviors in the outcomes and realized that the weather might not be predictable. Sensitivity to initial conditions became known as the "butterfly effect." Chaos was seen in the behavior of things as simple as a pendulum or a dripping faucet or a rising tendril of smoke, and researchers have applied it to everything from the beating of a heart to the fluctuations of the stock market.

Chaos is a fundamental property of planetary systems.[16] This doesn't mean everything is unpredictable; the orbits of most of the planets in the Solar System will be stable for billions of years. But it does mean that "rewinding the clock" on a planetary system to deduce the events that occurred or the initial conditions is unreliable. We can never see exactly "how it began."

HOME AWAY FROM HOME

In the years following the discovery of exoplanets, researchers refined their methods and pushed the limit of detection toward lower masses (Figure 3.5). The first exoplanets were Jupiter-mass or larger, but soon planets like Uranus and Neptune were being retrieved. Jupiter, Saturn, Uranus, and Neptune are 318, 95, 15, and 17 times, respectively, more massive than the Earth. This whetted everyone's appetite for finding terrestrial planets, and in particular planets like the Earth.

By 2010, astronomers had found 30 exoplanets less than 10 times the Earth's mass by using the Doppler method. So far, the smallest confirmed exoplanet weighs in at just under twice the mass of the Earth. Astronomers expect there to be many terrestrial planets orbiting stars like the Sun because the mass distribution of exoplanets rises steeply toward the limit of detection. Suppose you're catching fish using a net

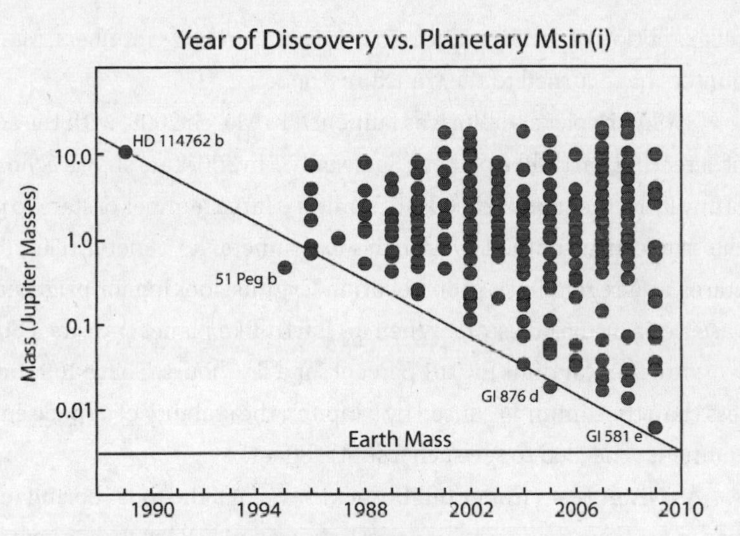

FIGURE 3.5. *The mass of known exoplanets as a function of the year of discovery, showing steady progress toward Earth-like objects. Most of the exoplanets were discovered using the Doppler technique, but many have been followed up with a detection of transits. The Kepler satellite was designed to detect low-mass planets using the transit or "eclipse" method and it has succeeded in detecting dozens of Earth-like planets.*

with 1-inch holes. If you inspect your net and find a few fish longer than a foot, a bunch slightly smaller than a foot long, and lots ranging from an inch up to a couple of inches long, you shouldn't conclude that fish smaller than an inch don't exist. Your best conclusion is that they exist in large numbers but you need a net with smaller holes to catch them.

Astronomers are trying hard to make a better net. Both pioneering groups in exoplanet detection are working toward detecting a Doppler shift of only 0.5 meter per second. This is about 1 mph and sensitive enough to measure the tiny tug caused by an Earth-like planet orbiting a Sun-like star. At some level, turbulent motions in the atmosphere of the host star may impose a limit on the mass that's detectable with a Doppler measurement, since those motions will be confused with the reflex motion of the exoplanet. To bag earths in large numbers, planet hunters have turned to the transit method.

NASA's Kepler satellite was launched in March 2009, with the goal of detecting Earth-like planets. Since only 1 in 300 stars shows eclipses of any kind, the satellite needs to monitor a large number of stars to get the true census of the lowest-mass exoplanets. Its 1-meter telescope stares at a region in the constellation Cygnus, looking for brightness variations in 156,000 stars. When an Earth-like planet transits a Sun-like star, the star dims by 0.01 percent for a few hours. That's 100 times less than the dimming caused by a Jupiter; the stability of a space environment is needed to see such a subtle effect.

As NASA team leader Bill Borucki says, "It's the most boring mission ever, doing the same thing every six seconds." Borucki has kept his eye on the prize for decades, enduring skepticism through the years when exoplanet detection was considered a pipe dream. His calm and persistent demeanor kept him going when NASA shelved the mission or sent him back to the drawing board four times. Just out of college, he was so sure he wanted to explore space that he only applied for one job, at NASA. He had published a proof of concept of the Kepler mission years before exoplanets were detected. History has finally caught up with him and this mild-mannered pioneer now runs one of the most exciting missions in NASA's portfolio.

Kepler has dramatically advanced the planet-hunting game. In 2010 it released its first batch of candidates and a small number of confirmed planets (Figure 3.6). *Kepler*'s data release in early 2011 included 1235 exoplanet candidates, most of which are expected to be confirmed by future data.[17] The haul includes a stunning 68 Earth-sized planets, 288 a few times the size of Earth, and 662 the size of Neptune. Of the new planet candidates, 54 are in the habitable zones of their stars, and five of those are Earth-like.

In less than a year the satellite has trebled a haul of exoplanets that took nearly 15 years to acquire and increased the number of earths

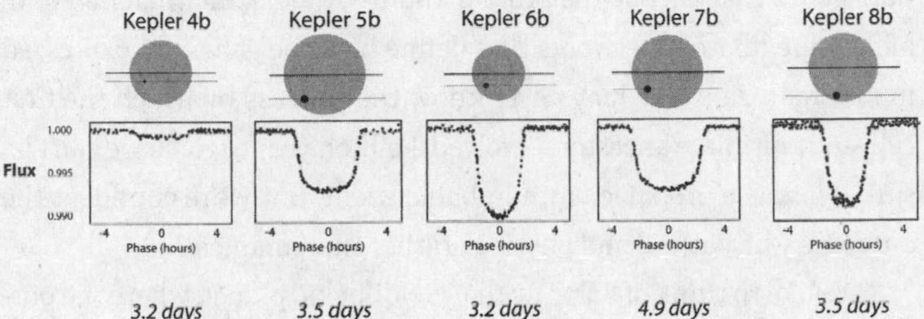

FIGURE 3.6. *The light curves of the first five exoplanet discoveries with the Kepler satellite show that it can easily detect the 1 percent eclipse depth typical of a Jupiter-like planet. The data quality is so high that eclipses of only 0.01 percent depth can be detected, enough to detect an Earth like planet. Most of the exoplanets announced by the Kepler team so far are "candidates," where the transit hasn't yet repeated in a way that would confirm the planet. It will take several years to reliably confirm an Earth-like orbit of its star.*

from zero to 68. *Kepler*'s early data is only sensitive to periods of a few days up to a few weeks. It will be two more years before *Kepler* can detect Earth-like planets on Earth-like orbits. The slightly larger "super-Earths" might also host water, and computer simulations of planet formation suggest that terrestrial planets routinely acquire as much water as in all the Earth's oceans.

Which leads to—pregnant pause—the matter of life. Life is matter, and that's the strangeness of it. Modern biology has left very little room for vitalism or a "ghost in the machine."[18] Many of the steps by which simple, and universally available, chemical ingredients assembled themselves into the first primitive cell have been studied in the lab. The revolution in modern biology is the realization that life is a branch of digital information. A four-letter chemical alphabet forms the genetic code, and the 3 billion letters in our genome are sufficient to imbue us with the profound and subtle attributes of intelligence and sentience. The genome codes two distinct types of digital information: genes that encode the protein and RNA molecular machinery of life, and regulatory networks that define how the genes are expressed in an organism.[19] We may never know the process by which the first cell—our ultimate ancestor—evolved 4 billion years ago, so we can't be sure if it was a probable or improbable event. But we're confident the outcome was natural and physical, rather than magical.

We like to think of Earth as the "Goldilocks" planet, where the porridge is a perfect temperature, and the music is to our liking, and the pillow is plumped up with a mint on it. But only humans and other large animals have such a fussy set of requirements for life. We share the planet with unfamiliar microbes that can tolerate temperatures above the boiling point of water and below its freezing point, pH values that range from drain cleaner to battery acid, pressures hundreds of times higher and lower than ground level air, and toxic chemicals at levels that would be fatal to us. Extremophiles, as they are known, indicate that the physical bounds on our form of biology are very broad.

It would violate the Copernican principle to presume that life is unique to this planet, given its quick emergence and widespread distribution here, and the ubiquity of its ingredients beyond. A minimal set of preconditions for life would be organic material, a local source of energy and water, and sufficient time for chemical reactions to transition into an organism governed by natural selection. These conditions probably exist in a dozen locations in the Solar System and as many as a billion spots in the Milky Way galaxy. Venturing into the realm of stars, the closest is on our doorstep.

It may be a voyage we could take.

The Sun's closest neighbors are members of a triple-star system in the southeast corner of the constellation Centaurus. Proxima Centauri is a cool red dwarf too dim to be seen by the naked eye, 4.2 light-years distant. Alpha Centauri A and B are 4.4 light-years distant and so close to each other that they weren't recognized as a binary until 1752. Their separation is 24 times the Earth-Sun distance, large enough for terrestrial planets to be unaffected by the binary orbit. The red dwarf has a very slender habitable zone and emits flares that might be hazardous to biology but A and B could host Earth-like planets.[20] Three research groups are on the hunt for planets using the Doppler method, with results expected in a couple of years.

The Earth clones that *Kepler* hopes to find will be dozens or hundreds of light-years away. If the Alpha Centauri system has one or more Earth-like planets, we already have the technology to suppress the starlight and look for biomarkers in the spectrum of the planet light. This might lead to evidence of a photosynthetic metabolism at work, evidence of microbes. Intelligent life would be hard to detect—Earth seen from this large a distance would give no hint of our cities and factories. And then?

Then we would want to go. The simplest way is to send an unmanned probe. *Voyager 1*, which was based on 1970s technology, would take 80,000 years to get there if it was headed in the right direction. Now we could do much better using miniaturization to reduce the energy requirements. A nanobot traveling at 10 percent the speed of light could make the trip in less than 50 years. In practice, the best strategy is to send a fleet of nanobots in waves, for redundancy, and so the advance probes could beam data and images back to Earth, boosting the signal at each wave. In this way the signal can return efficiently, analogous to firemen passing water in a bucket line, without suffering the dilution a signal sent from that distance would normally be subject to. A more ambitious approach is to send people. Science fiction writers and space visionaries have talked about vast "worldships" with multigenerational colonies living and dying onboard.[21] That idea may take centuries or millennia to execute. Voyagers with less patience may choose to strap themselves into a cosmic-ray hardened metal coffin, enter suspended animation, and be propelled into the unknown.

What would we find? In the movie *Avatar*, James Cameron fleshed out the imagined world of Pandora, a moon orbiting a gas giant near Alpha Centauri A. Both stars in the binary system are 5.7 billion years old, so life there could have a billion-year head start on us. They could be to us as we are to bacteria on Earth. Our imaginations may not be fertile enough to imagine the possibilities of life beyond Earth.

I seek comfort in the night sky and find it in familiar constellations. Sirius is blazing bright and high in the sky, but it seems too close to Orion. Nearby, I locate Procyon in Canis Minor and it too is disconcertingly far from its usual place, sparkling near the foot of one of the Gemini twins. Farther south, I see the summer triangle and Altair is also dislodged, from the head of the eagle to the tail of the swan. Everything is recognizable but skewed, as if distorted in a funhouse mirror.

Suddenly it makes sense. The dim star that cast its vermilion glow on the landscape must be Proxima Centauri. And the double star must be Alpha Centauri, a binary in the same system. I'm trillions of miles from home.

Like a parent searching for a lost child, I scour the star fields. Finally I spot a bright star that doesn't belong. The Sun. I close my eyes and with an act of will still my tumbling thoughts and try to break the spell that brought me to this alien world. Eyes still closed, the world comes into focus. I'm in a car, driving to work. But it's a car I sold several years ago, and a glance at the headlines of the newspaper on the passenger seat confirms my fear. It's my world, but it's my world as it was four years ago. The time it would take for light to streak across the void to Proxima Centauri. It seems impossible but I've stepped into the same part of a moving river for a second time. And I'm powerless to alter the events that will play out in that seam of time.

I open my eyes. The starlit scene is unchanged and obdurately weird. Then, I catch movement in my peripheral vision and look down. The small pebbles and stones underfoot are moving slightly. At first I think it's a trick of the dim light, but when I squat down I confirm my initial impression: the ground is in motion. Not just here, but 10 yards away, and as I walk farther, 100 yards away, and even a mile away.

Tiny wormlike creatures are working in concert to shift the soil for some unknown purpose. I pick one up and it squirms on my hand, blind and translucent. The static landscape is an illusion. This world is seething with life but it's an ecosystem of one. I'm surrounded by geology and biology on an epic scale. What form of sentience could be so small and helpless and yet sculpt a planet?

4

STELLAR NURSERY

IT'S BREATHTAKINGLY BEAUTIFUL. *Everywhere I look there are curtains and filaments of pale light. They pulse like glowworms mounted on black velvet. The nebulosity extends above my head and below my feet; I'm enclosed as if in a womb. But with no gravity I feel no sense of up or down. For a moment the disorientation sends a rising wave of panic through me. I flail my arms, with no effect. Then I relax. The vast hidden spider of the night can do what it wishes with me; I'm trapped in a diaphanous web of electric light.*

Now I notice the subtle colors. Their meaning comes back to me. Pink is the soft hue of hydrogen, the most abundant element. That green tinge is oxygen, etched into shells of hot gas around the youngest stars in the nebula. There's not enough to breathe, and no metal around for it to rust. Sulfur is pulsing warm orange. I anticipate its noxious odor but it's too thinly spread to detect. I can see veins of intense red from neon—the stars gaudily advertising their birth.

To be close to the familiar can make it unfamiliar, but I realize that I am in the lap of Orion, the great hunter of Homer's Odyssey. *Ahead of me is the nebula, with a trapezium of bright stars nestled at its center, swaddled infants surrounded by afterbirth. Holding out my hand, I see my body casts a shadow. Looking back over my shoulder, the blue-white supergiant Alnilam is so intense I can't look at it directly. In the distance beyond, halfway back to the Earth, Rigel and Betelgeuse stand like sentinels. I look nervously for attack ships, but see none.*

—

COSMIC CAULDRON

Since prehistoric time, people have understood that the Sun sustained life on Earth, but they could scarcely imagine the source of the energy. The Greek philosopher Anaxagoras was born in 499 BC to a wealthy family.[1] As a young man he moved to Athens as it became the seat of the emerging Greek empire, and he's thought to have taught Pericles, Euripides, and Socrates. He made a decisive break with ancient ways of thinking by claiming that the Sun was a red-hot stone larger than the Peloponnese peninsula. He imagined that the stars were similarly incandescent objects so remote that we didn't feel their heat.

This bold thinking got Anaxagoras in hot water. He was charged with impiety for daring to suggest that the Sun was a substantial physical object rather than a god, and he was sentenced to death. Luckily, his influential patron Pericles intervened and got the sentence commuted. Anaxagoras died in exile, but his clever speculations put in motion ways of thinking about nature that were the precursors for modern science.

A true sense of the prodigious energy released by the Sun requires us to know its distance. In the first century AD, Ptolemy gave the Sun's distance as 1200 times the Earth's radius, which underestimates the true value by a factor of 20. The Sun's distance and size were not accurately measured until 1672, when Giovanni Cassini and his young assistant Jean Richter made simultaneous observations of Mars from Paris and French Guiana. By measuring the parallax they calculated the distance from the Earth to Mars, and by applying Kepler's laws, the distance from the Earth to the Sun. The Sun is 1 million miles across and 100 million miles away.

Put an ice cube on the ground on a mild spring day and it will melt in about 40 minutes. That would be the fate of ice at any point in space 100 million miles from the Sun. So we can imagine a 1-inch-thick sphere of ice 200 million miles in diameter and it would be completely melted by the Sun in the same time. Shrinking that shell down to the surface of the Sun, it would become 1500 feet thick and 10,000 times the area of the Earth, like a vast, impenetrable glacier, and the Sun would still melt through it in 40 minutes!

The Sun doesn't vary noticeably in brightness from year to year or from generation to generation. A million earths can fit inside it. What could provide so much energy? In the mid-nineteenth century, the obvious answer was coal or gas or oil, the chemical fuels that were driving the Industrial Revolution. It seems bizarre to imagine the Sun as a million-mile-wide sphere of sloshing petroleum or lump of coal, but scientists took the idea seriously enough to do the calculations. The result: even if it were burning in an atmosphere of pure oxygen, a Sun composed of fossil fuel would snuff out in just 5000 years.

William Thomson threw himself into the challenge of explaining the energy source of the Sun. He was the most illustrious scientist of his time, the author of over 600 papers, and a prolific inventor. Elected as president of the Royal Society in London five times, he had also been knighted by Queen Victoria and then given the title Lord Kelvin, the version of his name used to brand the "absolute" temperature scale.[2] Kelvin extended earlier work by the German physicist Hermann von Helmholtz, who recognized that gravity was a source of heat. After all, a contracting cloud of gas and dust had led to the creation of the Sun. What was more natural than imagining that the Sun's energy came from its continuing gravitational contraction? Kelvin calculated how a shrinking Sun would convert gravity into heat. To match the rate at which energy reached the Earth, the Sun had to be collapsing at a rate that would reduce it to a point in only 30 million years. Since we don't see any evidence of the "Incredible Shrinking Sun," that number must be an upper bound to the age of the Sun.

It seemed—to put it mildly—a serious problem.

Not long before, Charles Darwin had written *On the Origin of Species*, where he estimated that hundreds of millions or billions of years were needed to evolve the observed diversity of species from a simple early ancestor. Darwin also deduced how long it would take for erosion to wash away the valley between the North and South Downs in southern England; he came up with 300 million years. And geologists agreed that the layering of the Earth spoke of a very ancient planet.[3] Kelvin contributed to the problem as well, by estimating that it would take the initially molten Earth 200 million years to cool to its current temperature. Darwin was so unnerved by the age controversy that he removed all mention of times scales from later editions of his book.

It took several decades for the issue to be resolved. At the turn of the twentieth century, physicists realized that there was an energy source more powerful than chemical burning or even the conversion of gravitational energy into heat. Radioactivity proved that atoms were not inviolable and could release phenomenal energy. Ernest Rutherford, who figured out that atoms were mostly empty space with the mass concentrated in a tiny fortress-like nucleus, calculated the energy released by heavy radioactive elements in normal rock. It was enough to greatly extend the age estimate of a cooling Earth, to billions of years.

A second piece of the puzzle appeared in 1905, when Albert Einstein derived the famous equation $E = mc^2$ as a consequence of his special theory of relativity. The speed of light squared is a vast number so his equation showed that a miniscule amount of mass could be converted into a large amount of energy. At the time, nobody saw how this insight was related to the Sun.

Then, in 1920 physicist Francis Ashton made the stunning discovery that heavy elements are less than the sum of their parts. If you add 1-pound weights to a set of scales, the total increases by simple addition: two 1-pound weights add up to 2 pounds, three add up to 3 pounds, and so on. It's common sense. But atoms don't work that way. A helium nucleus weighs slightly less than the two protons and two neutrons it's composed of. In the arithmetic of the nucleus, one plus one plus one plus one is slightly less than four. That same year, the famous English astrophysicist Sir Arthur Eddington realized that the "missing" mass is lost as energy according to Einstein's equation. So the fusion of hydrogen into helium can produce energy efficiently enough to keep the Sun shining for tens of billions of years. He saw with foresight the great promise of fusion and also the dangers it posed for the future of humanity: "If, indeed, the subatomic energy is being freely used to maintain their great furnaces, it seems to bring a little nearer to fulfillment our dream of controlling this latent power for the well-being of the human race—or for its suicide."[4]

The final ingredient in the story: a dash of quantum strangeness. In classical physics, the positive electric charges of two protons form an insuperable barrier that stops them from merging. The atomic nucleus really is an impregnable fortress. But in quantum theory there is a finite chance that the protons will get close enough to fuse. All the details were worked out by Hans Bethe, working at Cornell just before World War II. Bethe was the acknowledged master of nuclear physics and his set of three papers on the subject is known as "Bethe's Bible." In the inner 2 percent of the Sun's volume, where the temperature is 15 million degrees, 4×10^{26} watts is generated by a three-step fusion reaction (Figure 4.1). Every second, the Sun converts 20 skyscrapers'

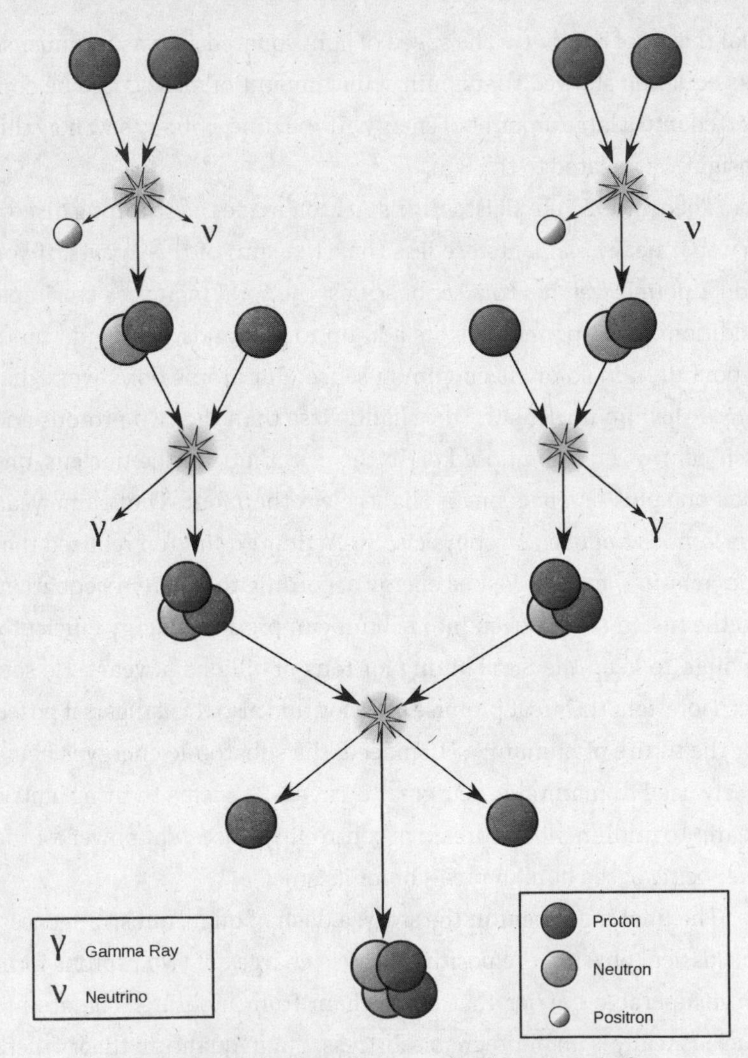

FIGURE 4.1. *The three-stage nuclear reaction that powers the Sun, called the proton-proton chain. In the first step* (top)*, two protons fuse to make deuterium, or "heavy" hydrogen. In the second step* (middle)*, a proton is fused with deuterium to make a light isotope of helium. In the last step* (bottom)*, two light helium nuclei merge to make a single normal helium nucleus, with two proton releases to facilitate further reactions. Energy is released at every step.*

worth of mass into radiant energy.[5] We don't see the cauldron where the gamma rays are created. These photons spend about 100,000 years in a drunkard's random walk, working their way out from the center while down-shifting in energy. Then, at the region we call the surface, the photons streak to the Earth in eight minutes.[6]

We take the Sun for granted, but it's actually a very strange beast. In the center, the state of matter is that of a high-temperature gas, yet the density is 150 times that of water. Overall, the energy production is equivalent to 100 billion megatons of TNT exploding every second. Yet the Sun isn't a bomb—it's thermostatically controlled and stable from year to year and millennium to millennium. At the center, the power generation is just 280 watts per cubic meter, the same as a compost heap. The Sun's prodigious energy isn't due to high power per unit volume; it's due to the huge overall size.

A STAR IS BORN

"We had the sky up there, all speckled with stars, and we used to lay on our backs and look up at them, and discuss about whether they was made, or only just happened." So said Mark Twain's Huck Finn, and he had a good point. When you look up at a sky full of stars, they're so artfully arranged and crisply delineated, it seems like aliens or a superior intelligence must have placed them there. In our Solar System, we've seen that a nudge from a nearby dying star may have been the trigger for a large and diffuse gas cloud to collapse into the Sun and eight orbiting planets. But how sure are we that this story applies generally? Physicists don't tell stories nearly as well as Mark Twain, but they're partial to a story about a spherical cow. It seems a dairy farm was having trouble with milk production, so they called in expertise from the local university. Unfortunately, a theoretical physicist was put in charge of the team. The team gathered data for weeks and the physicist assembled their data into a report. When the farmer received the report his heart fell as he read the first line, "Consider a spherical cow in a vacuum . . ."

The point: physicists or astrophysicists will often reduce a problem to its simplest form to make calculations more tractable even though the correspondence to reality may be greatly reduced. Often, it's the only way to make progress. Cows aren't spherical and neither are regions where stars form. Images from the Hubble Space Telescope show the nearest star formation regions to be chaotic tangles of gas and dust, and no amount of wishful thinking can conjure up anything remotely resembling a sphere. To see how far we are from a spherical cow let's peer deep into the Orion Nebula.

But first we need infrared vision.

In 1800, the astronomer William Herschel placed a thermometer in the different colors of a rainbow of dispersed sunlight and was surprised to notice that the temperature increase was greatest beyond the reddest red his eye could see.[7] The Sun's radiation creates power of a kilowatt per square meter at the Earth's surface, and just over half of that is at invisibly long wavelengths. Herschel showed that these "infrared rays" can be reflected, transmitted, and absorbed like visible light. It took a century before Ernest Fox Nichols, a physicist who served as president of Dartmouth and MIT, detected infrared radiation from another star. The new field of infrared astronomy developed slowly until the 1970s, when new detector technology allowed many astronomical objects to be detected in the infrared.

Just as night vision goggles "see" objects in the dark by their infrared radiation, so astronomers use infrared cameras to see regions too cool to glow in visible light.[8] Star-forming regions contain microscopic dust particles mixed with gas, and the dust is more efficient at scattering short wavelengths than long wavelengths of radiation (a similar effect explains why the sky is blue and the setting Sun is red). We see Orion through a smog of gas and dust. Only one in two waves of visible light can escape, making it look twice as faint as it would without the filmy gas. The deepest and darkest regions of the nebula are so opaque that only 1 in 1000 visible waves escapes; it's like looking through a fog bank. But infrared radiation is much less affected, allowing us to see deep into the nebula. Suitably endowed with 20/20 infrared vision, what would we see in Orion where stars are forming?

Let's take the actual trip imagined in the opening vignette. Approaching Orion, we're at the edge of one of the Milky Way's great spiral arms, so the density is slightly higher than average, about one atom per cubic centimeter. That's 100 million times sparser than the best manmade vacuum on Earth, which is in turn a trillion times less dense than the air on Earth at sea level. Even the busiest regions of space are pretty darn empty. We pass through a delicate veil of pale, glowing gas and enter the nebula, the forty-second object cataloged by the comet hunter Charles Messier. The entire star-forming region is 100 light-years across but we're homing in on a sparkling cluster of 2000 stars in the central 20 light-years. The temperature is a frosty 30 degrees above absolute zero.

We're now in a giant molecular cloud. The density is hundreds of times higher than when we started, but it's still an almost perfect vacuum. The higher density and extreme cold mean that fragile molecules can survive intact. Apart from hydrogen molecules, carbon dioxide, a noxious whiff of ammonia, and the silent killer—carbon monoxide—are present. In addition, over 130 different species of molecule occur at lower concentrations. The largest are buckyballs, discovered in 2010 and made of 60 carbon atoms. Other examples include formic acid, benzene, ethylene glycol, methane, the amino acid glycine, and the extremely nasty chemicals acetone and hydrogen cyanide (Figure 4.2). On the bright side, there's booze here—lots of it. The best estimate is 10^{28} shots of ethyl alcohol at 200 proof.[9] That's enough to keep everyone on the Earth drunk for a few trillion years. Just as well the bar is 1300 light-years away.

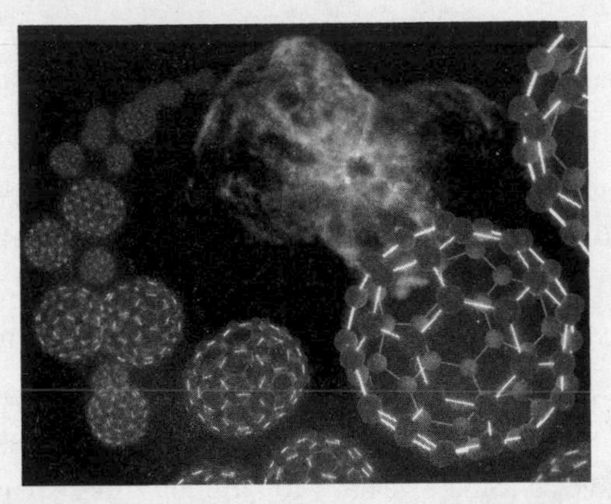

FIGURE 4.2. *In 2010, the Spitzer Space Telescope discovered buckyballs in space. These soccer-ball-shaped carbon molecules were found emerging from the region around a dying star. This image is a visualization of these large molecules, which are indicative of the complexity that can be found in cold interstellar space.*

As we head to the dense core of the molecular cloud, there's mayhem all around. Runaway stars that fled the nebula millions of year ago are moving at hundreds of thousands of miles per hour. Intense ultraviolet radiation from newborn stars has ripped the electrons off most of the hydrogen atoms. There's constant buffeting from all the turbulent gas. Supersonic "bullets" of gas 10 times the size of the Solar System are piercing the dense hydrogen clouds, their iron-rich tips glowing blue. Intense microwave emission is concentrated in narrow spectral lines—the telltale sign of masers. Humans didn't figure out how to engineer coherent radiation in the lab until 60 years ago; nature has been at it for billions of years.

Now the ride gets bumpy. As we pass deeper into the molecular cloud the density is millions of time denser than the region around the Sun, but still far thinner than the best terrestrial vacuum. The dust content gradually causes surrounding stars to fade from view. Before they go entirely, we notice small completely dark patches as black as ink, with ragged edges. They're dense and opaque regions a light-year across where a single star or a handful of stars are forming. The turbulence increases; it's like being inside the mother of all thunderstorms. Gas moves at supersonic speeds, threaded by invisible magnetic fields. In this huge cloud, with a mass of 10,000 suns, internal pressure can't support the weight, so it's collapsing in a gravitational free fall.[10] The cloud is fragmenting into hundreds of smaller regions, each one forming a plausible facsimile of a spherical cow that is both shrinking and rotating. They're stellar embryos.

We follow one of the collapsing knots of gas. At the center of the cloud rising temperature halts the collapse and shock-heats gas crashing onto the core. Even with infrared vision, we have trouble seeing through the murk. The core reaches 2000 Kelvin and glows a dull red. Molecules are ripped apart by the heat and electrons are torn from the hydrogen and helium atoms. These processes absorb the energy of the contraction and allow it to continue. A central object forms with mass similar to the Sun's. The protostar phase lasts 100,000 years.

The ball of plasma continues to contract but it's not a star yet. All the energy comes from gravitational contraction. This phase lasts 100 million years, so Kelvin wasn't too far off in his estimate. The object emits enough radiation to clear out nearby gas, and it sends jets of fast-moving gas along the poles of its spin axis. Meanwhile, gas and dust in the equatorial plane congeal into a thin disk where planets will someday form. The Hubble Space Telescope has seen hundreds of protoplanetary disks, or proplyds, in the Orion Nebula. These placental regions for the subsequent birth of planets are 10 to 20 times the size of our Solar System.

As the gas ball shrinks, its temperature rises, gathering itself for the rigors of the main sequence. When the central temperature reaches a million degrees, it gets a taste for fusion by burning a little deuterium and lithium. When the core hits 10 million degrees, the three-step jig that fuses helium from hydrogen takes center stage. The lights go on. A star is born.

TOURING THE ZOO

Nature loves power laws. In the natural world, it's very rare for every item in a class of objects to have the same size or every incidence of a phenomenon to have the same strength. There's often variation. With a power law the incidence of the phenomenon varies logarithmically with an attribute of the phenomenon.[11] Nature also likes little things or weak phenomena. So there are usually many more small or weak things than big or strong things.

That's terribly abstract. Let's get specific. In physical science, power laws describe the size of craters on the Moon, the strength of solar flares, and (luckily for us) earthquakes and hurricanes.[12] In biology, growth rates, life spans, and metabolic rates are all distributed like power laws. Even in the everyday world, power laws rule. The size distribution of power outages and traffic delays, the population of cities, the length of English words, the frequency of family names, popularity of books and music, and the daily variations of the stock market—these are all power laws.

So it is with stars. There are many more puny stars than huge stars. The power law for stars says that for each star 100 times the mass of the Sun, there are roughly 200 stars 10 times the Sun's mass, about 40,000 stars like the Sun, and a staggering 2.5 million stars one-tenth the mass of the Sun. It's not really surprising, since gravity is a power law force, so objects spawned by gravity will span many scales and the force weakens with distance so large objects will be less abundant than small objects. The chaos and turbulence of a fragmenting molecular cloud somehow coughs up the simplicity of a power law.

If stars form as a power law distribution of mass, why doesn't it go on forever? In other words, why aren't there stars as big as a galaxy or small enough to fit in your pocket?

The answer is that nature provides bounds to either end of the range of masses. If a gas cloud is more than about 120 times the mass of the Sun—the exact bound isn't well-determined observationally—it forms by gravitational free fall, but the collapse is so violent that the object blows itself apart and doesn't form a stable star. Even if it somehow could collapse more gently, the energy output would be so fierce it would drive off the outer envelope of the star.

Massive stars are profligate. They blaze ferociously and emit millions of times more light than the Sun. Intuitively, a star with 100 times the "gas tank" of the Sun should last 100 times as long, but in practice it races through its fuel so fast its lifetime is thousands of times shorter than the Sun's, just a few million years. Some of these stars could have experienced birth and death since our ancestor *Homo habilis* wandered in the Great Rift Valley in East Africa. Massive stars are extremely rare and the nearest examples are remote so it's hard to be sure which one might be the most massive.

Eta Carinae, a binary star system 7500 light-years away, where the more massive star is a luminous, blue variable 100 times the mass of the Sun, is one good candidate. Edmund Halley cataloged this "superstar." In the mid-nineteenth century, Eta Carinae was the second brightest star in the sky,[13] but it disappeared from view in the first half of the twentieth century. More recently it became visible to the naked eye. This massive star ejects gas at supersonic speeds. The gas is heated up to 60 million degrees, a temperature at which it glows with X-rays. Eta Carinae will die soon, but models of the star aren't good enough to say just when. Astronomers are keeping a wary eye on it.

The low-mass boundary for stars comes when the core temperature isn't high enough for hydrogen fusion to work. This boundary is 7.5 to 8 percent the mass of the Sun, or 75 to 80 Jupiter masses. The surface temperature of the puniest stars is a "cool" 3500 Kelvin so they're called red dwarfs. Astronomers speculated for years about failed stars, or brown dwarfs, but their extreme faintness made them very difficult to detect.[14] Brown dwarfs don't get the respect of exoplanets but the story of their emergence is similar. The first evidence of brown dwarfs came in 1995, and now we know of hundreds. A failed star will slowly fade until it's completely invisible.

Low-mass stars are misers. Their gas tanks can be 10 times smaller than the Sun's so you might think they'd fizzle out quickly. But they shine at 0.01 percent of the Sun's power so they live thousands of times as long. That means a red dwarf forming now will still be eking out a life as a card-carrying, hydrogen-fusing star 10 trillion years from now.

Welcome to the stellar bestiary.

In 1910, Enjar Hertzsprung and Henry Norris Russell were trying to understand how stars shine. They made a graph of luminosity or absolute brightness against surface temperature for a set of stars. Expecting a scatter diagram, they were surprised to see the properties of the stars concentrate in some areas of the graph and avoid others. By far the strongest feature of the diagram was a set of stars that undulated like a serpent from bright and hot to dim and cool. They called this trend line the "main sequence." Clumps of stars with other properties appeared in different parts of the diagram (Figure 4.3). Hertzsprung and Russell had made a map of the "zoo," not in the sense of showing where the animals live in space, but in the sense of diagnosing the relationship of their physical properties.

The diagram didn't lead to understanding directly, but Eddington used it to move toward a physical theory of energy generation in stars. We now know that the main sequence is the set of all stars fusing helium from hydrogen, including the Sun. These beasts vary in size and mass by as much as minnows do from sturgeon. Massive stars may be loud and showy, but they're rare and they flame out quickly. The low-mass stars dominate numerically and by virtue of their long lives. The story of stars in the universe is really a slightly dull tale of plodding red dwarfs.

Standing in front of the fish tank watching fish swim around is fine, but eventually you want to learn about other animals. The HR diagram contained stars with properties that placed them far off the main

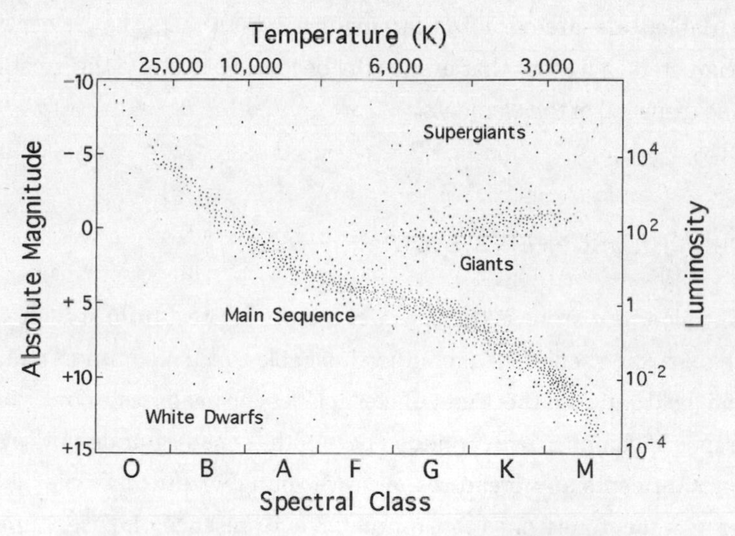

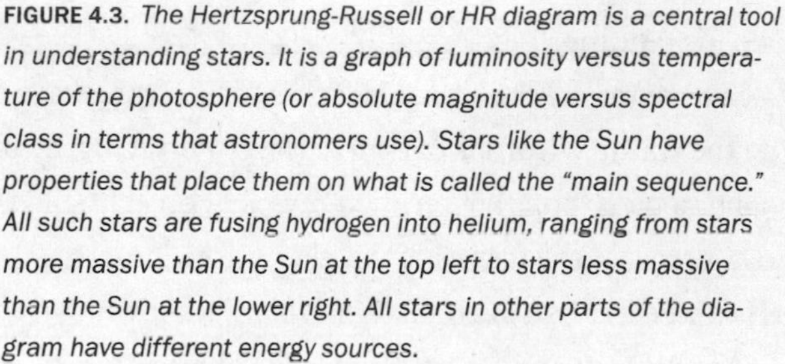

FIGURE 4.3. *The Hertzsprung-Russell or HR diagram is a central tool in understanding stars. It is a graph of luminosity versus temperature of the photosphere (or absolute magnitude versus spectral class in terms that astronomers use). Stars like the Sun have properties that place them on what is called the "main sequence." All such stars are fusing hydrogen into helium, ranging from stars more massive than the Sun at the top left to stars less massive than the Sun at the lower right. All stars in other parts of the diagram have different energy sources.*

sequence. Large objects cool faster than small objects, so bright and cool stars are much bigger than hot and dim stars. To use a domestic analogy, a huge hot plate glowing dull red puts out more energy than a tiny hot plate glowing a hotter shade of yellow. There are cool, luminous stars hundreds or even thousands of times bigger than the Sun, and hot, dim stars hundreds or thousands of times smaller than the Sun. Eddington knew from radiation physics that these stars couldn't make their energy the same way the Sun did.

It took decades for the zoo to be fully explored and understood. The exotic species that aren't like the Sun—the white dwarfs, red giants, supergiants, novae, supernovae, and more than two dozen types of variable star—are of more than just casual interest. They're central characters in a story that eventually becomes our story: the creation and dispersal of the elements.

THE PHILOSOPHER'S STONE

The blood-red stone is the object of intense interest from students at the boarding school. It's reputed to be able to turn common metals into gold and hold the secret of eternal life. The headmaster hides it in a special chamber in a forbidden part of the school, guarded by seven enchantments and creatures. A brave young wizard risks his life to protect the stone from a consummately evil presence who uses human hosts to do his bidding.

The headmaster is Albus Dumbledore. The evil presence is Voldemort. And the young wizard is of course Harry Potter. In the first book of the phenomenally popular series, a mystical object called the Philosopher's Stone is at the center of the plot.[15]

The Harry Potter books are fictional, but J. K. Rowling used a deep knowledge of history and culture to weave alchemy throughout the narrative. Alchemy is a pursuit aimed at turning common metals into gold, and generating the "elixir of longevity" and perhaps immortality. It's both a practical craft and a web of philosophical systems spanning all of the world's major cultures and 2500 years of history. History and mythology intersect in Nicolas Flamel, Dumbledore's partner in alchemy at Hogwarts School of Witchcraft and Wizardry.

Nicolas Flamel was also a real-life alchemist born in Paris in 1330. He worked as a bookseller in the shadow of a large cathedral, where he copied and "illuminated," or illustrated, books. One night, an angel appeared to him in a dream and showed him a special, gorgeous book. Amazingly, not long afterward a stranger came to Flamel's shop and, desperate for money, sold him a book just like the one in his dream.

Flamel spent the next 20 years trying to translate and understand its peculiar diagrams and symbols. He may have succeeded because word got around that he'd transformed half a pound of mercury into silver and then into pure gold. He became rich around the same time, and used his wealth for charitable works. Flamel supposedly lived to 88, a very long span in the fourteenth century.

Fact and fiction blur again at this point. Flamel's library passed down through his descendants for hundreds of years, and in the early seventeenth century one named Dubois allegedly used surviving traces of Flamel's Philosopher's Stone to turn lead balls into gold in front of King Louis XIII. Cardinal Richelieu, the king's ambitious first minister, coveted the power contained in Flamel's book so he imprisoned Dubois and later sentenced him to death and seized all his property. Richelieu built a special alchemical laboratory in his Chateau at Rueil, but died before he could crack the code. The book was never seen again.

Alchemy was not just for crackpots and mystics. As Louis XIII died, Isaac Newton was born. When Newton died, and to safeguard his reputation, the Royal Society deemed his alchemical writings "not fit for print."[16] He prepared his own personal translation of the *Emerald Tablet*, a Greek text in the Hermetic tradition that purports to reveal the secret of the primordial substance and its transmutations. Newton did experiments in alchemy at night and swore his servants to absolute secrecy. It was a shock and a sensation when trunks of Newton's unpublished writings were auctioned by Sotheby's in 1936, and a third of them turned out to be about alchemy. He in fact wrote more about alchemy than about light or gravity. The economist John Maynard Keynes purchased many of the documents, and after reading them he said that "Newton was not the first of the Age of Reason; he was the last of the magicians."

The modern view emerged slowly. Even the great Robert Boyle, who was called the "father of chemistry," clung to a theoretical framework based on alchemy. Then Antoine Lavoisier showed that matter could change shape and form but its mass always remained the same, and John Dalton demonstrated that matter is made of atoms that can be rearranged in different proportions. By 1800, all evidence suggested atoms were immutable. Elements obdurately retained their identities; all attempts to convert one to another failed. The dream of alchemy seemed dead. Until scientists found the Philosopher's Stone inside every star.

With stars, there's much more than meets the eye. Hidden from our gaze, their cores are conjuring up all the ingredients of the material world. Simple and spherical in form perhaps, but stars are protean in their ability to create new elements.

The story of the birth of elements is framed by the simple question, what is the universe made of? Let's start with ourselves. In terms of mass, we're roughly 60 percent oxygen, 18 percent carbon, 10 percent hydrogen, 3 percent nitrogen, and trace amounts of all other elements like calcium and phosphorus. Most of our mass is in the form of water. If you reach down and pick up a rock, the chemical composition is quite different. The most common elements in Earth's crust are oxygen, silicon, aluminum, sodium, hydrogen, iron, calcium, potassium, and magnesium. But chemical processes in the geologically active Earth have altered the proportions of elements, so a fairer sample of the material that makes up planets and moons comes by looking at the composition of primitive meteorites. That's combined with chemical analysis of the atmosphere of the Sun, which has 99 percent of the Solar System's mass. The universe is made of mostly "star stuff," and the best version of cosmic abundance comes from an average across a larger region of the Milky Way galaxy.

Imagine atoms in the universe scattered like playing cards. Instead of the normal suits and numbers, these cards are labeled as elements of the periodic table. The dominant elements are hydrogen and helium; everything else is amazingly rare. In terms of number of atoms, the universe is made up of 88 percent hydrogen, 12 percent helium, 0.060

percent oxygen, 0.026 percent carbon, 0.025 percent neon, and less than 1 part in 10,000 of all other elements. In the analogy of playing cards it's as if we marked all aces and two of the kings as helium atoms and the rest as hydrogen. We'd have to search through 32 decks of cards to find anything other than those two elements. If we had the patience to search 240 decks, or 12,500 cards, we'd find one nitrogen card, three neon cards, three carbon cards, and five oxygen cards. Everything else is hydrogen or helium. To find a single iron card, you'd need to sift through 190,000 cards or 3600 decks. Precious metals like gold, silver, and platinum are incredibly rare, cosmically speaking. Just 1 in 10 billion atoms, so gold at the current price is a steal. Imagine a huge Walmart store, emptied of goods and filled from floor to ceiling with decks of cards. You'd have to go through all of them to be sure of finding one gold atom.

The graph of the cosmic abundance of elements has several striking features (Figure 4.4). The Solar System has a large amount of helium relative to hydrogen. Helium is made in main sequence stars like the Sun, but there's too much to be explained by stellar fusion; we'll get back to that story later.[17] There's a deep trough near beryllium, a second peak at elements that are central to biology (carbon, nitrogen, and oxygen), a third peak at iron, exponentially increasing scarcity for atoms heavier than iron, and then a sawtooth pattern overlaid on the whole graph.[18] All of these aspects were explained in the middle of the twentieth century, thanks to a deeper understanding of stellar fusion and stellar evolution.

As biological creatures, we're partial to carbon. Some esoteric nuclear physics explains why carbon is as rare as it is, and why it isn't so rare that life couldn't exist. Two helium nuclei can fuse to make a beryllium nucleus, but it's radioactive with a half-life of a hundred-trillionth of a second. A star needs a central temperature of 100 million degrees to fuse the two nuclei together faster than they fall apart naturally. Only stars more massive than the Sun ever reach central

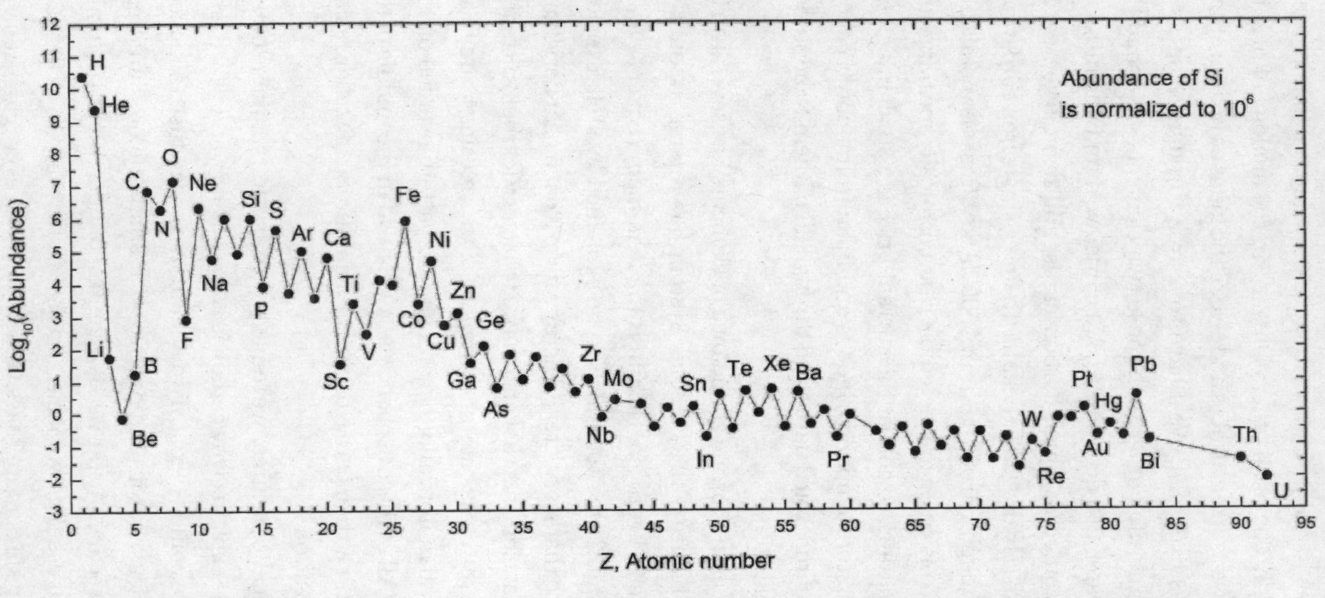

FIGURE 4.4. *The cosmic abundance of the elements, as measured in the Solar System. The vertical axis is a logarithmic scale, so heavy elements are enormously more rare than light elements. All of the features of the diagram are understood in terms of the big bang for very light elements and stellar nucleosynthesis for all the others. The vertical scale is arbitrary and normalized to silicon.*

temperatures this high. In the 1950s, astrophysicist Fred Hoyle realized that a special nuclear resonance greatly increased the probability of a third helium nucleus fusing with beryllium, and so forming carbon.[19] The fact that carbon exists at all is essentially a coincidence of nuclear physics!

To make even heavier elements, higher temperatures and higher mass stars are needed, and they're rare, which explains the rapidly declining graph of cosmic abundances. Fusion gets harder and harder because it has to overcome the increasing electrical repulsion between protons in the steadily growing atomic nuclei. The most abundant building blocks are hydrogen and helium, and since helium boosts the atomic number by two, the abundance curve has a jagged shape. In the most massive stars, carbon fuses with helium to make oxygen. But the star doesn't stop to breathe; it fuses oxygen with helium to make neon. And still it doesn't stop to advertise the fact; it fuses neon with helium to make magnesium. Larger units can combine too. Carbon fuses with oxygen to make silicon, two oxygens fuse to make sulfur, and two silicons fuse to make iron.[20]

It's like playing with Legos, except that the blocks are invisibly small and they're moving at the speed of light in a room with a temperature of a billion degrees.

For a massive star, this all must end in tears. The frenzy of alchemy takes place in a crescendo, with successive stages taking less time—carbon fusion takes about a millennium, oxygen fusion takes a year, and the final stage of silicon fusion takes only a day! The star ends with an "onion" structure, where raw hydrogen and helium are on the outside and layers of heavier and heavier elements are on the inside, nestled around an iron core. The core is a bizarre state of matter; its iron is hundreds of times denser than solid iron, yet it's a 3-billion-degree gas. Iron is a bottleneck because energy is no longer released by fusing toward heavier elements.[21] With no energy to keep the star puffed up, it collapses.

All the heavy elements beyond iron in the periodic table are created by massive stars. About half of those atoms are made in the atmospheres of those stars by stealth as neutrons are captured by the heavy nuclei. This process only goes up to bismuth, the heaviest stable element. The other half are created in the paroxysm of the star's death. Collapse is followed by an explosion called a supernova, and elements as massive as radioactive uranium and plutonium can be created in seconds in the billion-degree blast wave. Those heavy elements include gold.

The transmutation of lead into gold is possible. In 1972, physicists at a nuclear research facility near Lake Baikal in Russia found that some of the lead shielding at an experimental reactor had changed to gold, and in 1980, the U.S. chemist Glenn Seaborg used a reactor to turn a few thousand lead atoms into gold. These aren't quick paths to wealth; it's far easier to rely on stars. From the hoard of Croesus to the 4600 tons currently sitting in a vault at Fort Knox, humans have been harvesting the bounty of billions of years of stellar nucleosynthesis.

Stars are engaged in a vast recycling program. The universe would be a dull place if stars created elements and held on to them for eternity. Luckily, all large stars return a substantial fraction of their mass to the interstellar medium where it can become part of a new generation of stars and planets. Many atoms are removed from circulation in various stellar corpses: white dwarfs, neutron stars, and black holes. But a lot are sent back into space to be part of a new set of stories.

When you were conceived it was a special moment. From a tiny fertilized egg grew a person who can hold their existence and the entire universe inside their head. For your atoms, the significance is less profound. You represent one particular configuration of atoms that have experienced numerous permutations in cosmic history and will experience countless more in future eons. Each of us is impregnated with atoms forged in the centers of stars and cycled through multiple generations of stars. Their diverse journeys led them to converge on a patch of space where the Earth would form and we would animate the atoms with our dreams. We don't have to visit the stars; the stars have paid us a visit already.

It must have been the salsa.

I woke up one night in a cold sweat. I'd lectured my class that day on the cosmic abundance of elements, and then gone out for Mexican food with some friends. As always, I'd explained to the students how almost every atom in their bodies had once been part of a different star, their stories stretching back in time to long before the Earth formed. I was probably on autopilot, since the information was so familiar, telling them a fantastic story without really considering it deeply myself.

As I broke the surface of consciousness by tossing and turning I had been dreaming of atoms. I lay still and through one lidded eye saw a single blurry lash. My attention focused on the tip of the eyelash and suddenly I was down among the atoms, an incredible shrinking man.

Keratin. A protein that's the main structural element of hair and skin. Half of the atoms are carbon. The one nearest me was cooked up in a Sun-like star 6 billion years ago and blown off in a smoke ring where it languished in interstellar space before being swept up in the forming Earth. The next one over had a tumultuous history, churning through half a dozen stars before being interred on the planet to become food that would one day become part of me. There's a hydrogen atom. It's been unaltered since the birth of the universe 13.7 billion years ago. That's sulfur, forged inside a massive star and delivered to me by a titanic explosion 5 billion years ago.

And on and on. A trillion stories in one eyelash alone, a billion billion billion in my whole body. As Whitman once said: "I am large, I contain multitudes." My mind reeling, I submerged again into fitful sleep.

The Trapezium lies ahead, filling my field of view. Its four bright stars are embedded in smooth pink gas, overlaid with a sparse web of tendrils and filaments that stand out like veins on the back of a hand. I can see that three of the stars are in fact binaries, and the brightest one—Theta Orionis—has carved out a blister of hot blue gas with its intense ultraviolet radiation. The stars steadily shift. I realize my perspective is changing because I'm moving. Then I pass through sheets of dusty nebulosity, and an infant star whizzes by in a bubble of ionized gas shaped like a bullet's wake. I'm traveling swiftly through the canyons of space.

My destination seems to be a dark knot of gas. Something within it is glowing dull red like an ember of coal. I'm falling by gravity onto a newborn star, still shrouded in placental dust. What used to be vacuum now feels like a fine dry mist. I catch a smell of sweet and sour: benzene and ammonia. It may be my overactive imagination, but I recall that interstellar space has potassium cyanide, and I think I sense a whiff of bitter almonds.

Looking back along the path I've traveled, the stars are milky. I'm already well within the cocoon of the nebula and dust is obscuring more distant stars. With some difficulty I locate the solar neighborhood and the unassuming yellow star that shelters my home.

I'm inexorably falling into a newly ignited inferno. I concentrate hard and try to project myself back along the invisible tether that connects me to more familiar space. I descend toward the Earth. But my city isn't where it should be, and the landscape looks unfamiliar and virgin. I speed across an ocean, hoping for something recognizable. People are all on foot, draped in crude garments. Soldiers are on the move, in small bands, across a wooded continent. I see a great city in ruins. Tendrils of smoke rise into the sky.

The fall of Rome. The end of a vast empire and the beginning of the Dark Ages. This is the Earth as it was nearly 1600 orbits of the Sun ago. It's not my time—I don't exist. But how is it possible? If I were in Orion, I'd be seeing the Earth in 1600-year-old light. But I can't be in Orion. I must be at home. But home when? I've completely lost my sense of now.

5

THE EDGE OF DARKNESS

I'M IN A CITY OF GLITTERING LIGHTS. *They sparkle like diamonds flung onto black velvet; not the demure night sky of my childhood, but a wanton display of power and transmutation. It's bright enough that I could read a book. I set out what seems like a long time ago, toward Sagittarius. From a quiet suburb near the outer reaches of the galaxy I streaked through star fields, passing through the chaos of three spiral arms. Then I moved into the great bulge of old stars that travel on plunging orbits through the galactic disk. I feel like a country boy in the metropolis as I gawk at the light show.*

Ahead of me is the gravitational epicenter of 400 billion stars. Behind me and far away, the Sun is just a loyal follower, like a night watchman doing his rounds, moving lockstep with his neighbors on a quarter-billion-year orbit of the galactic center. Stars near me are moving faster now, stirred by some unseen force. The dense star cluster in front of me is so bright I have to avert my gaze. But I also sense a dark, impassive presence.

It's difficult to see Sol through the bustle of stars but I know where to find it, a warm yellow blob of light. And there nearby, the Pale Blue Dot. It swims into view, wild and unfamiliar again. I scan for traces of civilization but see nothing.

Then, almost imperceptible in the vast landscape, I notice a few small nomadic groups. These hairless apes are hard to distinguish from herds of other animals but they move steadily and with apparent purpose. Light has taken 27,000 years to reach my eyes. I'm seeing the Earth as it was when we were just beginning to flex our muscles as the alpha species. We've moved out of Africa and radiated throughout Europe and Asia. We've seen off the Neanderthals but haven't yet crossed the land bridge to the Americas. Just a million strong, we tread lightly on the planet. I'm wistful for that species—so young and raw and full of infinite promise.

CITY OF LIGHT

I'm groggy as I bundle up in warm pants and a down jacket. The jet lag involved in getting to Chile is a bitch. From Los Angeles, the route goes across to Miami, followed by a long night flight down the spine of the Andes. I had just enough time to freshen up in the guest house in Santiago before heading to the regional airport for the one-hour hop to La Serena. Then I made a four-hour drive on dirt roads across the barren, lunar landscape of the southern Atacama Desert to the Las Campanas Observatory. I checked into my room, grabbed a few hours of fitful sleep, and am preparing to go to the telescope. Tomorrow is my first night so I want to see how everything works and be well-prepared.

The dormitory is down the ridge from the telescope, which sits on a promontory overlooking the desert floor. A paved path snakes through the dirt and desert scrub. There are no lights; lights are the enemy at any major observatory. I open the door. The cold night air makes me catch my breath. I've left a balmy summer LA for the high Andes in midwinter. I pull on my gloves, turn up the collar on my jacket, and cautiously start walking.

To my left is the down slope of the range, ridges descending like a crumpled blanket until they disappear under a pale gray marine layer. To my right is the Cordillera. It takes 20 minutes to become fully dark-adapted, and as I walk I see more and more stars in a brilliant swathe that defines the jagged silhouette of the Andes.[1] I stop for a minute to catch my breath. At 8000 feet the air is markedly thinner than at sea level. It's absolutely still and utterly quiet. Looking down, my gloved hands are casting shadows on the path. I look up.

I'm stunned. The Milky Way passes directly overhead. The vault of night is ablaze (Figure 5.1). The galaxy arches over my head from horizon to horizon—ragged curtains of light straddle a seam as black as coal. I follow the tail of the swan to Altair, then the bright stars at the

FIGURE 5.1. *City of Light. A U.S. Park Service employee took this spectacular image of the galaxy we live in, a sight many people will never see, at Racetrack Playa in Death Valley in 2005. The full 360-degree panorama artificially curves the Milky Way, which defines the plane of our galaxy. The ragged dark features are due to obscuring dust rather than the absence of stars.*

head of the eagle, the vertex of the summer triangle, and on to Sagittarius, the Archer. His taut bow is poised to send an arrow into the vast pile of stars. I feel that if gravity loosened its grip slightly I would be flung headlong to the center of the galaxy.

We live in a city of light. To grasp the vastness, we need a scale model. We can get there in two steps. In the first step, we'll shrink objects and space by a factor of 10 million. The Earth is an apricot, and the Moon a pea at arm's length. This cozy arrangement is the full extent of human voyaging. On this scale, the Sun is a 10-foot globe 100 yards away, while the nearest star is 30,000 miles away. With this scale model, most stars are still unimaginably far, so we shrink space again, this time by a factor of 100 million. Stars are now the size of atoms, and a solar system is contained within a region of space the size of a small grain of sand. The typical distance between stars is 30 feet and the Milky Way galaxy is the size of the continental United States. This city of light holds many billions of stars. A legend on this map would read $1{:}10^{15}$, or 1 inch equals 1 light-month. It would probably add: objects are farther away than they appear, so stock up well before traveling in the Milky Way.

This modern sense of scale was not easily reached. To most ancient Greeks, the chaotic appearance of the Milky Way was so far removed from the Pythagorean perfection of the crystalline spheres that they assumed it was an atmospheric phenomenon. However, Anaxagoras and Democritus, bold thinkers in so many ways, thought that it might be made of distant stars. Several Arab astronomers had the same idea during medieval times,[2] but it wasn't verified until Galileo's telescopic observations in 1610.

Mapping the Milky Way took a long time. Imagine a tribe of hunter-gatherers living in a valley. They can explore their valley and maybe climb to get a view of more distant terrain, but they have no way of understanding the continent they live on. It's beyond comprehension. In the late eighteenth century, the musician-turned-astronomer

Sir William Herschel "strip-mined" the night sky with his telescope, counting the stars in different directions and using the Earth's rotation to sweep out swathes of sky. He was correct in deducing that we live in a disklike distribution of stars but wrong in thinking that we're at the center of the disk. He wasn't looking deep enough. Out to 1000 light-years the disk looks more or less the same in all directions. Also, a view to the center of the galaxy is precluded by the absorbing dust in the space between stars and in star-forming regions.

In 1917 the young astronomer Harlow Shapley, who was working at the Mount Wilson Observatory, resolved the issue. Shapley mapped the distances to large agglomerations of stars called globular clusters that move through space above and below the plane of the galaxy. He showed that their orbits are centered on a point remote from the Sun, placing us far from the center of the galaxy. However, he misjudged the size of the entire system of stars.[3]

The true size of and detailed structure of the Milky Way didn't become clear until radio astronomy matured in the 1950s. Cold hydrogen gas has a spectral line at a wavelength of 21 centimeters (1421 MHz if you are searching for the hiss on your radio). Radio waves are unaffected by dust so the 21 centimeter line was used to observe the far side of the Milky Way and map the rotation of the entire disk. Harvard physicist Ed Purcell and his graduate student Harold Ewen first detected the line in 1951. Working weekends and evenings, Ewen built a radio detector and a horn-shaped antenna that he pointed out the window of his lab. There were hazards; during heavy rain the horn funneled water and flooded the lab, and in winter Harvard students tossed snowballs into it. The birth of radio astronomy had come 18 years earlier, when Bell Labs engineer Karl Jansky detected radio waves from the constellation Sagittarius. Since normal stars don't emit radio waves, this was a sign that something extraordinary was going on in the center of our galaxy.

BLACK HOLES

The Milky Way is full of beginnings and endings. Around us in the Milky Way we can see stellar nurseries like the Orion Nebula and evidence of the titanic explosions that occur when massive stars die. Most of the several hundred billion stars in the Milky Way are timeless. They have such low mass that regardless of how long ago they formed, they will eke out a living converting hydrogen into helium for billions of years to come. When they die, they'll collapse to a concentrated state called a white dwarf, and slowly radiate their residual energy until they fade to black. For stars much heftier than the Sun, the beginnings and endings are linked. Such stars create the heavy elements that become part of a new generation of stars and planets, and a dying star can trigger the formation of a new star from a nebulous gas cloud.

The death of massive stars can lead to the formation of the most exotic object known in nature: a black hole. When all nuclear fuels have been exhausted, the core of the star collapses in gravitational free fall. Most of the mass of any massive star will be sloughed off gently during its evolution or violently in a supernova explosion at the end. The resulting compact object forms when the mass that hasn't been ejected undergoes core collapse. Stars more than about eight times the Sun's mass will end as supernovae. From 8 up to about 25 times the mass of the Sun, the core collapses to a bizarre object called a neutron star (Figure 5.2). In the crushing gravity, protons and electrons merge to form neutrons and the entire dead star has the density of an atomic nucleus. Imagine all humans squashed into the volume of a cube of sugar. Neutron stars were hypothesized in the 1930s and confirmed with the discovery of pulsars in 1967.[4]

If the collapsed core is less than about three times the Sun's mass, it will form a neutron star. If it's more than that, it will continue to shrink to an even more condensed state, from which nothing can escape. The

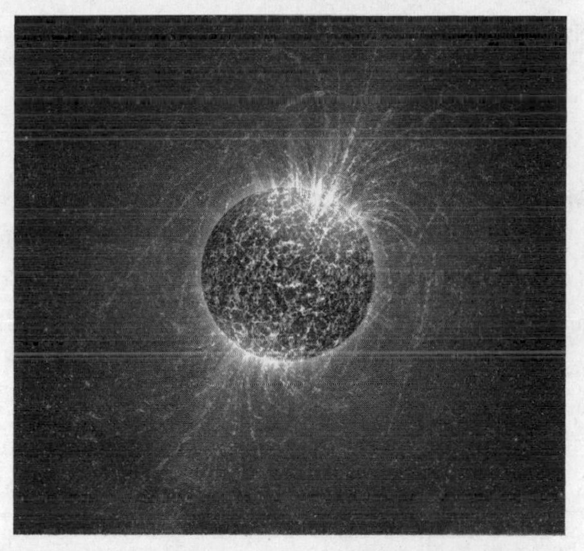

FIGURE 5.2. *Neutron stars are the end states of massive stars where the collapsed core is between one and a half and three times the mass of the Sun. With no electrical repulsion operating, the star is as dense as an atomic nucleus with an atomic number of 10^{57}. The star is about 5 miles across and threaded with a super-strong magnetic field. There are roughly 300 million neutron stars in the Milky Way.*

death of the most massive stars signifies the birth of something about which we can know almost nothing.

Let's eavesdrop on the gods, who are arguing about the best jewel in the heavens.

There's no greater star in the firmament than the Sun and those like her. Such stars are liquid gold. The speaker is Apollo, his voice strong and confident. My son, your paean is very pretty, says Zeus, reclining nearby. Who will speak against him? Zeus looks around expectantly. I will, says Artemis. She's Apollo's twin, equally confident and unwilling to cede him the argument. My choice is the white dwarf—opalescent like the Moon and a veritable diamond in the sky. No precious stone is more abundant.

Ares speaks next, his voice an ominous rumble. Not true. The battle is won by those whose bloodlust burns the longest. Give me an army of such warriors and there's no land I cannot conquer. The red dwarfs outnumber all other soldiers. They never quit. To you they might just be a dull glow in the night, but to me they have the sheen of rubies. Zeus listens tolerantly to his oldest son and nods.

Midas is nearby, listening. He interjects. You are all mistaken. When the most massive stars die they forge gold. There can be no greater gift to the gods. A supernova is the pinnacle of the stellar realm. Zeus is feeling contrary. Not so, he says. The gift is paltry, almost an insult. The gold is flung carelessly into space. Not enough would reach me to gild my brow. Just as a single water molecule isn't wet, a single gold atom has no luster. Be gone, Zeus says dismissively, and take your donkey ears with you.

Hades shuffles over, his three-headed dog Cerebrus snarling beside him. What do you want, snaps Zeus? You're here on our sufferance. Hades returns his gaze evenly. Brother, you forget your origins. It's only the luck of the draw that put you here in Olympus. You do not respect what you cannot see, he says. But the dark world completely eclipses the sunlit world. The most magnificent stars in the firmament are those that wear a helmet of darkness when they die. Black holes. They are the black pearls of the night.

What exactly is a black hole? They're variously referred to as pinched off regions of space-time, the most condensed forms of matter known, information membranes, places where space and time have no meaning, objects with no hair, the mark of the abyss, and this from the comedian and actor Stephen Wright: "Black holes are where God divided by zero."

The formal definition of a black hole: it's a region of space-time that's not in the causal past of the infinite future. Say what? That's what you get if you ask a general relativist, assuming you can pry their attention away from the consuming pleasures of tensors and manifolds. A more accessible definition would be a region with gravity so intense that not even light can escape. But that definition begs a few questions. What's the state of matter in that region? What happens to stuff that falls in? Doesn't light travel at a constant speed, so how can it be trapped by gravity?

To answer these and other questions about black holes, we must come to grips with gravity and, particularly, the general theory of relativity.[5] If Newtonian gravity is a stately minuet, the gravity of Einstein is a wild improvisational tango.

Here's Newton's view. Space and time are infinite, absolute, and linear. Objects move in space and show behavior over time, and the two can be cleanly separated. Mass and energy are also obviously distinct, as different as stone and water. Mass is intrinsic and unalterable, while the energy of something can vary. Gravity is a force that acts instantly over the vacuum of space. Mass tells gravity how much force to exert, and force then tells the mass how to move.

Here's Einstein's radical revision. Space and time are linked and are interchangeable, part of a four-dimensional construct called space-time. Space-time may be finite or infinite, linear or curved. It is definitely not absolute, because its properties change according to the "local" gravity. Mass and energy are also interchangeable, and are linked by the famous equation $E = mc^2$. Gravity is a force that acts at light speed, not instantaneously. In the words of John Wheeler, a physicist who also coined the term *black hole*, "Mass-energy tells space-time how to curve; curved space-time tells mass-energy how to move."[6]

In 1905, Einstein came up with the special theory of relativity, which says that the laws of physics are the same for all observers in constant relative motion. This applies to light as well, which is observed to have a constant speed regardless of relative motion. Bizarre consequences are the result: rapidly moving objects contract or are squashed in the direction of motion, their mass grows, and their time slows down or is stretched. This suppleness of space and time shocked physicists at the time, and they're still a little unnerved, as if physics was hip and edgy when they would have preferred it to remain dull and regular. Einstein wondered if he could extend his ideas to nonuniform motion, or acceleration. The clue that he could was the equivalence principle. Galileo noticed that falling objects fell at the same increasing rate regardless of their mass or composition. This is really strange because it means that two apparently quite different masses—the mass of an object as given by its resistance to any change in its motion, and the mass of an object given by its motion under the force of gravity—are the same.

Not many people get pleasure thinking of someone plunging to their demise. But while sitting in his patent office in Bern in 1907, Einstein realized that someone free-falling in gravity, like in an elevator whose cable had broken, wouldn't feel their own weight. He later called this his "happiest moment."

To see the equivalence, imagine you're in an elevator floating in deep space, far from any star. You'd be weightless because it's a situation of zero gravity. That's indistinguishable from the more ominous situation of being in a plunging elevator.[7] Then Einstein considered another pair of situations. Suppose you're in an elevator being accelerated by a rocket at 9.8 meters per second per second. That's indistinguishable from sitting in a stationary elevator at the surface of the Earth. One scenario involves gravity and the other has acceleration but no gravity, yet the two are equivalent. Einstein saw that there was nothing special about changing motion by the force of gravity as opposed to changing motion due to any other force.

Thinking like a physicist means trying to see the unity in all physical phenomena. Gravity being a "special" force with its own mechanism is inelegant; it's better if all forces are on equal footing.

Einstein's audacious move was to reformulate gravity as a geometric theory. Instead of Newton's absolute and linear space and

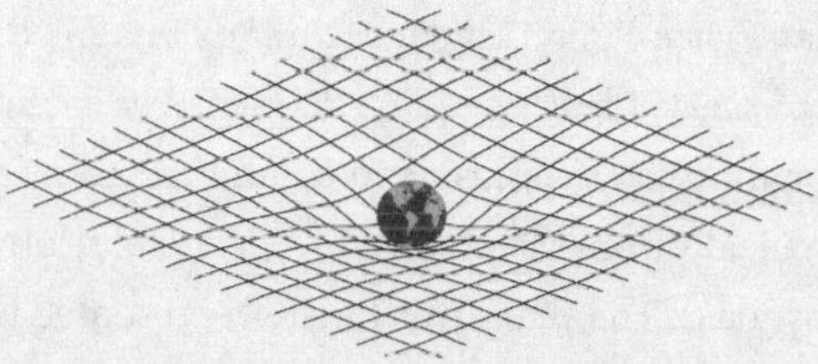

FIGURE 5.3. *The core of Einstein's formulation of gravity is the idea of curved space-time. Newtonian gravity acts in linear space but in the theory of general relativity, mass and energy create curvature of the space-time, and the curved space-time causes particles and radiation to follow curved trajectories.*

time, space and time in general relativity are malleable and curved (Figure 5.3). His "field" equations relate the curvature of space-time to the energy and momentum of mass and energy within it. General relativity is utterly different from Newton's concept and has profound implications.

Let's go back to the elevators. Suppose you shine a flashlight across the elevator. If it's an elevator with no changing motion or no gravity the beam of light will go straight across. But if the elevator is being accelerated by a rocket at 9.8 meters per second per second the light will follow a curved path. By the equivalence principle, this isn't any different from the situation of an elevator at rest on the Earth's surface. So that beam of light must also curve downward. Gravity bends light! Or more correctly, the mass of the Earth distorts space-time and light follows the curvature of space-time.[8]

Armed with a geometric theory of gravity, we can contemplate objects with portals signposted like the entry to hell in Dante's *Divine Comedy*: "Abandon hope all ye who enter here."

In 1783, English geologist John Michell speculated that the size of the Sun could be scaled down to create a situation where the escape velocity would reach the speed of light.[9] A decade later, the French mathematician Pierre-Simon Laplace similarly imagined the existence of "dark stars." These ideas were ignored for over a century because light was thought to be massless and so unaffected by gravity. Black holes follow logically from the prediction of general relativity that light is deflected by mass. The year after Einstein presented the formalism of general relativity, Karl Schwarzschild solved the equations for the situation of a nonrotating, spherical body, calculating the size of an object that would be so dense that it would be a black hole. Schwarzschild did this work while serving in the German Army in World War I. Tragically, he developed a skin disease and became gravely ill. Einstein presented his pioneering ideas just months before Schwarzschild died. In 1939, Robert Oppenheimer took the next step just before he started work on the Manhattan Project, by predicting the collapse of stars above 3 solar masses (which started their lives above 25 solar masses) to a state where matter and radiation are trapped.

Once formed, is the beginning of a black hole the end of everything? Does it consume everything around it like a cosmic vacuum cleaner? To borrow Oppenheimer's words from another context, if a black hole could speak, would it say, like the Indian deity Shiva, "I am become death, destroyer of worlds"?

No. Far from a black hole, the gravity is mild and is indistinguishable from Newtonian gravity. That's as it should be; general relativity is a theory that supersedes Newton's theory but approaches it when gravity is weak. When gravity is strong, general relativity must be used. If an evil alien empire crushed the Sun to a radius of 3 kilometers, it would become a black hole. However, the Earth would continue undisturbed in its orbit. (The loss of sunlight eight minutes later would, however, be more than an inconvenience.) Nobody should obsess about death by black holes. We should, however, be curious—black holes have some very unusual properties indeed.[10]

LOST HORIZONS

Since a black hole is a purely theoretical construct, we have to ask if it actually exists in nature and how many there might be? The most obvious (but not the only) way to make a black hole is when the core of a dying massive star collapses under gravity. Massive stars are rare, but the galaxy is vast so lots of them should have died and left behind black holes.[11] As a rough estimate, the supernova rate is 1 per 30 years, or about 300 million in the 10-billion-year life of the Milky Way. The mass distribution of stars is very steep, in the sense that the most massive stars are scarce, so we'd expect most of the remnants to be neutron stars. Twenty to 30 million black holes should have formed in our galaxy. The challenge is to find something as black as pitch.

An isolated black hole is invisible. But many stars are part of binary systems, which opens up the possibility of detecting the black hole through its interactions with a visible companion. The evidence is indirect because rather than see the black hole, we see matter being stressed as it falls toward the black hole. As often in astronomy, the evidence came from a surprising direction.

In 1948, researchers launched an instrument designed to detect X-rays with a rocket developed by the Germans in World War II. X-rays have high energy and short wavelength and can't reach the ground from space. The instrument detected X-rays from the Sun but showed that they're a million times weaker than the visible light; stars seemed to be very boring X-ray sources. But in 1962 a similar rocket discovered X-rays coming from the constellation Scorpius. The source, named Sco X-1, emitted billions of times more X-rays than the Sun. The true birth of X-ray astronomy came with the launch of the Uhuru satellite in 1970. It found hundreds of sources—the sky is alive with X-rays.

Most of these X-rays come from binary systems where a normal star siphons gas onto a dark companion (Figure 5.4). Two stars are locked in a tight gravitational embrace. Gas is ripped from the atmo-

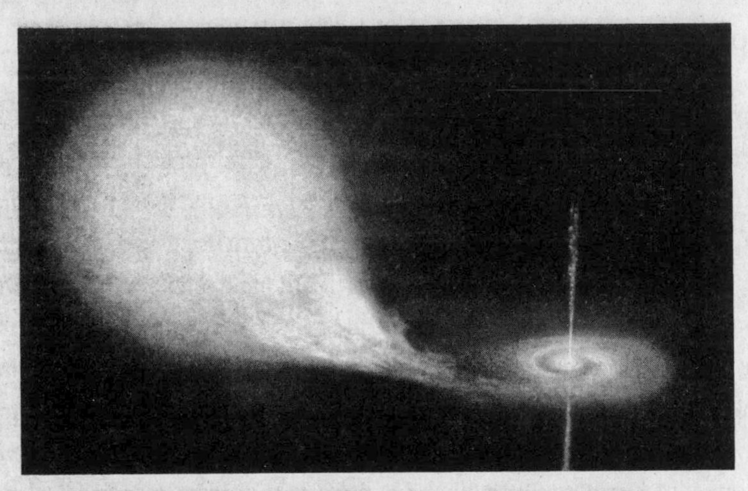

FIGURE 5.4. *Black holes are black, since no radiation can escape the event horizon. However, a black hole in a binary system can suck in gas from its normal companion. The gas forms a disk in the plane of the equator of the spinning black hole, and jets of plasma are beamed out along the poles. X-ray satellites detect this high-energy radiation.*

sphere of the big star and it spirals onto the dark star, where it forms a disk of hot plasma. Only a supercompact object has the gravity to attract a lot of gas and heat it such that it emits copious high-energy radiation while falling in. These binaries are gravity engines and their exhaust comes out as twin jets of X-rays along the polar axis of the spinning dark star.[12] The visible star in the binary system is observed to track the orbit and determine the masses of the bright star and the dark companion. Most compact companions prove to be neutron stars, including Sco X-1, but the brightest X-ray source in the constellation Cygnus, called Cyg X-1, is a good candidate for a black hole because its inferred mass is more than three times the mass of the Sun. There are 20 candidates known and for half of them the inferred mass is so high that the evidence is compelling.[13] All are more than a few thousand light-years from the Earth.

We are warily stalking our prey. Black holes are firmly predicted as a consequence of stellar evolution but an isolated dark, compact object is undetectable. As a result, the best evidence involves binary systems where a normal star has a dark companion that's massive enough that it must be a black hole. The dark companion is identified by a disk of extremely hot gas and by particle acceleration caused by the intense gravity. The X-rays come from the vicinity of the black hole, not the black hole itself.

Black holes have many extraordinary properties, but their defining attribute is an event horizon. An event horizon isn't a physical barrier. It's the boundary between knowledge and ignorance. Think of it as a point of no return. Matter and radiation can reach us from near the event horizon, although space-time near it is heavily distorted. Within the horizon, all matter and radiation are trapped forever and we can know nothing about their fate. Schwarzschild calculated the size of a black hole, the radius at which the escape velocity is the speed of light and nothing can escape. It's 30 kilometers for a stellar remnant 10 times the mass of the Sun. All the intense action and space-time curvature happens in a region the size of a small city. The distance to the event horizon scales with mass, and in principle it's possible to have black holes of any mass or size! We mentioned that if an evil alien genius crushed the Sun down to 3 kilometers in radius, it would become a black hole. If they also squashed the Earth to the size of a shelled peanut, it would also be a black hole. Let's hope this is very unlikely.

Here's what would happen if you fell into a black hole. Let's assume you left your friends in a spaceship orbiting at a safe distance and then ventured down in a small probe. As observed by your friends, time for you would slow down as you approached the event horizon while the image of you got redder and dimmer. Assuming they were patient, your friends would see you take an infinite amount of time to reach the horizon and as you did, you'd become too dim to be seen. You, meanwhile, would see nothing strange going on with the clock of your space probe. You'd fall through the event horizon in a finite time, but would be unable to say exactly when you crossed it, or where it was.

Actually, that description is purely hypothetical, because even if there were a black hole near enough to reach in a spaceship, you wouldn't survive the journey in. The gravity near a black hole is so strong that if you fell in feet first you'd be subjected to brutal tidal forces—stretching you lengthwise and compressing you sideways. You'd be spaghettified. It's even more unpleasant than being pulled apart like pizza dough. If you approached the event horizon you'd be ripped to smithereens on many different scales at once: limbs, muscle fibers, cells, even DNA. When Einstein was having his mischievous thoughts about people in falling elevators, he'd no idea of the tortures that gravity can inflict.

The second essential attribute of a black hole is a singularity. At the center of the black hole is a region of infinite density and space-time curvature. Physicists were uncomfortable with the implications of singularities, but hard work over several decades showed that they're unavoidable.[14] General relativity can't be used to calculate a state of infinite density so it's been said that the theory contains the seeds of its own demise. Space and time as we know them cease to exist at the singularity; quantum gravity is needed to make sense of it. No such theory exists but the physicists drinking superstring Kool-Aid think they're on the right track.

Schwarzschild had calculated a solution for a static black hole, and it was anticipated that black holes would share the rotation of the stars that gave rise to them. In 1963, nearly 50 years later, Roy Kerr found the solution for a rotating black hole. (It's one sign of the difficulty of general relativity that only four exact solutions to the field equations have been found in a century of effort.) Spinning black holes reach a whole new level of weirdness.

Due to the phenomenon of frame-dragging, space itself swirls around the spinning black hole like water exiting a bathtub. Rotation as rapid as 99 percent the speed of light has been observed. A spinning black hole has two event horizons, and the faster the black hole spins the closer together they come.[15] Between two event horizons space and time are interchanged. Beyond the outer event horizon lies an ellipsoidal region called the ergosphere. Particles and radiation within are dragged along by the rotation, but they can escape. Roger Penrose hypothesized that it may be possible to extract a third of the black hole's mass energy from the ergosphere, so it might one day have practical applications. Inside the inner event horizon is the singularity. The singularity for a spinning black hole is a ring rather than a point—a circle of infinite gravitational forces. But it's not as dangerous as the nonspinning point singularity because it's possible to avoid it. In fact, except along the equator, it repels rather than attracts.

The ring singularity of a rotating black hole offers two extraordinary options: time travel and escape. In theory, a voyager could use the ring to visit anyone within the inner horizon arbitrarily far forward or backward in time.[16] You could arrange a poker game, where each of the players is you at a different age. However, you couldn't visit a celebrity like John Lennon before he died, or yourself before you traveled to the black hole; those people are on the far side of two impassable event horizons. In theory, you can also leave the black hole by crossing the singularity, but that leads to a place with the quality of "negative space." Nobody knows if it's a portal to another universe or just a mathematical abstraction.

Black holes are tantalizing. If you could venture inside one you'd see amazing sights—gravity at its most Baroque. But you'd never return to tell the tale or even get the information out. Distant observers remain ignorant of anything inside the event horizon. Black holes only have mass, spin, and charge (the last is not likely to apply in practice). This state is so simple it's been said that black holes have "no hair," where hair is a metaphor for information about what went into the black hole.

It doesn't matter how the black hole was made—by consuming matter or antimatter or being fed encyclopedias or candy bars or lots of small rodents—the result is the same as seen from the outside.

The "amnesia" of a black hole leads to a serious problem. Entropy is the disorder of a system, or the number of equivalent physical states. The second law of thermodynamics says that entropy must increase but the entropy of anything falling into a black hole seems to be lost. Jacob Beckenstein argued that black holes have more entropy than any object of the same volume. (The huge number of ways that a black hole could have reached its apparently simple state can be seen as another way of conveying its large entropy.) Steven Hawking pursued the analogy with thermodynamics by recognizing that an object with entropy must have a temperature, and if it has a temperature it must radiate energy.

Black holes aren't black! Energy leaks out at a feeble rate as particle-antiparticle pairs are created near the event horizon and one falls in while the other escapes. Black holes aren't eternal either. The release of "Hawking radiation" means that black holes slowly evaporate. These effects are subtle; for a massive star that ends as a black hole the temperature is 10^{-8} Kelvin and the time for evaporation is 10^{68} years. That's a phenomenal time compared to the age of the universe so far and probably not worth waiting around for. These effects are also hypothetical; they've never been observed.

Black hole evaporation led to a related problem, the information paradox, which has flummoxed a lot of physicists over the years. Whether a black hole is made of encyclopedias or simpler forms of matter, information seems to have disappeared. Many physical states evolve into just one state. As the black hole evaporates, the radiation that comes out is mute about the states of matter that built the black hole. This was obnoxious to many physicists since a hallowed principle says laws of physics are reversible. Quantum theory doesn't work if the laws of physics aren't reversible. So, in 1997, Stephen Hawking's colleague John Preskill was upset enough to bet an encyclopedia that information *isn't* lost inside a black hole. In 2004, Hawking conceded the bet, as he and others had figured out ways that information about the states of matter inside a black hole might be "coded" on the event horizon, in the same way two-dimensional holograms code information about three-dimensional objects. These formulations typically involve the esoteric equations of quantum gravity and string theory.[17]

Stephen Hawking put it this way: "If you jump into a black hole, your mass-energy will be returned to the universe, but in a mangled form which contains the information about what you were like but in a state where it cannot be easily recognized. It's like burning an encyclopedia. Information isn't lost, if one keeps the smoke and ashes. But it's very difficult to read."[18]

Black holes remain enigmatic. Their existence is indicated beyond a reasonable doubt by massive, dark companions in binary systems but it's difficult to measure any properties other than mass. Nobody has demonstrated the existence of an event horizon or a singularity. Most of the bizarre phenomena caused by intense gravity have not yet been observed. Black holes will not give up their secrets easily.

BIG BEAST

Black holes are strange and ominous beasts. Perhaps it's good news that the nearest examples in the Milky Way are several thousand light-years away. But there's a big beast at the center of the galaxy that's a million times bigger than any stellar remnant.

The Milky Way was thought to be a fairly dull galaxy until 1974, when the strongest radio source in the direction of Sagittarius was found to be very compact. Later observations showed that the region of emission was smaller than the Solar System and was poised motionless at the center of the galaxy. The Sagittarius radio source was more compact and more energetic than any in the galaxy. Learning more was hard because of the dust between us and the galactic center. Only 1 in 10 billion visible photons escape to reach us.

Infrared observations were used to see through the dust. There's a dense star cluster centered on the strange radio source. Stars in the cluster are so densely packed that if we were located there the night sky would be as bright as the daytime sky and some stars would be close enough to roam into the Solar System, scattering planets like tenpins. Over a million stars are packed into a space 3 light-years across; near the Sun such a region would contain only a handful of stars. In the late 1990s, infrared imaging and methods for removing atmospheric blurring let large telescopes resolve individual stars in the central star cluster for the first time (Figure 5.5). The cameras can see a quarter at a distance of 5000 miles. Competing research groups in Germany and California patiently tracked the space movements of stars from year to year and after 20 years of work they've seen complete orbits for several of them.[19]

The result is compelling evidence of a dark object 4 million times the mass of the Sun. The star motions are used to measure the mass causing the motions, and even though there's a stellar pile-up in the central region, the excess mass can't be explained by normal stars. The concentration of millions of solar masses in a space smaller than the Solar System must be a black hole. Stars swarm like angry bees around the dark object. Half a light-year away they're moving at 100 miles per second and even closer in they whizz around at 500 miles per second. A star called S1, which approaches within a light-day of the beast, is moving at a scorching 900 miles per second or 3 million miles per hour.

This gigantic dark object 27,000 light-years away is the best example of a black hole. Measurement of dozens of orbiting stars rather than just one to diagnose the mass means the evidence is better than for any of the stellar-mass black holes.

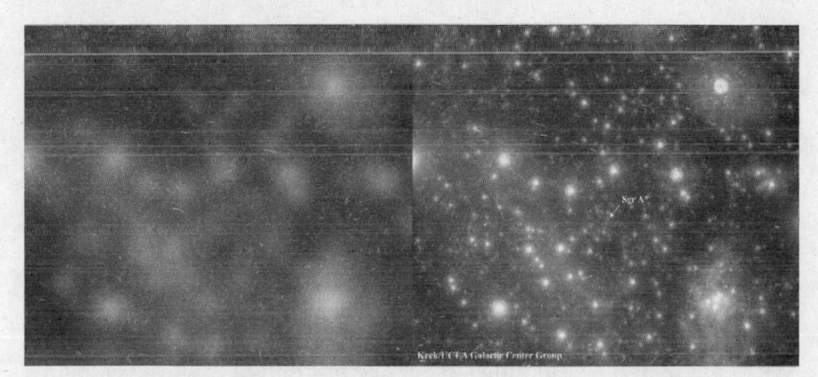

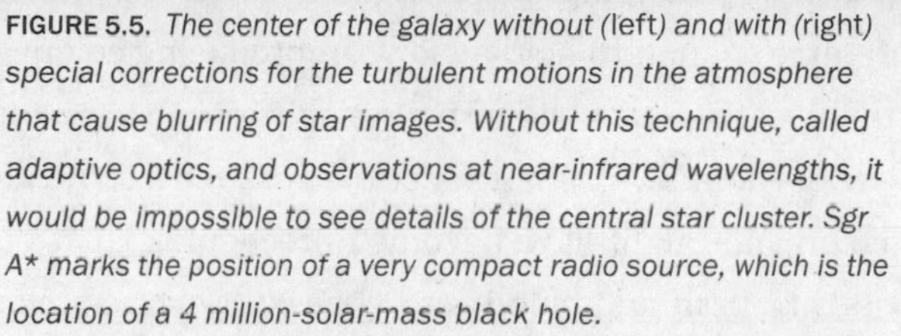

FIGURE 5.5. *The center of the galaxy without (left) and with (right) special corrections for the turbulent motions in the atmosphere that cause blurring of star images. Without this technique, called adaptive optics, and observations at near-infrared wavelengths, it would be impossible to see details of the central star cluster. Sgr A* marks the position of a very compact radio source, which is the location of a 4 million-solar-mass black hole.*

With the black hole proven beyond a reasonable doubt, the research has moved on to using this object as a test of general relativity and theories of black holes. The current state-of-the-art of radio imaging can resolve a scale just three times the event horizon size.[20] It may soon be possible to see signs of space-time being "dragged" around the black hole or watch the death throes of matter falling toward the event horizon. Even more exciting would be to observe the "shadow" caused by the event horizon or test the "no hair" theorem that says black holes are truly monolithic in their properties.

Finally, a supermassive black hole sounds remarkable, but it's not as exotic a form of matter as you might expect. As black holes get more massive, the size of the event horizon grows proportional to the mass but the volume grows at a faster rate so the density is lower. Average density inside the event horizon of the galactic center black hole is a trillion times lower than the density inside a more prosaic stellar black hole, or about 100 times the density of lead. The gravity is strong but the large size of the black hole means that the tidal stretching force is a lot lower. Spared from spaghettification, a voyager is not doomed as he or she approaches the event horizon. This is a trip we could make one day, and see what happens inside nature's time machine.

The journey to the center of the galaxy is so daunting it's rarely even contemplated in science fiction. Our space-faring capabilities are still immature—the nearby stars are far beyond reach. Perhaps one day we'll touch the stars but not before we've invented new technologies. We might do it by perfecting suspended animation or by developing a spacecraft large enough to hold a viable, reproducing colony.[21] Either way, the umbilical to Earth will be irrevocably broken by the trip, with no way to rejoin families that had since died or even a civilization that had moved on and might have become unrecognizable. It's easier to send robots than flesh-and-blood across the gulfs of interstellar space. If we're not alone, the emissaries we meet may be machines.

—

So we end the first part of our journey through space and time. We're humbled by the inaccessibility of the stars. But while the distances are vast, the realm in terms of light travel is proximate. We see planets as they were hours ago, nearby stars are they were decades or centuries ago, and the center of our galaxy as it was a few tens of millennia ago. Looking to the supermassive black hole at the galactic center, gravity has trapped some of the light forever. The light we see left the scene when we were little more than savages, and not sophisticated enough to comprehend the cipher of gravity. The Milky Way is still our cosmic backyard. Greater strangeness awaits.

—

I've been so entranced by looking back in time to the Neolithic Earth that I've failed to notice my accelerating motion. In vacuum there's no sense of speed but I've moved toward the middle of the dense star cluster and the stars are shifting their relative positions second by second. Tendrils of hot gas thread the space between stars, giving the effect of a luminous spider's web covered with sunlit dew.

Below my feet is a swirling vortex of gas emitting flashes and bursts of energy. To me it's down, but I'm in free fall so there's no feeling of motion and no sense of danger. The radiant beauty of the scene pushes away all my anxiety. A few minutes later, there's an amazing sight: the vortex has a black bite taken out of it. The event horizon.

As I get closer and the black ellipse and its surrounding swirl of gas get bigger, I see starlight near the edge of the horizon sheared and distorted like in a funhouse mirror. There's no landmark to set the scale but I know that the blackness is 10 million miles across. It would dwarf the Sun and now it makes me feel like a piece of cosmic flotsam.

Suddenly I have a change of perspective. I'm not falling though space. Space is falling. A waterfall of space is coursing into the black hole and I'm carried along with it. A second realization follows the first, this one more unsettling. The arrow of time reverses and my thoughts unspool in perfect backward chronology. I've just crossed the outer event horizon and time and space have swapped. Moments ago, I had a sense of the freedom of my movement and the inevitability of time passing. Now I've gained freedom over time but my trajectory is irrevocably locked onto the gravitational pit below. My throat tightens. I can still see out to the star fields beyond but nobody can see me. I cannot escape. I've crossed the Rubicon.

I'm in the chaos of space-time between the outer and inner horizons. Space flowing in at near the speed of light meets space flowing out equally fast. The extreme distortions of space are confusing. The black hole seems smaller though I know I'm rushing toward it at greater and greater speed. The outside universe has shrunk to a half-dome of sheared light above my head. I can feel an unmistakable tidal force now. In mere seconds I'll reach the singularity. I release myself into the experience, my fear morphing into anticipation, for what on Earth could compare to this?

PART II
REMOTE

6

ISLAND UNIVERSE

THE PINWHEEL IS ENORMOUS, REACHING THE EDGE OF MY VISION IN EVERY DIRECTION. *Its white tentacles curl out in all directions, their silky cumulus dotted with glittering diamonds. At the periphery each arm fades into the black of night; there's no discernable edge, the light diminishes like a voice lost on the wind. At the center they join in an embrace of yellow yolk. I'm not directly above the center, so from my perspective the overall shape is oval. The central bulge is bisected by a ragged dust lane—a mud track meandering through a field of marigolds.*

I'm hypnotized by the sight. After a while I get the distinct impression of motion. The great spiral is very slowly rotating. But wait, it's too fast, an entire circuit should take hundreds of millions of years. And it's going in the wrong direction; the arms should swirl backward, not forward. Ah. Perhaps it's me that's spinning. I smile at the relativity of the situation.

The galaxy is larger than it was a moment ago. I'm falling in. I've no sense of speed until halo stars start flicking by, moving so quickly each one is a blur. Then a spherical cloud of stars looms in the distance. They're pale yellow-orange, a mustard tinge. The cloud grows. Its core is so dense that I can't see through it and I'm headed directly for it. I brace for impact.

But no impact comes. The swarm of stars spreads out and suddenly I'm inside it. Stars streak by on all sides but the spaces between them are so vast I'm in no danger. For a moment I'm embedded in a halo of dancing light, like a cloud of fireflies. Then back into dark space. I pass through the globular cluster unscathed and descend onto the great spiral, heading for a location two-thirds of the way out from the center, at the edge of a blue-white spiral arm. It looks strangely familiar.

—

NATURE OF THE NEBULAE

Edwin Hubble looked up at the moonless night sky. It was a perfect night. Behind him, the 100-inch Hooker telescope moved smoothly to the next target on mercury bearings and the sound of its drive motors reached him through the warm summer air. He'd driven up to Mount Wilson through orange groves and horse ranches, and the drive gave him time to clear his head and plan the night's observing.

Looking out from the summit across the mostly dark valley, he could see lights from the sleepy town of Pasadena. Beyond and to the left was the city of Los Angeles. He stayed well away from it. Los Angeles was a den of thieves, bootleggers, and brothels; most of the cops were crooked and the mayor's top aide had just been indicted for running a protection racket. Beyond and to the right was the town of Hollywood. He could see the place where the new "Hollywoodland" sign was being erected, and some of the studios were illuminated for night shoots. On the horizon, oil fires flickered off the coast of Long

Beach. He wrinkled his brow involuntarily. Commerce was crass and these lights polluted the darkness of the night sky.

It was 1923. The population of Los Angeles was half a million, but it was a boomtown and would double in size by 1930. The big movies of the year were Cecile B. DeMille's *The Ten Commandments* and *Robin Hood* with Douglas Fairbanks. The four Warner brothers had just set up a studio, and Roy and Walt Disney had started doing animation in the back of a Realty office. Talking pictures, the Oscars, and the stars outside Grauman's Chinese Theater were all in the future. In Pasadena the small campus of the California Institute of Technology was starting to be well-known. One of its founders, George Ellery Hale, had raised the money to build the 100-inch telescope on Mount Wilson. Another, Robert Millikan, had just won the Nobel Prize in Physics.

Ambition was in the air. Hubble felt it tugging at him. But he never suspected that one faint smudge of light would profoundly change our view of the universe.

It was late October and Hubble was inspecting a photographic plate he'd taken earlier in the month. He'd written "N" next to three smudges that had brightened, indicating they were novae, stars that undergo eruptions where they brighten for weeks or months at a time.[1] But comparisons with other plates in his observing sequence convinced him that one of the stars had regular, periodic variations, so he excitedly crossed out the "N" and relabeled that star "VAR!" to indicate it was a regular variable.

The distinction was very significant. Different novae vary in absolute brightness by a factor of 3 or more, so the apparent brightness is a poor guide to the distance. However, Cepheid variables pulsate in a predictable way, and in 1908 Henrietta Leavitt at the Harvard College Observatory found that the intrinsically brighter Cepheids had longer periods of variation and the relationship was tight and well-defined. It meant that the apparent brightness and the period together gave the distance to the star. The 100-inch was the first telescope with enough light grasp to see individual stars at very large distances, and Hubble was determined to use it to probe the size of the universe. He set his sights on an enigmatic class of celestial objects called nebulae.

Nebula comes from the Latin word for cloud, and it had been known since the invention of the telescope that there were fuzzy patches of light that seemed to be associated with regions of star formation. The comet hunter Charles Messier cataloged 103 nebulae in 1781. His list of stationary targets was actually a "reject" list for comet hunters to avoid. Within 20 years William and Caroline Herschel boosted the number to 2500. In the mid-nineteenth century William Parsons, aka the Third Earl of Rosse, turned his big new telescope to the nebulae. It was located in rain-sodden southern Ireland and nicknamed the "Leviathan of Parsonstown." He identified 14 nebulae with spiral structure and claimed that he could resolve the smooth nebulosity of some of them into the pinpoint lights of myriad individual stars. Parson's Leviathan was the largest telescope in the world for over 60 years, until it was eclipsed by the glass that Hubble used in the sunnier skies of southern California.[2]

The stage was set for Hubble's observations by a formal debate three years earlier in Washington, DC. On one side was the brilliant young Harlow Shapley, who was on the staff at Mount Wilson Observatory, until he was appointed director of the Harvard College Observatory. Shapley overlapped for two years with the even younger and equally brilliant Hubble, who was also from the Show Me state. They had a civil but often frosty relationship. In the debate, Shapley chose to defend the viewpoint that the Milky Way was large and the nebulae were gas clouds near its periphery. This was the nebular hypothesis. To Shapley, the Milky Way *was* the universe.

On the other side was the older, and much less flashy, Heber Curtis, who would later become the director of the University of Michigan observatories. Curtis was respected and rigorous, holding to a high standard of evidence before he'd be convinced of a viewpoint. Curtis

FIGURE 6.1. *The Andromeda galaxy (M31) in a view from a modern telescope. In the early twentieth century it wasn't clear if spiral nebulae like this were star-forming regions in the Milky Way or "island universes" distinct from our own galaxy. Hubble identified and measured bright variable stars in the outskirts of the nebula to make his measurement of distance.*

defended the idea that Andromeda (or M31) and other spiral nebulae were "island universes," systems of stars remote from the Milky Way. This idea was first put forward by English astronomer Thomas Wright, who wrote, in a book that also explains the appearance of the Milky Way, "the many cloudy spots, just perceivable by us . . . in which no one star or particular constituent body can possibly be distinguished; those in all likelihood may be external creation, bordering upon the known one, too remote for even our telescopes to reach."[3] The island universe hypothesis was developed and popularized by the influential German philosopher Immanuel Kant, who acknowledged his debt to Wright (Figure 6.1).

Who won: the rising star or the cautious pro? It was a draw, as each protagonist won on some points and lost on others.

Shapley presented arguments that the Milky Way was 300,000

light-years across, 10 times larger than previously thought. He also used the distribution of globular clusters to deduce that we were offset from the center of the galaxy. In both suppositions, he was right. However, he also used bad distance measurements to argue that the Andromeda nebula was contained within the Milky Way, and he accepted at face value a questionable measurement of the rapid rotation of the spiral nebula M101 to indicate that it also had to be in the Milky Way.[4]

For his part, Curtis was correct to assert that the spiral nebulae were external galaxies, even though the decisive evidence was not yet in hand. He pointed out that nova distances put them beyond the Milky Way, and spectroscopy showed they had radial velocities so fast they would escape the Milky Way. He also noted that they were often bisected with obscuring material, like our galaxy. However, he unfairly impugned Shapley's use of the Cepheid variables to measure distances and he was wrong about the size of our galaxy.

Hubble resolved the debate emphatically. He is unquestionably a titan of twentieth-century astronomy, but he had a needless tendency to amplify his own legend. At high school he excelled in the shot put, high jump, basketball, and he was an amateur boxer. His only bad grades were in spelling. In the later telling, boxing promoters tried to persuade him to turn pro and intended to groom him for a shot at a world title belt. He also liked to talk of a duel with a German naval officer whose wife had flirted with him. Hubble studied, but never practiced, law, and worked as a schoolteacher until he felt the calling to astronomy. He served in the army at the end of World War I, emerging with the rank of major. For years afterward, he introduced himself as Major Hubble. He'd been a Rhodes Scholar, and he retained the accent and mannerisms of an Oxford don throughout his life, to the amusement of his colleagues.

Armed with the 100-inch telescope and dark California skies, Hubble devoted himself to studying the nebulae. Within a year, he'd found a dozen more Cepheids and used Leavitt's period-luminosity relation to measure the distance to the Andromeda nebula. It was a staggering million light-years away, well beyond the outer extremity of the Milky Way. He measured the distances of several dozen other spiral nebulae and all were millions of light-years away. Hubble expanded the size of the known universe by a factor of 100.[5]

Hubble is the hero of this story, a giant in the history of cosmology. His story is well sung, in part because one of his many gifts was self-promotion.

The unsung hero is Henrietta Leavitt. All of the distances that Hubble and Shapley measured depended on her discovery that the Cepheid variables can be used as "standard candles." Leavitt went to Radcliffe and became interested in astronomy, but her studies were interrupted by an illness that left her profoundly deaf.[6] She and a number of other women were hired as "computers" by Harvard College Observatory to laboriously scan and measure images on photographic plates. During this work she discovered the relationship between period and absolute brightness of Cepheids. She wanted to pursue her discovery, but her lowly rank didn't permit it. Shapley promoted her after he took over the Observatory, but unfortunately she died of cancer and didn't ever see Hubble use her discovery to recast our view of the universe.

Shapley became an esteemed astronomer, but the discovery of the nature of the nebulae was a blow for him. When Hubble wrote to tell him of the distance to Andromeda, Shapley brandished the letter and said, "Here is the letter that destroyed my universe!"[7] The biggest irony is that he could have made the discovery two years earlier. He handed some plates to Milton Humason, one of the Mount Wilson assistants, to look for rotation of the nebula. Humason didn't see rotation but he saw stars that he thought might be Cepheid variables and he marked them with a pen on the back of the plate. Shapley was sure they couldn't be Cepheids, since he believed spiral nebulae were gas clouds in the Milky Way. So he removed the ink marks with his handkerchief.

SPIRAL ARCHITECTURE

Hubble continued to work on spiral nebulae—now called galaxies[8] to recognize them as systems of stars distinct from the Milky Way—for the rest of his life. He invented a classification scheme based on their appearance that's still in use today.[9] Some of his other ideas, such as the belief that the different types represent an evolutionary sequence, proved to be incorrect. Hubble was the first to use George Ellery Hale's masterpiece, the 200-inch reflector on Mount Palomar. With it he could see far enough to estimate that there are millions of spiral galaxies. Using the telescope in space named after him we now know there are billions.

Since we live in a spiral galaxy, isn't ours the one we know best? Yes and no. We all grow up in families, and think we know them well. But our own family can be an imperfect guide to the properties of families in general. We can certainly see the ingredients of the galaxy we live in, and study its star formation in great detail. But in one sense we're too close. Living in the disk means our view of the galactic center and the far side of the galaxy is obscured by dust. And we'll never get to gaze down on the gorgeous spiral arms. For that, we have to turn to the nearest examples: Andromeda (M31) at a distance of 2.5 million light-years, the slightly more distant Triangulum (M33) 3 million light-years away, M81 in Ursa Major at a distance of 12 million light-years, and the Whirlpool (M51) and Pinwheel (M101) spirals 23 million light-years from the Milky Way.[10]

Even with its fleet feet, light could be excused for being weary after traveling so far. We see the nearest galaxies as they were millions of years ago. Light collected by our telescopes today left M31 and M33 when early humans had brains three times smaller than ours and they were unexceptional among the beasts of Africa. When the light we see left M51, human and apes hadn't yet diverged in the tree of life. And when the light we see left M101, the largest apes were no bigger than a dog. We can't know what these galaxies are doing now. We'll know that in millions of years. We're always stuck with ancient light.

Galaxies take us from the realm of voyages to the realm of history. They're too far for travel or communication, except in our imaginations. But the finite speed of light means we can use them to peer back in time. History unfolds through the observation of galaxies.

Scientific understanding often begins with classification. We didn't catch these butterflies, but they're arranged on black felt all around us, ready for our inspection. Caught in midflight, pinned in a frozen moment of time, spirals have random orientations. Some are face-on, their symmetry in full view (Figure 6.2). Some are edge-on, their disks riven by a coal-like seam of darkness (Figure 6.3). Many appear at oblique angles, like Andromeda, with its spiral pattern foreshortened.

FIGURE 6.2. *The Whirlpool galaxy (M51), a face-on spiral galaxy giving a perfect view of the spiral arms. This image is taken with the Hubble Space Telescope, showing how the bright knots of star formation are found along the spiral arms. M51 has a relatively small central bulge.*

FIGURE 6.3. *This edge-on spiral galaxy, NGC 4565, is also nicknamed the "Needle galaxy" and it was discovered in 1785 by William Herschel. It's 30 million light-years away. Obscuring dust occupies the midplane of the disk and the peanut-shaped bulge is also clearly visible.*

They're not as vividly colorful as butterflies. There's no cobalt blue and no vermilion, just a slender slice of the color wheel centered on pale yellow. However, the patterns are subtle and exquisite. Hubble gave spirals a classification based on the visibility and shape of the spiral arms and the prominence of the central smooth bulge of stars. The Hubble sequence runs from "a," with smooth, tightly wound arms and a large, bright bulge, to "d," with fragmentary, loosely wound arms and a puny bulge.

The Milky Way and other galaxies in the middle of the sequence are called grand design spirals. They have two arms wrapping the galaxy in an elegant embrace, and the arms are decorated with knots of star formation and dusty clouds. Only 1 in 10 spirals has this classical appearance. Just over half have multiple arms, including some with odd numbers of arms and short spurs that branch off the main arm. Looking at these mongrels makes us glad we live with a galaxy with pedigree. Most spirals lack perfect symmetry. A final third have structure so chaotic that the spiral pattern is not visible or is barely discernable; they're called flocculent spirals. Half of all spirals, including the Milky Way, have bars. Bars are linear features that emerge from the bulge and tether the arms. Others have rings of young blue stars and many have dust lanes that thread the nuclear regions.[11] There's plenty here to divert the avid butterfly collector.

The structural components of a spiral galaxy are distinguished by the type of stars they're made of and their orbits. Disks have young and luminous stars in circular orbits; the most vigorous star formation occurs in the spiral arms. A disk is much thinner than it is wide, with the geometry of a large, thin crust pizza. A bulge is like two fried eggs back to back. It's a compact region of older stars moving on elliptical orbits in three dimensions. A halo is a spherical component that spans the disk and extends beyond it; its larger mass belies the fact that it hardly seems to be there. Halos are made of huge globular clusters and individual old stars, all moving on looping elliptical orbits. There's also a supermassive black hole at the center, which can be difficult to detect unless the galaxy is nearby.

The twisting arms that define a spiral galaxy turn out to be an optical illusion. To see this, we have to look at how these galaxies rotate. If spirals turned like solid objects, such as an old-fashioned LP, the rotation speed would increase linearly going out from the center and features at different distances would stay lined up. But spirals have a rotation speed that's almost constant with distance from the center. So outer regions take longer to complete a circuit than inner regions and the galaxy must "wind up," just what's needed to get the spiral pattern. But there's a problem: it happens too fast! It only takes a couple of rotations to make a nice spiral, or about half a billion years, and galaxies like the Milky Way have had time to rotate two dozen times. They should be so tightly wrapped the arms wouldn't be visible.

What else could make a spiral? Imagine driving down a freeway and encountering a lot of older women wearing black leathers on Harleys. Let's call them "Hell's Grannies." They're cautious so they're driving at 40 mph. You approach them at 70 mph and have to slow down to get through the jam. Then you speed up to 70 mph again. Everyone else does the same thing. As seen from overhead in a helicopter, the traffic jam moves at 40 mph but almost all the cars are moving at 70 mph. We can think of this as a "density wave," a density enhancement that moves at a different speed from the objects within it.

In a rotating galaxy, Hell's Grannies form a spiral pattern by marking the places where the density is higher than average. In the inner part of a galaxy, the stars move faster than the wave and they overtake it (like you passing the grannies). In the outer part, stars move slower than the wave and are overtaken by it (the grannies overtaking you). So the spiral pattern can move slower than many of the stars and it avoids getting wound up. It's a transient feature associated with an ever-changing set of stars.

Recall that spiral arms are marked by lots of hot young stars and sites of active star formation. When a star approaches the wave, it speeds up due to the stronger gravity. When it leaves the wave it slows down. So it lingers near the wave, which is why we observe it as a region of higher density—a spiral arm. Increased density causes compression, which causes gas clouds to collapse and form stars. The young stars trace out the spiral arms but they will eventually age and disperse, by which time the wave has moved on to form stars in a different place. Spiral arms are created by roving density waves.[12]

Spirals come in a range of masses and sizes, from less than a billion solar masses and 10,000 light-years across to a trillion solar masses and 300,000 light-years across. The Milky Way is typical of a middleweight spiral so we can use its numbers as illustrative. A spiral may have 400 billion stars weighing 200 billion solar masses—which is a sign that most stars are lower mass than the Sun. The stars divide roughly 80 percent in the disk, 20 percent in the bulge, and 1 percent in the halo. Also, the disk has about 10 billion solar masses of gas, the raw material for forming stars, and it forms about five new stars a year. That sounds like a pathetic rate of star formation, but it would consume the gas in 2 billion years and most spirals are older than that so gas must get replenished or "topped up" from intergalactic space.

There's also a mystery ingredient.

Sometimes an observation is so weird that nobody knows what to do with it. That was the case back in 1933 when Caltech astronomer Fritz Zwicky measured the motions of galaxies in the Coma Cluster and saw them moving far too fast to be held in the cluster by the gravity of the galaxies there. We'll return to his work in the next chapter. But 40 years later, researchers studying the way spiral galaxies rotate were faced with an unpalatable choice: either we don't understand gravity, or most of the mass is a mysterious substance that emits no light.[13]

Here's what they did. Around 1970, Ken Freeman and his colleagues in Australia were using radio measurements of the gas in spiral galaxies to map the rotation of the disks. Meanwhile, Vera Rubin and Kent Ford at the Carnegie Institution in the United States were doing the same thing with optical spectroscopy.[14] Both groups saw rotation speed stay constant all the way to the edge of the galaxies. To see why that's surprising, think of the Solar System. The planets move in orbits that are driven by the Sun's gravity and their speeds are given by Kepler's third law. The orbital speed goes down with the square root of the distance. So the Earth orbits at 30 kilometers per second while Jupiter, five times farther out, orbits at 13 kilometers per second. Mass in spiral galaxies is centrally concentrated so the rotation speeds should start declining well before the edge (Figure 6.4). But they don't.

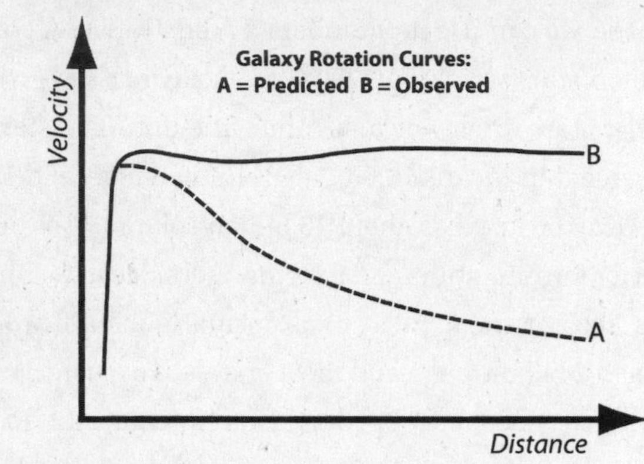

FIGURE 6.4. *A graph of galaxy rotation speed as a function of distance from the center is called a rotation curve. If the mass is mostly in the central regions, the speed of stars and gas in the disk should steadily decline moving out from the center (A). Instead, what's observed is a flat rotation curve (B) and dark matter in a large halo is invoked to explain the rapid motions.*

This was not the first time Vera Rubin had dealt with skepticism (or worse) from her colleagues. She'd earned a BS from Vassar and a master's from Cornell, but her master's thesis presented data on large-scale motions of galaxies in the local universe and that went against standard views of cosmology at the time. It got a frosty reception and negative publicity at a large astronomy meeting. She naively wrote to Princeton for a graduate catalog, but got no answer; Princeton didn't accept female students for the PhD program until 1975. Her PhD thesis at Georgetown dealt with the clustering of galaxies, but that field didn't become mainstream for 15 years so her work was again ignored. She was the first woman ever to observe at Mount Palomar but was not allowed to stay in the observatory's men-only dormitory, nicknamed the "Monastery."

So it was familiar terrain in the 1970s when she began presenting the "rotation curves" of spiral galaxies, and was met with incomprehension and shaking heads. But nature doesn't exist to make us feel good and the data were robust. Rotation has since been measured for thousands of spiral galaxies, and in all cases the rotation speeds are too fast to be explained by the visible matter. Astronomers gradually accepted their unpalatable choice. Since tossing out Newton's highly successful gravity theory was anathema, they got used to living in a universe with a form of matter that exhibited gravity but had no interactions with light. Since the Milky Way also has constant rotation to its outer edge, we're surrounded by dark matter too. Models of spiral galaxies gained a new component: an extended dark matter "halo" with six or seven times as much mass as the entire stellar mass.[15]

The Milky Way is surrounded and pervaded by a trillion solar masses of mystery "stuff"! As Vera Rubin has said, "In a spiral galaxy, the ratio of dark-to-light matter is about a factor of 10. That's probably a good number for the ratio of our ignorance-to-knowledge. We're out of kindergarten, but only in about third grade."[16]

GALAXY ASSEMBLY

Spiral galaxies are magisterial. It's hard to believe that the sumptuous textures and delicately wound arms arose from gravity acting on an amorphous gas cloud. We can hear a faint echo of the design argument of William Paley, a philosopher and theologian of late-eighteenth-century England, who argued that if you found a watch by the side of the road you'd have to presume that it was the work of an intelligent designer, for how could unguided forces lead to something so complex and detailed?

A spiral galaxy isn't a watch, but it's pretty impressive, and it seems too complicated to result from the action of a single long-range force. What can we say about how the leopard got its spots or, rather, how the galaxy got its arms? And assuming spiral galaxies haven't been around since the origin of the universe, how do they arise?

The good thing about being an observer (I'm one) is that nature is bountiful. You can always go out and find more stuff. With any large telescope you can take deep images of the sky and find thousands of galaxies and count and classify them to your heart's content. Theorists don't have it so easy. They must explain why things are the way they are. While nature is prolific, it's also subtle, and the current universe doesn't always leave neatly arranged evidence of origins. It has taken decades of research to have a good idea of how galaxies formed.

In the early 1960s, theorists proposed that a spiral galaxy like the Milky Way formed by the collapse of a large gas cloud.[17] This picture has been called "top-down" structure formation because large things form directly from the primeval gas. Halo stars and globular clusters are frozen "relics" of the early formation epoch, while the gas settles into a rapidly rotating disk and continues to form stars from new gas raining in from space. The top-down theory fell into disfavor when we learned that globular clusters in the halo have a wide range of ages, from 13 billion years old, or nearly primeval, to a relatively youthful 3 or 4 billion years old.

The alternative theory is called "bottom-up" structure formation. In this scenario the Milky Way and other spirals formed by the assembly of smaller pieces over cosmic time.[18] The bottom-up idea is motivated by the fact that structure forms from initially smooth conditions in the presence of a sea of weakly interacting dark matter.

Cooking a spiral galaxy requires a delicate baker's touch. Computer simulations have become a major tool in understanding how galaxies form and evolve. To simulate a galaxy, put dark matter, stars, and gas in a computer (metaphorically rather than literally), switch on gravity and the underlying cosmic expansion, and come back in a few billion years. Dark matter is the container within which the structure forms; it's nonstick since it doesn't interact with radiation. For a long time, simulators had trouble producing spiral disks as neat and thin as we see. When small galaxies merged sequentially over time, the result was always a huge bulge. To abuse the cooking analogy, it was as if you tried to make a thin pizza from little bits of dough and you ended up with a big wad in the middle.[19] Flour—dark matter—is everywhere, covering everything. It's a total mess.

In the last few years, computer power has increased, and simulators have incorporated sophisticated algorithms to handle both hot (stars forming) and cold (gas falling in from deep space) ingredients. Here's the current view of how spirals form and grow. The galaxy is created from a combination of smaller dwarf galaxies and free-floating gas. Disks are built steadily when cold gas flows smoothly onto a galaxy from intergalactic space, bulges are built by mergers, and bars are created by the close passage of a companion galaxy. Forming spiral patterns is a tricky balancing act. Merging with a smaller galaxy can disrupt spiral arms, while the same small galaxy passing close by can induce their formation. New gas arrives from space but the disk also recycles and ejects gas due to the life cycles of massive stars. Spiral arms are transient features and the entire disk may form and reform several times over the history of the galaxy.[20]

Do spiral galaxies show evidence of this complex history? Yes! All the ingredients can be identified. If big galaxies like the Milky Way grew by mergers with smaller galaxies, there should be lots of dwarf galaxies around. We live in a loose agglomeration of galaxies called the Local Group. It spans 10 million light-years and includes us, M31, and M33. Apart from the three large spirals there are three dozen dwarfs, ranging down to galaxies with a millionth the mass of the Milky Way. There should also be lots of gas available. A careful census shows four or five times more gas in intergalactic space than has been consumed to make all stars in all galaxies; the untapped gas reservoir is vast.[21]

Gravity does the rest. Our galaxy has companions that are prominent features of the southern sky: the Large and Small Magellanic Clouds. The Milky Way's disk is warped like a fedora and models indicate that the distortion was caused by a close passage of the dwarf

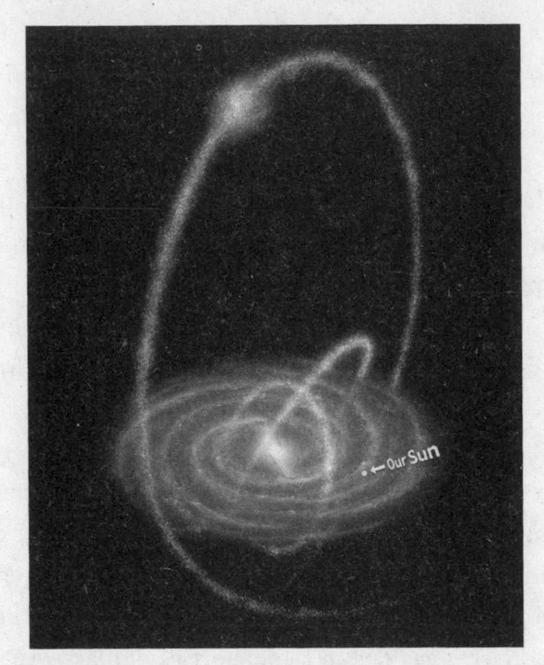

FIGURE 6.5. *This schematic view of the Milky Way shows the location of three "streams" of stars found by NASA's Spitzer Space Telescope. The inner two streams are likely to be disrupted globular clusters while the outer one is made of stars ripped away from a dwarf companion galaxy and strewn along its orbit. The streams are signs that the Milky Way was assembled from smaller units over time.*

galaxies. In 1994, the Sagittarius dwarf galaxy was discovered on the far side of the galactic center. It's on a polar orbit of the Milky Way and seems currently to be plunging into the disk. It's being ripped up and "eaten" by gravity but our galaxy is a messy diner so stars are stretched out into a stream that traverses the halo.

Over the past decade several of these tidal streams have been found in the halo of the Milky Way (Figure 6.5). The observational signatures of a galaxy eating dwarf companions are subtle since the star streams have low density and they disperse over time. Large new spectroscopic surveys of the halo are able to find them by taking spectra of all the stars in a large area of sky. In 2010, two tidal streams were found in the halo of the Andromeda galaxy, so galactic cannibalism is a general phenomenon, and all galaxies seem to grow by assembly from smaller pieces.[22] The gourmand isn't always eating, but splatter marks on his bib and dribbles of gravy on his face show he's being well fed.

IS ANYONE OUT THERE?

I walk in darkness through a stand of pine trees to a rocky lip at the edge of the San Gabriel Mountains, where hang gliders take off. The city glitters below me. A grid of light has been cast like a fishing net from San Bernardino to Santa Monica.

The population of Los Angeles County is about 10 million. I imagine each person burning 10 lights, in their cars and houses, for a total of 100 million. If each light stood in for a thousand stars it would be just like the disk of the Milky Way galaxy. Staring at the city of light, and imagining the stories that accompany each light, I get a hint of the grandeur of a galaxy. Looking up, there's not much to see. There are several dozen bright stars, which barely provide enough of a pattern to recognize the constellations. I can imagine Edwin Hubble preparing for a night of observation with a dark sky. The Square of Pegasus is just visible but even though I look hard in the right direction I've no hope of seeing the Andromeda galaxy.

When I did see Andromeda for the first time through a small telescope, I had a single thought: is anyone looking back?

In 1961, the young radio astronomer Frank Drake wrote an equation on the board at a meeting in Green Bank, West Virginia, and he was the first to establish the Search for Extraterrestrial Intelligence (SETI) as a real scientific discipline. Drake merely intended for his equation to frame the discussion; he didn't realize that it would enter the popular culture. The Drake equation provides an estimate of the current number of intelligent, communicable civilizations in the Milky Way galaxy, and it's the product of seven numerical factors. They are: the rate of stars forming per year, the fraction of those stars that have planets, the average number of planets in any system where planets exist, the fraction of those that develop life at some point, the fraction of those that actually develop intelligent life, the fraction of those that are able to communicate in space, and the length of time they survive in a communicable state. The product of these factors is labeled N, the number of potential pen pals.

The first three factors are increasingly well-known through astronomy research, but the last four are completely unknown. Drake nailed his colors to the mast with his license plate, which says $N = L$. That is, the number of human-peer, space-faring civilizations is equal to their time spent in the technological phase. If we optimistically imagine that time to be thousands of years or more, we have plenty of companionship. Others are less sanguine, and think that N may be much smaller, to the point where we're alone. The issue will be decided by looking, not by rumination or debate, so SETI researchers refuse to be pessimists.

When I think of the vast amount of real estate in the galaxy—a billion habitable moons and planets, give or take a few—it's hard for me to believe they're all stillborn, or that none of them spawned a species full of piss and vinegar like us. It doesn't matter whether N is only a few, or 10, or 100. If there are any intelligent civilizations out there, it's an electrifying thought.

Not only are we unlikely to be alone, we're unlikely to be the first. Our galaxy has been making stars and their attendant planets for 11 billion years. We can look at our example—a habitable planet where it took nearly 4 billion years to evolve a species with technology—and conclude that such an outcome is rare or unlikely. But we could just as well say that since it did happen here it could happen elsewhere, and there are Earth-like planets out there where life could have gotten a 6- to 7-billion-year head start on us. Those creatures might be so advanced as to be unrecognizable.

However many there are in our galaxy, there will be a similar number in our twin, M31. Quite likely, someone or some*thing* is looking back across the gulfs of space. If I actually saw a flashing signal from an alien civilization in M31, it would be a 2.5-million-year-old message.[23] It's unlikely that civilization exists now. Perhaps it only signaled for a short time and that shell of information happens to sweep across Earth 2.5 million years later while I'm alive. Of course, the same is true for me. A signal that I send to Andromeda would take 2.5 million years to get there, by which time I'll be long dead and humans probably won't exist. It makes my head spin to think about it.

We'd like to find kinship in our backyard, on an Earth-like planet a few dozen light-years away. But if some life-forms have a culture that lasts for millions of years, they'll be able to communicate among galaxies. Andromeda is bound to us by gravity and is heading in our direction at a nippy 200,000 mph. In about 3 billion years, these two pinwheels of stars and their potential galactic "federations" will merge. No species will be harmed because the spaces between stars are so large that they almost never collide, even in a merger. However, the approach of another galaxy so that it takes over the sky will be the topic of conversation around whatever passes for the water cooler eons from now.

As I approach the disk, I have the distinct impression I'm falling upward, that the spiral is above me, pulling me with invisible threads. It makes no sense—in space all sense of orientation comes from local gravity. Whatever long journey I've been on, I'm coming home. The disk grows and fills my field of view and the bulge moves to the side, like a great cloud of angry yellow jackets. I'm slipping between two spiral arms, so I look for familiar landmarks like the stellar nurseries of Orion and Taurus to orient myself. Before I see anything familiar I'm in the disk with stars passing me on either side.

There's a pale yellow star dead ahead, with a milky planet off to the side. I look at it expectantly, but it's wrong. It's not Earth. Where am I?

I wheel around and look back in the direction I've been traveling. A spiral galaxy is poised in the distance, impossibly far away. The Milky Way. Fighting a rising tide of panic I reel myself in, focusing my thoughts inward to where I really am, and away from this projection. In my mind's eye it is the Earth. It looks familiar, unchanged. But swooping down through cloud layers there's nothing resembling a human. It's taken light 2.5 million years to reach where I think I am. The apelike creatures on the savannah have learned to use primitive stone tools but their brains are three times smaller than mine; the genus Homo *has not yet emerged.*

I turn back toward Andromeda. It's a mirror image of the Milky Way. But similar isn't as comforting as the same. If this is a doppelganger of my galaxy, is the planet below me a doppelganger of the Earth? And what kind of life would I find there, if I'm brave enough to look?

7

COSMIC ARCHITECTURE

IF EACH STAR IS A STORY, I'M FACING AN UNIMAGINABLE NUMBER OF STORIES. *The cluster in front of me has a huge elliptical galaxy as its tether, at the center of a swarm of hundreds of galaxies. Most of them are smooth and featureless ellipticals; the spirals stand out because they're so rare. There are also thousands of dwarf galaxies, each no more than a smudge of light yet each made of 100 million stars. That must be 1000 trillion stars in total, and I'm stunned just thinking of it.*

Yet I know there's something else defining this space. Something with enough heft to easily outweigh all those stars. Perhaps it's just my imagination but I feel it infiltrating me, subtly permeating me. It has no more shape or form than space itself. It just is.

Then something magical. Something I didn't notice at first, my eyes were so drawn to the giant galaxies. The cluster is laced with the light of hundreds of faint-blue galaxies. They're knotted with star formation and elongated, but the elongations aren't random. Blue galaxies are aligned parallel to their neighbors and the overall appearance is like iron filings drawn into loops around the central galaxy. Could some great and remote intelligence have arranged them like fragments of concentric circles? It looks too intentional and neat to be the result of chance and the mindless force of gravity.

UNIVERSAL EXPANSION

We should take a moment every day to give thanks that the universe is lumpy. In this sense: without the sculpting force exerted by gravity, we wouldn't be here. The universe started life as a hot gas. If it had stayed smooth and gaseous it would be a simpler but duller place. To understand how the universe has evolved, we return to Edwin Hubble.

When we last met Hubble, he was redefining our ignominy by showing that the Milky Way is one stellar system in a universe of thousands of stellar systems spread over tens of millions of light-years. That alone would etch him in the history books. But he wasn't done. His second great discovery set the stage for modern cosmology.

Hubble was a player in the discovery of the expanding universe, and perhaps not the most important one. To adapt a common sporting metaphor—appropriately, given that he was a noted athlete—people saw Hubble on third base and just assumed he had hit a triple.[1]

In 1912, Vesto Slipher was a young researcher at Lowell Observatory and just three years out of graduate school when he started taking spectra of the spiral nebulae. He found that the Andromeda nebula's light had a blueshift—it was approaching us. However, almost all the other spiral nebulae were receding, and at prodigious speeds. By 1915 Slipher had observed redshifts in 11 out of 15 spiral nebulae and when he presented his results at the American Astronomical Society's annual meeting, he got a standing ovation.

Two years later, he had 17 nebulae with redshifts and the average speed of recession was a nose-bleed-inducing 1,500,000 mph (700 kilometers per second). These speeds were so much faster than the speed of any star in the Milky Way that it seemed unreasonable that the nebulae were contained in the galaxy. He wrote, "It has for a long time been suggested that the spiral nebulae are stellar systems seen at great distances... This theory, it seems to me, gains favor in the present observations."[2] Slipher wrote this eight years before Hubble found variable stars in Andromeda and resolved the island universe question.

Slipher later published a catalog of his spectra for 44 spirals and the implications of the redshifts were widely discussed through the 1920s. In 1924, Karl Lundmark assumed that galaxies were standard objects and used their size and brightness to infer distances. When he plotted redshift against distance, he thought there might be a relationship but not a very definite one. Hubble made the next important step, using Slipher's redshifts but estimating more reliable distances with Cepheid variable stars in the galaxies. He found a clear correlation, or a linear relationship between radial velocity and distance, for 24 spirals.[3] This graph, called a Hubble diagram, is one of the iconic images of modern cosmology. The linear relationship is called Hubble's law (Figure 7.1).

Meanwhile, in a parallel progression that mostly took place in Europe, theoretical physicists were grappling with the implications of a radical new theory of gravity proposed by Albert Einstein. General relativity made a connection between the density of mass and energy and the curvature of space-time. The theory was confirmed by a 1919 eclipse expedition, where the Sun was observed to bend light from a distant star by the predicted amount. Einstein and others quickly realized the theory could be used not just for localized gravity but to describe the gravity of the entire universe. Just as space-time near a compact star like a black hole was curved, so the space-time of the whole universe might be curved due to the action of all the matter it contained.

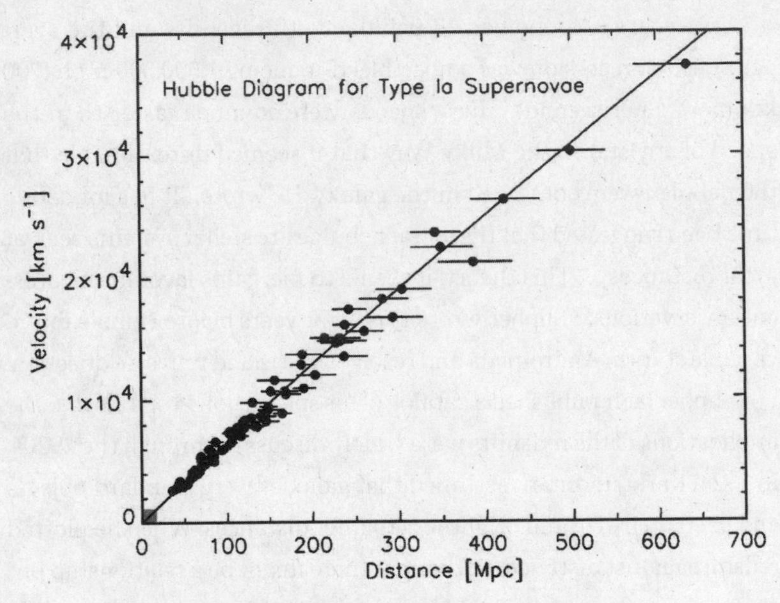

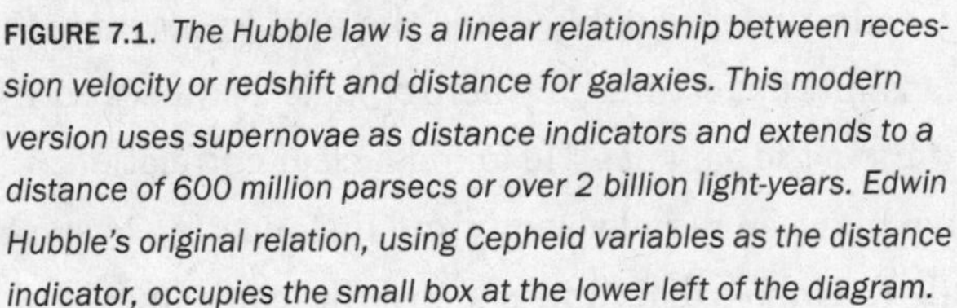

FIGURE 7.1. *The Hubble law is a linear relationship between recession velocity or redshift and distance for galaxies. This modern version uses supernovae as distance indicators and extends to a distance of 600 million parsecs or over 2 billion light-years. Edwin Hubble's original relation, using Cepheid variables as the distance indicator, occupies the small box at the lower left of the diagram.*

Einstein solved his equations for the universe as a whole, but at the time astronomers thought the universe was a single large system of stars with no overall motion. The equations of general relativity are intrinsically dynamic—the solutions want to take flight with expansion or contraction—and to suppress that tendency and produce a static solution Einstein had added a term called the "cosmological constant" to his solution. Because of this he missed predicting the expansion of the universe. He later called this the "biggest blunder" of his life.[4] In the 1920s, both Alexander Friedmann and Georges Lemaître found expanding solutions to the equations of general relativity. By the time of Hubble's 1929 paper, there was a clear theoretical context for the linear relation between redshift and distance.

We live in an expanding universe.

It took a while for observers to fully understand the theory. Even after Einstein visited Mount Wilson Observatory in 1931 to thank Hubble for providing the observational basis for cosmology, Hubble was reticent about accepting the implications of general relativity, writing in 1936: "expanding models are a forced interpretation of the observational results."[5] It was hard for Hubble or anyone else to grasp that galaxy redshifts are not Doppler shifts.

Let's start with the familiar experience of standing by the side of the road while a fire engine or a police car passes by. The pitch of a siren rises as it approaches us and then recedes once it passes us. A pitch is a frequency, so the frequency increases and then decreases. Equivalently, as a source of sound waves approaches us, the wavelength gets shorter, and as the source recedes, the wavelength gets longer. The phenomenon was first described and explained by Christian Doppler in 1842.[6] It works the same way with light waves. A light source approaching us has a shorter wavelength than if it were at rest—a blueshift. A light source receding has a longer wavelength—a redshift.

Intuitively, it works like this. When a source of waves approaches you it "catches up" with its own waves, scrunching them in the direction of motion so shortening the wavelength. When it goes away from you, it "races away" from its waves, stretching them out and increasing the wavelength.[7]

Astronomers were familiar with the Doppler effect and had been using it to map the motions of stars in the Milky Way since the second half of the nineteenth century. However, galaxy redshifts are fundamentally different since they're caused by the expansion of space. The Milky Way is held together by its internal gravity so cosmological redshifts don't become apparent until we enter the realm of the galaxies.

FIGURE 7.2. *Cosmological redshift is distinct from the Doppler effect. The expansion of space-time carries all galaxies away from each other, while stretching or redshifting the wavelength of light waves traveling through the universe. This analogy is in one dimension, but in space the expansion is in three dimensions, which might be curved according to general relativity.*

In general relativity, the cosmological redshift is caused by the expansion of space *itself.* No reference to any object or a particular location is needed to define this kind of redshift. Galaxies are all carried away from each other by the expansion. As light waves travel though expanding space, their waves are stretched by the expansion. The longer and farther they travel, the more they're redshifted (Figure 7.2).

Intuitively, as well as literally, it's a stretch. This may help. Think of the galaxies as beads glued to the surface of a balloon. Blow up the balloon and each bead will move away from every other bead. If you made measurements of the expansion rate, it would have the linear relation between distance and expansion rate that Hubble measured. Now draw a wiggly line on the balloon to represent a wave of light. If you inflate the balloon the wavelength of the waves gets longer or stretches. That's a cosmological redshift.[8] Don't think of the galaxies as moving *through* space; it's not a ballistic situation. Space expands and carries the galaxies with it like flotsam.

At first glance, the Hubble diagram seems to subvert a core principle of astronomy since the time of Copernicus: there's nothing special about our position in space. Surely if all galaxies are moving away from us then we're the center of the universe? Not necessarily. All Hubble's observation shows is that galaxies are redshifted and the farther away the galaxy is, the larger the redshift. General relativity says that redshift is caused by expanding space and the expansion is global, not local. If we could hypothetically transport ourselves to a distant galaxy, we could measure distances and redshifts of a set of galaxies, including the Milky Way, and if we made a graph we'd find the same linear relation that Hubble did. Maybe a distant alien has actually done it, and has proudly announced a law that bears their (probably unpronounceable) name.

The universe was considered timeless in most human cultures,[9] and in the scientific tradition until the twentieth century. But if every galaxy moves away from every other galaxy, it suggests a time when they were all much closer together. We can imagine "rewinding the clock" and going back in time to a smaller, denser universe. A simple calculation based on projecting the current expansion back in time to when all galaxies were on top of one another gives a rough estimate of the age of the universe. It's 14 billion years, about three times the age of the Earth.

If the universe had a beginning, it's a legitimate scientific question to ask how it began. Be patient, and time travel will take us almost all of the way there.

LARGE-SCALE STRUCTURE

We'll return to the question of origins later, but Hubble's work spurred a new discipline devoted to the study of galaxies. Astronomers began to count them to fainter levels, measure their motions, and attempt to understand their properties.

Cartographers in the Age of Exploration mapped out the world for the first time. William Herschel felt a similar sense of exploration as he mapped out the Milky Way in the late eighteenth century. Starting in the 1930s, astronomers explored a vast new terrain of galaxies. In studying this dynamic universe, they made the assumptions that we don't occupy an unusual or atypical location in space, that the region we see around us is a "fair sample," and that same physical laws apply throughout the universe. Earth's early cartographers were in the same boat. They assumed that we don't live on a white sand beach while the rest of the world is a jungle or a swamp. They assumed that their own country gave them good sampling of the range of terrain and features they might encounter elsewhere. They also had to assume that waves, clouds, and geological processes are the same everywhere.

This set of assumptions is called the cosmological principle. Formally: seen on a sufficiently large scale, the universe looks the same for all observers. Colloquially: the universe is the same whoever you are and wherever you are.

It's worth delving into this a bit because the cosmological principle has philosophical implications. At a base level, we assume the universe is knowable and follows rational physical laws. Science wouldn't work if this weren't the case.[10] We also need the concept of an "observer," an intelligent being capable of measuring and contemplating their larger environment. When an owl looks at the sky it sees more stars than we do, but apart from their use for navigation, they're just points of light. Dolphins are sentient, intelligent mammals, but since they live in an aqueous environment, they'll never be astronomers. We have large brains and technology, so we've been able to understand our place in the universe. Perhaps there are superobservers out there, aliens who can do general relativity while they sleep and zip from star to star just by concentrating. To be an observer, you don't have to be a "Master of the Universe," you just have to be "smart enough."

You also have to be in the right place. Arthur Eddington speculated that if we happened to live on a planet shrouded in dense clouds, like Venus, our knowledge of gravity would let us deduce the existence of stars we'd never seen. That's controversial, but even if you grant it there's no way physical intuition would have led us to predict that we live in a vast and expanding universe. There are other locations—the center of the galaxy, inside a globular cluster, near the event horizon of a black hole—where we could live without ever seeing galaxies. Is this our dumb luck? Is it naïve to think the universe wouldn't trick or mislead us?

But the cosmological principle isn't an article of faith. It makes testable predictions: isotropy and homogeneity. These are also attributes of the expanding universe solutions in general relativity.[11]

Isotropic means "the same in all directions." *Homogeneous* means "the same at all locations." Imagine two planned communities, each built using identical houses. One is built on a grid and so is homogeneous, because anywhere you happened to be in the community would look the same. But it's not isotropic because the grid is based on particular directions. The other community is built on streets that are concentric circles centered on a circular park. It's isotropic if you're standing in the central park, because all directions look the same, but it's not homogeneous because the curvature of the streets depends on your distance from the center. You can have homogeneity without isotropy and vice versa. Now imagine a third community, more like a shanty town, where houses are randomly scattered across a wasteland. What you see would be roughly the same at any location and looking in any direction. This most resembles the universe (in two dimensions), where the houses are like galaxies.

The real universe is approximately, but not perfectly, homogeneous and isotropic. We don't see exactly the same galaxies in one direction in the sky as another, but on average the numbers of bright and faint galaxies and galaxies of all types are similar. And our location near a spiral arm in the disk of a midsized spiral galaxy is quite particular, but not extraordinary in any way. If we were at an analogous location in M31 or any other galaxy, the universe would appear more or less the same. Homogeneity is harder to verify than isotropy. We can test isotropy by pointing telescopes in different directions. However, to test homogeneity we'd have to travel to remote galaxies to be sure the universe there looked and behaved the same. By analogy, you could stand in a large forest and say, "It looks pretty much the same in all directions," or you could do the hard work of hiking all over to be sure it's the same. Homogeneity and isotropy follow from the Copernican principle.

Cosmology rests on a bizarre premise. The measurement of invisible, expanding space-time depends on locating galaxies, which gather in clusters and are departures from the smoothness, so they violate the cosmological principle. The gravy is smooth if you ignore the lumps!

In the 1930s, Hubble and Shapley independently mapped the galaxy distribution, just as Herschel had mapped the distribution of stars in the Milky Way 150 years earlier. They both found regions of sky with larger concentrations, or "clouds," of galaxies than usual. There's a large and loose aggregation of galaxies in the direction of the Virgo constellation, a similar one in the southern sky near Fornax, and a fainter but very dense cluster in the direction of Coma. The galaxies were not uniformly or randomly distributed as a central premise of cosmology would demand.

In these photographic surveys, about 70 percent of the galaxies were spiral or irregular, and 30 percent were smooth, reddish, and round or elliptical in shape. Hubble and his longtime collaborator Milton Humason made the striking observation that the densest clouds or clusters had more of the elliptical galaxies, while spirals were more generally distributed, without a lot of close neighbors. Humason's personal story is unlikely and inspiring. A high school dropout, he loved mountains and so got a job as a mule driver taking materials up to Mount Wilson to build the new observatory. He stayed on to become a janitor and was so quick to learn the trade of observing that he became a night assistant. The director of the observatory, George Ellery Hale, recognized his talent and made him a staff member, over the objections of many others on the staff. Most of the observations that made Hubble famous relied on the meticulous contributions of the man with no high school diploma.

Once I got a tour of the plate vault at the Carnegie Observatories in Pasadena, where Hubble and Humason worked. As we inspected the photographic material from the 1930s and 1940s, each plate a few millimeters thick and the size of a record sleeve, the librarian pointed out features and interesting galaxies. The plates were negatives, with black galaxies and stars sprinkled on an otherwise transparent sheet of glass. Some plates had elongated images or were fogged or imperfect in some way. "Those are Hubble's," said my guide, with a wry smile. "He did his best work at the telescope when Humason was driving."

In 1948, the massive 200-inch Hale telescope saw first light on Mount Palomar. Alongside it was the less heralded 48-inch telescope, which had been designed to take images of large swathes of the sky. For a decade, the 48-inch mapped the whole northern sky in red and blue light, generating 1874 photographic plates. The National Geographic Institute, the organization that sent Peary to the North Pole and Byrd to the South Pole, funded the survey. Mapping the night sky was another epic milestone in the history of cartography.[12] Using this new and powerful resource, George Abell made a catalog of 4000 clusters of galaxies. Anyone staring at the plates could see rich textures and patterns in the galaxy distribution.

But the view was incomplete. Imaging surveys couldn't convey the depth of the third dimension.

You're inside a vast forest. Imagine the trees have been thinned out so you can see a long way. Looking around, the forest looks more or less the same in all directions and the trees right next to you are the same

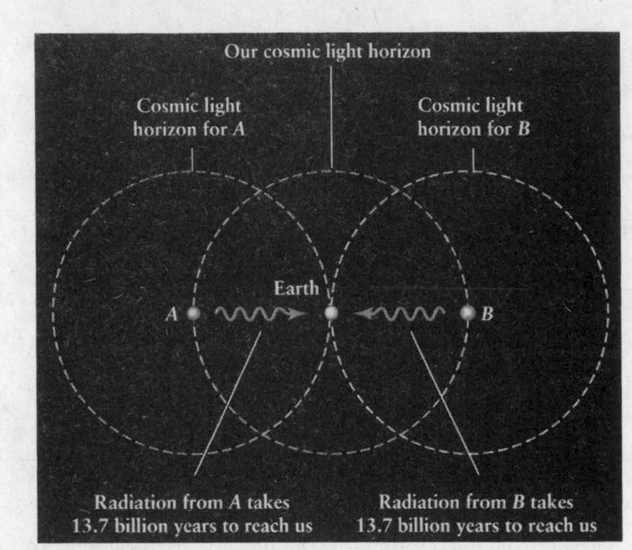

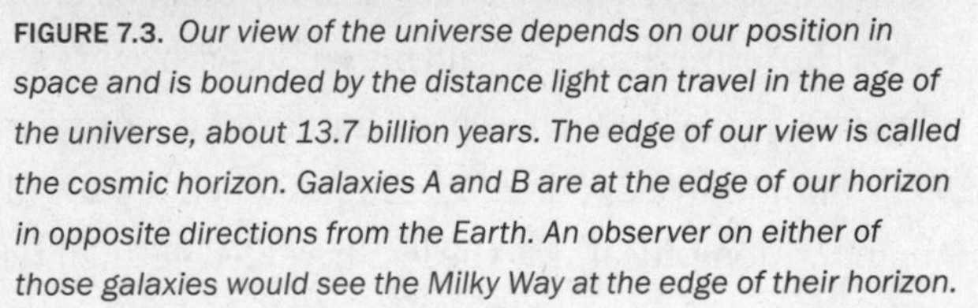
FIGURE 7.3. *Our view of the universe depends on our position in space and is bounded by the distance light can travel in the age of the universe, about 13.7 billion years. The edge of our view is called the cosmic horizon. Galaxies A and B are at the edge of our horizon in opposite directions from the Earth. An observer on either of those galaxies would see the Milky Way at the edge of their horizon.*

kinds you see farther away—the cosmological principle. But your view of distant regions of the forest is imperfect. More significantly you are limited to the single dimension of your circular horizon (Figure 7.3).

Now you're in a helicopter above the same forest. Its shape and size and variations in the density and species of trees are laid out for you like a map. If, for example, the trees had been planted in little circles throughout the forest, that would be impossible to detect standing on the ground, but readily visible from the air. The subtle architecture of the forest becomes clear with the addition of an extra dimension.

So it is with the universe. With no sense of depth everything becomes squashed onto the celestial sphere. Galaxies that are neighbors in the sky may be at very different distances, and the close juxtaposition of galaxies that are far apart in three dimensions may create the illusion of a physical association. (The same is true of constellations; stars in them are often widely separated in space.) Hubble pioneered the use of Cepheids to measure distances, but they can only be distinguished in the nearest few dozen galaxies.

The solution: use redshift as a proxy for distance. Cosmic expansion implies a linear relationship between distance and redshift. So if you measure the redshift of a galaxy—easily done with a spectrum—you can infer the distance. In the 1970s, astronomers started to gather galaxy redshifts and so define their locations in three dimensions.[13] It was too much work to measure redshifts for all galaxies in a region so surveys typically trawled a strip of sky that followed the rotation of the Earth. The results were "slices of the universe." At first, the maps were threadbare, with a few dozen galaxies tracing the bony skeleton of three-dimenstional structure. Improving telescopes, spectrographs and detectors allowed the sky to be strip-mined on an industrial scale; the Sloan Digital Sky Survey has recently produced nearly a million galaxy redshifts.[14]

Astronomers reached for metaphors to describe what they saw in the three-dimensional maps. The scales involved were so immense they headed for the comfort and familiarity of the kitchen. Just looking at the galaxy distribution on the plane of the sky, without redshifts, it seemed that the clusters were meatballs floating on a gravy of single galaxies. With redshifts added as an ingredient, the connectedness of the structures was apparent. Some researchers saw linear structures and these filaments looked to them like a tangle of spaghetti. With so much mess in the kitchen, others had the itch to clean up, and they pointed out that there were interconnected voids and structures with equal volumes, like a sponge. Some of the voids were very large and that suggested a foam of soap bubbles, where galaxies were confined to the intersecting sheets of soapy film.

Who was right? All the chefs were right at some level.[15] The topology of large-scale structure is complex and defies simple description.

The best metaphor used by researchers today is the "cosmic web." There are filaments and walls of galaxies, with voids between, and clusters are found where the structures intersect. Very few galaxies are truly isolated. One number—the fractal dimension—accurately describes the structure on scales from a few million light-years up to a few hundred million light-years. That number is 1.7, or partway between stringy, which is a fractal dimension of 1, and sheetlike, which is a fractal dimension of 2.[16]

Redshift surveys turned up some prodigious structures: a supercluster, or cluster of clusters, that's 550 million light-years across, a giant void a billion light-years across, and a wall of galaxies 1.4 billion light-years long.[17] As Jonathan Swift noted, "a flea has smaller fleas that on him prey, and these have smaller fleas that bite 'em, and so *ad infinitum*." Astronomers worried this progression might continue upward without end, like a fractal, but were reassured when the largest surveys found that the cosmological principle was valid on scales above 300 million light-years. In other words, any 300-million-light-year "chunk" of the universe looks about the same and has about the same number of galaxies as any other. To anyone with really blurred vision, such that they couldn't make out anything smaller than 300 million light-years, the universe is smooth and uniform (Figure 7.4).

Galaxies are flawed markers of expanding space-time. Gravity makes them gregarious; the most likely place to find a galaxy is near another galaxy. Clustering of galaxies can be seen in their motions as well as their positions because gravity forces faster motions when galaxies are concentrated in space.

When astronomers started using redshifts to map out the three-dimensional structure of the universe, they noticed departures from the Hubble law. Hubble's linear relationship between distance and radial velocity (or redshift) is really a statement about expanding space-time. Galaxies tug on their neighbors, and the action of gravity steadily gives them a component of motion distinct from the motion

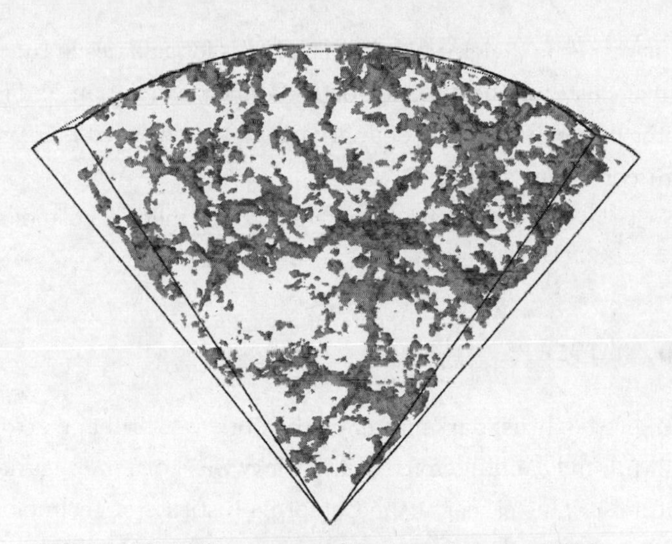

FIGURE 7.4. *Redshift survey map of the three-dimensional distribution of galaxies in a "slice" on the sky. In this picture, we are at the apex of the slice and it extends 2 billion years outward. The gray regions are smoothed representations of where the galaxies are found in this volume. The largest coherent structure, the Sloan Great Wall, runs across the slice roughly halfway out and is over a billion light-years long. (Data are from the Anglo-Australian Two-Degree Field survey and the Sloan redshift survey.)*

caused by the expanding universe. Astronomers call this a "peculiar" velocity, and we can imagine them turning their noses up at such a departure from pristine Hubble flow.

If the gravity within a region of space is strong enough, matter in that region is exempted from Hubble expansion. Galaxies (and everything within them) are held together by their own gravity so don't expand. We're rushing headlong into the embrace of Andromeda and thumbing our nose at the cosmic expansion that would otherwise drive us apart. Dense clusters also seem to be regions where galaxies are contained by their mutual gravity. If that's the case, the spread in their motions can be used to "weigh" the cluster. In 1933, Caltech astronomer Fritz Zwicky measured redshifts for galaxies in the nearest great cluster, in the direction of the Coma constellation, and found that they were like a cloud of angry insects, moving with surprising and disconcerting speed.

So it was that a big black fly entered the ointment of cosmology.

DARK MATTER

The man who thrust dark matter on the world was brilliant, arrogant, insightful, and cantankerous. Fritz Zwicky was born in Bulgaria and spent most of his career at the California Institute of Technology in southern California. He's the most famous astronomer you've never heard of.

Zwicky was pondering the mystery of extremely high energy particles from space—called cosmic rays—and decided they could only come from the explosive detonation of a dying star. He coined the term *supernova* and discovered more of them than any other astronomer. He thought a supernova would leave behind an ultradense core of pristine neutron material. Theorists scoffed at the idea, but later decided it might work and three decades later pulsars were discovered. He used the 48-inch telescope at Mount Palomar to produce a monumental catalog containing tens of thousands of galaxies, including hundreds of clusters, and he correctly speculated that there were many more dwarf galaxies than large ones. His judgment wasn't flawless—he had oddball ideas about gravity, the age of the universe, and redshifts.

I was a postdoc at Caltech, and even though Zwicky had been dead for years by that time, he was a legend. Senior astronomers wistfully told stories about him even though they must have suffered the lash of his tongue. He didn't suffer fools gladly and he bore epic grudges. One of his colleagues was afraid Zwicky might kill him. His department chair called him "vain and very self-centered." In the preface of his galaxy catalog he named colleagues and accused them of stealing his ideas, calling them "fawners" and "thieves." He once said, "Astronomers are spherical bastards. No matter how you look at them they are just bastards."[18]

Zwicky used radial velocities of the galaxies in Coma to measure the mass of the cluster, and compared it to the stellar mass implied by adding up the light of all the galaxies (Figure 7.5). He was stunned when the first number was 10 times larger than the second. He floated four explanations for the unexpectedly large mass of the cluster, all of which were unpalatable to astronomers. Perhaps the laws of physics were different in the Coma Cluster or it was made of unusual stars. Perhaps the cluster had not settled by gravity into its final configura-

FIGURE 7.5. *The Coma Cluster is 320 million light-years from Earth and contains tens of thousands of galaxies, most of which are dwarfs. There are a few bright spirals on the outskirts of the cluster, and two large ellipticals near the center. The spread of velocities or redshifts within the cluster is so large that it must be held together by an invisible form of matter.*

tion, so the velocities weren't a true reflection of the mass. Finally, 90 percent of the mass of the cluster might just be invisible. Zwicky called the last concept "dark matter."

Over the next few decades, the first three options melted away. No evidence came from any other direction for variation in the laws of physics or in the nature of stars, and the Coma Cluster was smooth and spherical so Zwicky's original mass calculation was correct. Then Vera Rubin and others showed that spiral galaxies rotate too quickly for them to be held together by visible matter, and Zwicky's original paper was dusted off and read. It's easy to understand why the topic of dark matter languished for 40 years. Astronomers didn't like the message and many of them wanted to shoot the messenger (perhaps literally in Zwicky's case).

By the 1980s, dark matter was part of the wallpaper of cosmology, even though nobody had any idea what it was. Spiral galaxies were embedded in massive dark halos, and with more difficulty the same was shown to be true of elliptical galaxies as well. Simulators had to feed it into their computers in the right proportions or they couldn't create a universe anything like the one we see.

Final vindication of Zwicky and his daring ideas came in 1979, five years after his death, with the discovery of gravitational lensing.[19] Zwicky realized that clusters were massive enough to bend light, in agreement with a core prediction of general relativity. The bending, distortion, and even magnification of light would be a confirmation of relativity and a new way to probe dark matter. Crisp images from the Hubble Space Telescope showed that old red galaxies in a rich cluster are often surrounded by tiny blue arcs of light, arranged like fragments of concentric circles around the cluster core. Each blue arc is a distorted and magnified image of a background galaxy. Light from the distant galaxy can take multiple paths around and through the cluster, so a single object can produce a mirage with different images straddling the cluster.[20] In some cases, hundreds of background galaxies are distorted in this way by gravity's fun-house mirror.

Light takes eons to trace out this optics experiment. The background blue galaxy may be 5 or 6 billion light-years away. Its light leaves before the Earth has formed. Along the way gravity makes the photons swerve and wiggle slightly. They travel in all directions and only a tiny fraction of them head toward where I will one day be. It's smooth and almost linear sailing for 5 billion years until the photons encounter the Coma Cluster. They follow the warping of space-time caused by the cluster's dark matter. Four of the trajectories veer around the cluster and afterward happen to aim at the Earth, where creatures have just crawled out of the oceans onto the land. After another few hundred million years the photons arrive at a telescope that astronomers have built just in time, which captures them and registers them as four distinct images of a single galaxy. Astronomers standing nearby will marvel at the mirage and mutter something about dark matter.

Lensing has been used to "weigh" dozens of clusters and in each case the mass is dominated by the invisible stuff. If all this talk about dark matter is making you suspicious, you're not alone. It's rather perverse to live in a universe where most matter is invisible and doesn't interact with radiation. All dark matter can feel is gravity. If you had a handful of it, it would pass through your hand and fall gently to the center of the Earth. The universe is mocking us with its secrets.

Since dark matter coexists with normal matter and is only revealed by a gravity calculation, theorists wondered if Newtonian gravity is wrong on large scales. Change the gravity force law slightly and you can do away with the need for dark matter. Luckily, nature provided a perfect place to test for the existence of dark matter (but not its fundamental nature). The Bullet Cluster consists of two clusters that collided a long while ago and passed through each other.[21] Most of the normal matter in each cluster is hot gas, not stars; when the clusters met, the gas piled in the middle like two tossed buckets of water sloshing together, while the galaxies and the dark matter crossed paths like ghosts in the night. Lensing maps prove the dark matter is displaced to either side of the normal matter. Modifying Newton's gravity law can't explain this observation because the direction of gravity has shifted away from the normal matter.

Dark matter is real.

MAKING ELLIPTICAL GALAXIES

I'll never forget the first time I witnessed the bounty of the universe. Trusted to solo for the first time, I gingerly moved the 12-inch-square plate from the tank with developer to the tank with fixer. As the timer completed the three-minute countdown, I moved the plate again to the water tank for final rinsing. I could feel the razor-sharp edges of the glass through my surgical gloves. The darkroom was bathed in the dull red light of a brothel. Normal procedure would be to grab a few hours of sleep and come back to inspect the plate and enter it in the catalog, but I was impatient to see my first Sky Survey plate so I put it in the air dryer and waited. After 10 interminable minutes I moved the plate onto the light table and flicked on the fluorescent light to illuminate it from behind. Eagerly I peered at it through a hand lens.

Dark blobs snapped into focus. From a distance the plate looked like it was covered with ink splatter, but up close many of the blobs resolved into rounded nebulosity and spiral arms. I scanned the plate and saw galaxy upon galaxy scattered among the foreground stars of the Milky Way. I'd seen galaxies before in pictures, like flies pinned to the page. But here were 10,000, caught in midflight just an hour before, and fixed immobile in the photographic emulsion, like flies suspended in custard.

I was twenty-one and just starting a PhD at the University of Edinburgh. I'd been sent to one of their remote facilities, the UK Schmidt Telescope, set in the Warrumbungle Mountains of New South Wales. I was there for an apprenticeship in astronomical photography in its waning days. Charge-coupled devices (CCDs) at the time were like Maseratis—high-performing, but liable to fail and end up as a pile of scrap—while photographic plates were like Ford trucks—durable, versatile, and able to cover a lot of terrain.

The apprentice system is excellent because it accommodates learning and it forgives mistakes. I'd already had a night where I didn't focus the telescope well enough and all the images were donuts not stars. And a night where the telescope lost its guide star so the images were ugly smears. And a night where I'd switched on the white light instead of the red light and spoiled a night's worth of plates. Not forgetting the night where I foolishly rested a plate on a chair and then sat on it—imagine the sound of 144 square inches of millimeter-thick glass splintering into small shards.

I was feeling clumsy and cursed, and I was hungry for success. So it was with delight that I saw the crisp galaxy images strewn over the plate. The UK Schmidt Telescope is a twin of the Mount Palomar 48-inch telescope at Palomar, extending Zwicky's pioneering survey to southern skies. I imagined the brilliant curmudgeon in a darkroom like mine, squinting at newly discovered clusters of galaxies and nodding in satisfaction. The plate I'd taken was 1 of 600 needed to tile the southern sky, just another brick in the wall of astronomy, but that didn't diminish my feeling of accomplishment.

I was dog tired. I had observed all night and processed the plates, so the Sun was high in the sky as I walked back to the dorm. The Siding Spring Observatory is located in a rolling mountain range at the edge of the Australian outback. The air was fragrant with eucalyptus, and kangaroos were grazing by the side of the road. Kookaburras cackled in the gum trees. But I didn't relax too much; these mountains are home to some of the world's most venomous snakes and spiders.

As a newcomer to astronomy, I was amazed that I could travel halfway across the world to trap light from galaxies halfway across the universe.

Our lives are so short compared to the lives of galaxies that galaxies seem timeless. They're ancient but not eternal. The archaeology of spirals tells us they were assembled from smaller galaxies. Galactic construction continues today, but at a much reduced pace because cosmic expansion reduces the availability of gas and smaller galaxy building blocks.

Hubble, Humason, and Zwicky realized in the 1930s that different galaxy types are not found with equal probability everywhere in

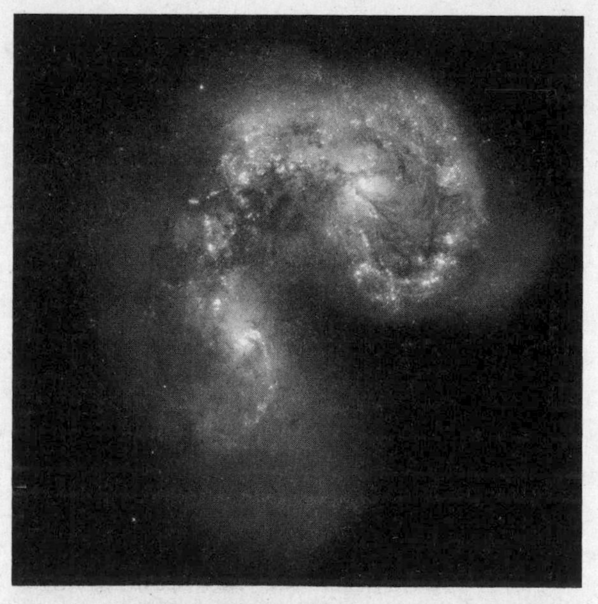

FIGURE 7.6. *The Antenna galaxies started to interact a few hundred million years ago, and within a billion years they will fully merge, the star formation will stop, most of the stars will be sent into elliptical orbits, and the galaxy will resemble an elliptical. Some ellipticals formed early in the universe, but others are formed by mergers like this at later times.*

the universe. Spirals prefer low-density environments like the one that contains the Milky Way and M31, while ellipticals prefer the denser environment of a cluster. Astronomers wondered: was this nature or nurture? Ellipticals have older stars and little current star formation. Perhaps they're frozen relics of the first wave of galaxies. But they're also inhabitants of regions with violent galaxy encounters, so maybe they're the result of interactions and mergers.

The verdict has come down on the side of mergers. Simulations show that galaxies with the look and feel of ellipticals can form either by a series of mergers of small gas-rich galaxies or by the epic collision of two large spirals like the Milky Way. This seems strange, as if vanilla plus vanilla could give chocolate. But the merger process scrambles the neat circular orbits of the stars in the two disks and scatters them into a near-spherical cloud (Figure 7.6). Any gas is either swept away or consumed in a burst of star formation, leaving a nearly gas-free galaxy.[22] Most of the major mergers occurred long ago when galaxy groups assembled and the universe was denser, so the stars we see now are old and red. And some ellipticals have faint outer shells and arms of stars, relics of their violent history. The Milky Way is already the result of mergers and acquisitions, but the grandest event lies in the future when we fall into the Andromeda galaxy in about 2 billion years. From our vantage point near the Sun, we'll have ringside seats for this merger. I have my ticket and a bag of popcorn ready.

The Coma Cluster reels me in. It seems that every galaxy I can see is in its thrall, but I'm the little fish that got away, so it tugs me with the mass of 10,000 trillion suns. These galaxies have turned their backs on the expanding universe and are looping around the cluster on great ellipses. I wonder if my path will send me skimming past the cluster or plunging into the 10-million-degree gas that glows with X-rays in the cluster core.

The light from the distorted blue galaxies has been squashed, sheared, and steered by gravity. With pleasure I note that some of these images have mirror-image counterparts on the opposite side of the cluster. It looks like the optics experiment of some madman from an alien super-race. A shame that Einstein couldn't live to see his masterwork brought so vividly to life. Gliding silently toward the cluster core, the blue arcs peel away from my field of view and a new set of images grows to take their place. Welcome to the fun house.

I feel the enormity of space and time. In either, I'm not even a drop in the ocean. Somewhere behind me—I almost dread to look—is a small world in a faraway galaxy. I see it now, and put my full attention there. By a lucky coincidence its image is lensed and magnified by intervening dark matter so that its details are visible even from this great distance.

Familiar but strange, it no longer looks like home. There are forests of giant ferns and crude trees. Primitive reptiles but no birds or mammals or even dinosaurs. The single supercontinent, Gondwanaland, is girdled by lush swamps near the equator. This ancient light is from the Carboniferous, the time of life's first great conquest of the land. There's nothing for me there. I cast my lot with the sea of galaxies and dark matter that's steadily drawing me closer.

8

NUCLEAR POWER

THE MASSIVE GALAXY SWIMS INTO VIEW BENEATH MY FEET. *At its center is a source of light so bright I have to shield my eyes. Gradually they adjust, and the slender light beam becomes visible. It glows a ghostly blue and reaches with unerring straightness from the center of the galaxy past me into deep space. This is 3C 273, the mother of all quasars.*

I feel the tug of something dark and immense but I try to resist it and move toward the gossamer thread of light. It looked small from a distance, but as I approach I can see it's a fat pipe of fast-moving glowing particles. Suddenly I'm inside it, pulsing light enveloping me, and I feel a surge of acceleration. It's a churning, coruscating river, coursing through the dark canyon of space. Stars outside streak by—their blurs tell me I'm moving incredibly fast. There's a metallic taste of fear in my mouth but there's also exhilaration. I arch my back, spread my arms out wide, and surf the radiant plasma.

The buffeting increases until I'm being shaken like a rag doll.

Then suddenly, and with no warning, I'm flung through the sheath of the glowing beam and back into the vacuum of space. My ride has carried me far above the galaxy; its nucleus no longer hurts to look at.

Scanning the sky, I have little hope of getting my bearings. I'm nearly 2.5 billion light-years from home. The Milky Way is just a middleweight spiral galaxy among thousands that dot the sky. But I have plenty of time, so I keep looking. Eventually I spot the Virgo Cluster and the Local Group at its periphery, and my stellar system and its sibling, Andromeda. With deep concentration I zero in on the Pale Blue Dot.

The planet is almost unrecognizable. There's a primeval continent with ocean surrounding it. It shows the scars of repeated deep glaciations and ice fields reach halfway to the equator. Earth is on the cusp of global climate change that will almost cover it with ice and take the biosphere to the brink of extinction. The exposed rocks have strata laced with red bands, times when oxygen was released into the atmosphere and rusted the rocks. The bands are markers of an epic struggle between bacteria that invented the trick of photosynthesis and primitive life-forms for whom the oxygen released in photosynthesis was deadly. I'm witnessing the biggest extinction event in the planet's history, but I'm grateful, because without it I'd have nothing to breathe and the Earth would be deadly to me. It's my planet but it's so strange it's hard to think of it as home.

MURMURS OF DISCONTENT

As Hamlet told his confidant Horatio, "There are more things in heaven and Earth than are dreamt of in your philosophy."

As the twentieth century unfolded, astronomers considered the universe to be made of stars. The space between them was a tenuous medium of gas and dust, the raw material for making new stars.

Galaxies were assemblies of stars, gas, and dust in different configurations. Yet even before the nature of the nebulae was understood, there were signs of strange activity in the centers of some of them. In 1908, Edward Fath at Lick Observatory took a spectrum of the spiral nebula M77 and saw prominent emission lines of hydrogen, nitrogen, and oxygen coming from its bright nucleus. Two decades later, Edwin Hubble measured similar lines in two other spiral nebulae.

This was intriguing but nothing to make a fuss over. Most spirals had spectra that looked like the superposition of the spectra from a lot of stars; there were sharp features of absorption but not emission. But gas near one or more hot young stars can be excited to a very high temperature by the stars' intense ultraviolet radiation, which creates narrow spectral features from the elements that make up the gas.[1] Hubble duly noted that the nebulae with emission lines had more hot blue stars than those without emission lines.

The mystery sharpened in 1943, when Carl Seyfert published a study of a dozen galaxies with strong emission lines, including the three first noted by Fath and Hubble.[2] Seyfert's beautiful new data showed that the lines came from a bright nuclear region, so compact that it looked starlike in the best photographs. He also noted that the emission lines were broad, spanning a much larger wavelength range than was seen in the gas around clusters of young stars. Some new mechanism had to be operating to make the gas move this fast and be so "hot."[3] Seyfert galaxies, as they are called, were consigned by astronomers to the category labeled "things we don't yet understand."

A second strand of evidence for exotic phenomena in galaxy nuclei came from such an unexpected direction that nobody put the pieces together for decades.

For a decade Grote Reber was the only radio astronomer in the world. He'd read about Karl Jansky's discovery of radio waves from the Milky Way in 1933 and wanted to learn more; he tried to work with Jansky at Bell Labs, but in the heart of the Great Depression there were no jobs available. So he decided to build a radio telescope in his yard. Reber was a ham radio expert, and his telescope was a sophisticated and steerable 30-foot dish—a phenomenal technical achievement for any individual, and much more advanced than Jansky's. In 1939, he discovered an intense radio source in the Cygnus constellation. The position of Cygnus A, as it is called, was too poorly determined to identify an optical counterpart, but it proved to be the first external galaxy detected in radio waves. Clearly something extraordinary was going on: the radio waves from Cygnus A were far more intense than the radio waves from the center of the Milky Way, just as Seyfert's galaxies had optical emission far brighter than anything coming from our galactic center (Figure 8.1).

Radio astronomy took off after World War II as many engineers who had been working on military radar returned to civilian life. Suddenly Grote Reber had a lot of company. In 1946, Martin Ryle showed for the first time that signals from widely separated radio receivers could be combined to simulate a large telescope with much higher angular resolution. This technique is called interferometry.[4] Higher resolution translates into sharper images and more accurate positions for radio sources. But better positions didn't immediately resolve the nature of the radio emission. By 1950, there were 72 radio sources across the sky, and most couldn't be identified with a known optical object. The theoretical prejudice was that they were "radio stars" relatively close to us in the Milky Way because nobody knew an emission mechanism that could make a distant galaxy so bright in radio waves.

Armed with better positions, optical astronomers revisited some of the strong radio sources. Caltech astronomers Walter Baade and Rudolph Minkowski showed that Cygnus A was associated with a train wreck of a galaxy, and argued with each other about its nature. Baade bet that an optical spectrum would show the signature of high excitation and very hot gas and noted with pleasure, "Last week, Minkowski

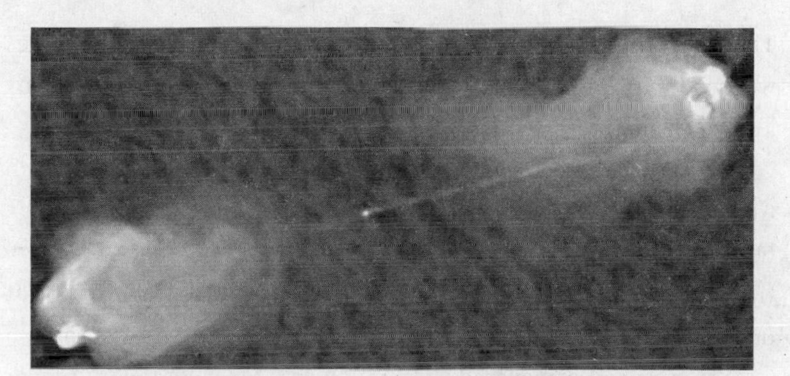

FIGURE 8.1. *Cygnus A is the brightest radio source in the direction of the Cygnus constellation. This radio map shows emission by electrons moving near the speed of light. The bright dot at the center is compact radio emission at the center of a chaotic-looking elliptical galaxy. Jets extend in opposite directions far beyond the galaxy, and end in lobes of diffuse radio emission. Cygnus A is 600 million light-years away.*

got the spectrum of the disputed object and promptly paid off the bet (one bottle of Scotch) which I had made with him."[5] They also noted that Virgo A, the brightest radio source in the Virgo Cluster, lines up with M87, a large elliptical galaxy with a very unusual optical jet coming from its center.

Evidence was accumulating of peculiar phenomena in the centers of some galaxies.

Radio astronomy was maturing rapidly but progress on the nature of the strong radio sources was slow. By 1958, there were over 2000 radio sources but only 7 had been identified with weird galaxies like the one in Cygnus. In the early 1960s, a new catalog of strong radio sources became available and interferometry gave positions of unprecedented accuracy. Maarten Schmidt, a young Caltech researcher, went to the 200-inch telescope at Mount Palomar and got a spectrum of the 273rd object from the third Cambridge radio catalog, called 3C 273.

The spectrum he took made no sense. Strong and broad emission lines were coming from the starlike object, but not at the wavelengths of any known element.[6]

Perhaps the object was made of entirely different substances from the rest of the universe? Nobody liked that suggestion very much. Then he noticed that the pattern of spectral lines exactly matched the spectrum of hydrogen, but only if the object was rushing away from us at one-sixth the speed of light and was 2.4 billion light-years distant!

Schmidt had discovered the first Quasi-Stellar Radio Source, or quasar.[7] Here was a distant galaxy, yet it didn't look like a galaxy. It had thousands of times the total optical light of the Milky Way, but so tightly concentrated that it appeared pointlike. It emitted millions of times more radio waves than the Milky Way. Astronomers worried that the redshift might not be due to cosmic expansion but might be caused by the object being ejected from the Milky Way.

The redshift of 3C 273 was unprecedented, which is why it took so long to figure out the spectrum. Spectral lines that have a particular wavelength in the lab are shifted to longer (redder) wavelengths in the quasar. Galaxies known at that time have much smaller redshifts. For example, the Coma Cluster has a recession velocity of 7000 kilometers per second, 2.3 percent of the speed of light. Spectral lines in Coma galaxies are therefore shifted by 2.3 percent to the red, a small effect. 3C 273 has a redshift 5 times bigger, and the second quasar discovered, 3C 48, has a redshift 16 times bigger.

These redshifts mean that active galaxies are receding very rapidly, and a cosmological model gives a way to convert redshift into lookback time. Light left 3C 273 when Earth had no multicelled organisms. It left 3C 48 when Earth was still a primeval planet, with frequent impacts as a reminder of the early chaos of planet-building.

Over the next decade, a more complete picture of quasars emerged. Deep imaging showed that the point sources live in normal galaxies so they are at the large distances indicated by their redshifts. Better radio maps showed that the emission is often concentrated in an unresolved point source, but other radio galaxies have twin jets and extended lobes of radio emission. Better spectroscopy showed that quasars are related to Seyfert galaxies, but at larger redshifts and with higher luminosities. Most strikingly, optical surveys for sources with strong ultraviolet emission found quasars with weak or absent radio emission. Radio emission was the exception not the rule. The term *active galaxies* was coined to bracket all these phenomena, from low-redshift, mildly luminous Seyfert galaxies to high-redshift, extremely luminous quasars.

Everything now hinged on a single question: what was the source of the phenomenal energy from active galaxies?

GRAVITY ENGINES

To visualize a quasar, imagine being high over a big city at night in a helicopter. We'll use Los Angeles as the example again; the 100 million lights represent a galaxy, where each light stands in for 1000 stars. Variations in the light from quasars show that the energy comes from a region not much larger than the Solar System.[8] If Los Angeles extends 50 miles in all directions, imagine light equal to 1000 times the lights of the entire city combined crammed into a 1-inch space downtown!

Continuing the analogy, if the central light was less intense and the hovering distance not too great, the city would be like a Seyfert galaxy, where the bright nucleus is embedded in a normal galaxy. But if we rise far above the Earth to a distance where the individual lights of the city have faded from view, the intense central light source would still be visible. This is like a quasar.

In 1969, the University of Cambridge theorist Donald Lynden Bell put forward the idea of "gravity engines" as the power source for quasars. He hypothesized that a supermassive black hole would naturally form in the deep gravity potential at the center of a galaxy. The black hole is dark because nothing can escape the event horizon. Strong gravity near the black hole draws gas in to a rapidly rotating disk. Gas at the inner edge of the disk is sucked into the event horizon, and the black hole grows. Gravity and friction heat the disk to tens of thousands of degrees and it glows intensely in the ultraviolet. From a distance, this intense emission swamps the starlight from the surrounding galaxy.[9] This idea is already familiar from solar-mass black holes, where the black hole itself is dark but gas falling in from a stellar companion is heated and accelerated enough to emit UV rays and X-rays. Lynden Bell's hypothesis was very daring since the concept of black holes was new and there wasn't even any evidence for "normal" black holes as a result of stellar evolution. And here he was proposing "beasts" millions or even billions of times more massive.

Why was Lynden Bell driven to this idea? Here's his reasoning. Suppose you try to explain a quasar by normal nuclear reactions in stars. A quasar emits a power of 10^{40} watts, an unimaginable number. (As a comparison, it takes 10^{13} watts to power Earth's civilization, so it would take 1 billion billion billion Earth-like civilizations to power one quasar.) By $E = mc^2$, that amount of energy weighs 10 million solar masses. But nuclear fusion can only convert mass into energy with 0.7 percent efficiency so the "waste" mass left behind in powering quasars by fusion is a billion solar masses. If we imagine compressing a billion stars into a volume the size of the Solar System, the gravitational potential energy would be 10 to 20 times larger than the fusion energy, so the nuclear energy from stars becomes irrelevant. Gravity power always wins.

Broad emission lines are another sign of a gravity engine at work. The velocity spread of the hot gas near the nucleus is thousands of kilometers per second. Gas and stars in the most massive galaxies don't move faster than 300 to 400 kilometers per second. Only an extremely massive and compact object can drive such rapid motion.

Lynden Bell and his colleague Martin Rees continued to develop the theory and they soon accounted for more of the exotic properties of quasars and radio galaxies.[10] It was almost inevitable that the stellar pile-up in the center of a galaxy would create a black hole millions or even billions of times the mass of the Sun. The rapidly spinning black hole had an accretion disk in its equatorial plane and along its poles it acted like a vast particle accelerator. Plasma could be accelerated to 99.9 percent of the speed of light and escape the galaxy, creating jets and lobes of radio emission. Lynden Bell predicted the radiation that would be seen from an accretion disk, that inactive black holes should exist at the centers of many galaxies, and that they could be detected by their gravitational effect on surrounding stars. All of these predictions were confirmed over the following 30 years.

Nature makes not only black holes a few times the mass of the Sun but also monsters several billion times the Sun's mass. Measuring the mass of the black holes requires a technique that maps the motion of gas clouds a few light-months away from the event horizon.[11] Kepler's law then gives a reliable mass estimate. The mildly active galaxy M87 in the Virgo Cluster is the nearest big black hole with a mass 3 billion times that of the Sun. Compared to this, the black hole in the center of the Milky Way is just a baby.

A VISIT TO THE ZOO

In a famous Indian parable, blind men approach an elephant to learn more about it and each one touches a different part of the elephant. The man who touches the leg says it's a pillar, the man who touches the tail says it's a rope, the man who touches the trunk says it's the branch of a tree, the man who touches the tusk says it's a pipe, and the man who touches the ear says it's a fan.

All are right, but none are right, because nobody has information on the whole elephant. Scientists are used to dealing with incomplete information but the problem is particularly acute in astronomy since we've routinely discovered aspects of the universe that were utterly unexpected. Nature seems to take delight in surprising us, which makes it fun to be an astronomer.

Our knowledge of the universe is bounded by our observational tools. The biggest boon to astronomy in the past century was the invention of detector and telescope technologies that worked beyond the visible spectrum. Radio astronomy came of age in the 1950s, followed by infrared astronomy in the 1960s and X-ray astronomy in the 1970s. By the 1980s, astronomers could gather and measure radiation from cosmic sources that spanned a factor of a trillion in wavelength, from meter-long radio waves to gamma rays the size of an atomic nucleus. Most of this wavelength range can't penetrate the Earth's atmosphere so the flowering of invisible astronomy also depended on innovations in rocketry and satellite technology. Prying open the electromagnetic spectrum has been crucial for understanding active galaxies.[12]

Having panchromatic detection capability is important because active galaxies have panchromatic emission. Stuff in the familiar universe—stars, gas, and dust—emits radiation with a characteristic wavelength that depends on temperature. It's called thermal radiation. Stars like the Sun have radiation peaking in the middle of the visible spectrum. The hottest stars emit most of their radiation in the ultraviolet, while cool objects like brown dwarfs emit most of their radiation in infrared light; but they all emit some energy as visible light.

By contrast, the nuclear emission from active galaxies includes many situations when most or all of the radiation is invisible. If particles are accelerated in a magnetic field, the spectrum of the resulting radiation is very broad, with no peak or characteristic wavelength. That's called nonthermal radiation. For example, the bright quasar 3C 273 emits nonthermal radiation spanning the entire electromagnetic

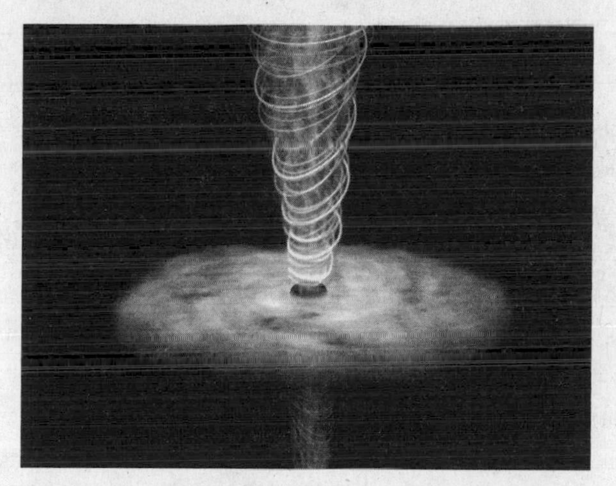

FIGURE 8.2. *Visualization of the phenomena occurring in the centers of many galaxies, where high-energy emission is driven by a gravitational engine in the form of a supermassive black hole. The black hole is fueled by a disk of very hot gas that forms in the equatorial plane of the surrounding galaxy and funnels gas onto the black hole. Gas is accelerated near the black hole and emerges along the poles of the black hole's spin axis, sending jets of plasma deep into space. The jets are contained by the magnetic field that also threads the black hole.*

spectrum, from radio waves through gamma rays.[13] The environs of a supermassive black hole can accelerate electrons to 99.999 percent of the speed of light, an energy much higher than reached by the Large Hadron Collider in Geneva. Spiraling electrons release a torrent of radiation. Physicists guessed this mechanism might be operating in cosmic radio sources back in 1950, and it was soon attributed to radio galaxy jets and the optical jets in M87 and 3C 273 (Figure 8.2).[14] As with the ultraviolet radiation from a hot accretion disk, gravity power from the black hole is the ultimate energy source.

And so to the zoo.

It's easier to spot game under the cover of darkness, so astronomers enjoyed early success using night-vision goggles, in this case working at long radio wavelengths. The radio sky is "quiet" because stars and normal galaxies are not significant radio emitters, so any strong radio source is potentially interesting.

The radio bestiary is fascinating. When radio astronomers perfected interferometry they were able to make maps a thousand times sharper than the images of optical astronomers. The distinguishing features of extragalactic radio sources are cores, jets, and lobes. The radio core is coincident with the center of a galaxy, almost always an elliptical. It's presumed to be the immediate environment of the supermassive black hole. Twin jets emerge from the poles of the black hole. They're made of relativistic plasma sheathed by magnetic field lines. The diffuse and hot gas moves at close to the speed of light out of the host galaxy and into the near-perfect vacuum between galaxies. Jets can be traced for hundreds of thousands or even millions of light-years. Sometimes they wiggle, as if an unsteady hand holds the hose. Sometimes precession of the black hole makes them trace twin spirals in the sky. Sometimes they curve back from the central source as it moves through the thin intergalactic gas, like windswept hair. And sometimes they're straight and unwavering, like death rays.

The plasma in radio jets often moves out at half the speed of light or more. When it ploughs into the space between galaxies it creates a glowing "hot spot," like the place where a blowtorch runs into metal. Then, over several million years, it creates a diffuse lobe of emission. Double-lobed radio galaxies are rare beasts; only 1 in 1000 elliptical galaxies has this kind of activity. The largest radio galaxies are about 15 million light-years across, 100 times the optical size of an elliptical galaxy.

Radio galaxies can't change their spots, but they move. In the 1980s, astronomers learned how to combine radio waves from telescopes on different continents, mimicking the angular resolution of a telescope thousands of miles across. This let them observe the base of the jet, where it emerges from the black hole. Hot spots moved out from the central engine and to their amazement, the hot spots in some radio sources seemed to move faster than light. Sources with this behavior usually have strong central emission and no double lobes. Relativity isn't being violated; the effect is an optical illusion. If the jet points almost directly at us, near-light-speed motion of blobs along the jet can appear as faster-than-light transverse motion. To create a similar effect, imagine shining a powerful flashlight at a distant surface. You could in principle pivot the flashlight and make the spot move across the surface at "warp" speed, but information isn't actually transmitted over the surface that fast.

Spotting active galaxies in the wild using visible light is hard because they're so well camouflaged, their ferocity masked in sheep's clothing. The first quasars were discovered by using radio emission to pinpoint them. Without the radio emission, they appear as innocuous points of light, indistinguishable from stars of the same brightness in the Milky Way. No quasar is visible to the naked eye; the brightest is 600 times too faint, and large numbers aren't found until a million times fainter than the eye can see. Even at that level, there are 100 stars in the Milky Way for each quasar in the distant universe.

How to find the wolf hiding among hundreds of sheep? Two methods proved successful. The first keyed off the nonthermal emission that makes a quasar energy distribution flatter than any star (Figure 8.3). The central engine is extremely hot, so the trick was to look for strong ultraviolet radiation. The second method uses images where the light from each object is spread into a miniature spectrum, and thousands of spectra can be harvested from a single image.[15] The quasars stand out due to their strong, broad emission lines. Using both methods, the success rate leapt from less than 1 percent to 50 percent. Tens of thousands of quasars were bagged. Astronomers filled their trophy cases. And it quickly became clear that the history was misleading: most quasars do not have strong radio emission

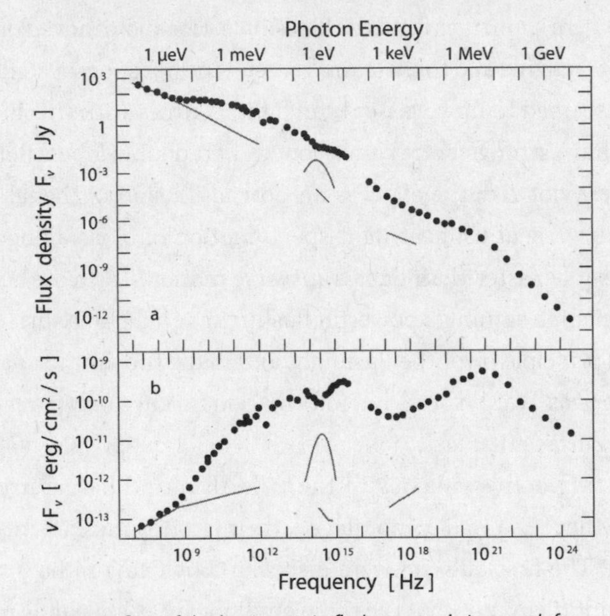

FIGURE 8.3. *One of the best ways to find quasars is to search for their nonthermal emission, which is spread over many orders of magnitude in wavelength, unlike stars or normal galaxies. This graph shows the energy from 3C 273 over a factor of 10^{18} in wavelength, from radio waves to gamma rays. The top panel shows flux and the lower panel shows energy per unit frequency, indicating that the energy budget of 3C 273 is dominated by ultraviolet and soft X-ray emissions. The curved solid line is the light from an elliptical host galaxy, typical of the thermal emission from any collection of stars, and flat lines show the contribution of the jet.*

The Sloan Digital Sky Survey is the current state of the art in optical surveys for quasars. From 2000 to 2008, a modest 2.5-meter telescope took deep multicolor CCD images of more than a quarter of the sky. First, the imaging was used to find a million quasar candidates based on the fact that their colors were unlike any star.[16] Next, spectra were taken of the candidates. The final haul was 120,000 quasars. It's hard to think of quasars as exotic fauna when there are so many in a single survey.

Hunting quasars by their light has been very successful, but if you want to see through obstacles you put on your X-ray glasses. The X-ray sky is as "quiet" as the radio sky so anything that emits X-rays is indicative of high-energy phenomena. A typical quasar emits most of its energy at ultraviolet and X-ray wavelengths so X-ray surveys are the most efficient way to find them. The Chandra X-Ray Observatory can find 1000 quasars in a patch of sky the size of the Moon. X-ray surveys have revealed new populations of beasts. Radio surveys were missing 90 percent of the quasars found in optical surveys, but optical surveys were missing 75 percent of the quasars found in X-ray surveys. The high-energy radiation comes from the inner edge of the accretion disk, only 5 or 10 times the size of the event horizon. This is as close to the dark engine as observations can take us. Although fewer than 1 in 100 galaxies harbors a quasar, they blaze so intensely that their light makes up 10 to 20 percent of the radiation in the universe.

What about the lairs of quasars? The sharp, deep images of the Hubble Space Telescope have shown that radio emitters prefer large elliptical galaxies as nests, while quasars with optical and X-ray emission dominating are more eclectic, living in a mix of spirals and ellipticals (Figure 8.4). Many quasars have close companions that help feed them in two ways. The gravity of the companion can send gas to the inner regions where it fuels the black hole, like it's "stirring their food." And sometimes the small companion galaxy is swallowed whole, which provides a feast for the hungry black hole.

I make no claim to be a big game hunter, but once I traveled to the Caucasus in Russia to try to trap some elusive prey. There's a type of active galaxy called a blazar that has always intrigued me. Blazars are strong, compact radio sources with variable emission. They're thought to be quasars where the view from Earth is down the throat of the relativistic jet. I'd heard of a source called OJ 287, which is the fifth brightest active galaxy in the sky. It was known to vary by a factor of 100, and there were rumors of variations on timescales as quick as 20

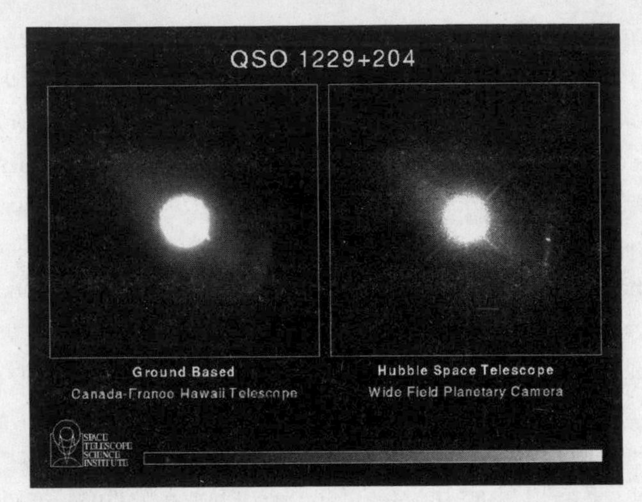

FIGURE 8.4. *Quasar host galaxies as seen with the superior imaging capability of the Hubble Space Telescope. Quasars with strong radio emission tend to be in elliptical or disturbed galaxies, while quasars without strong radio emission are found in galaxies of all types. Just a small amount of fuel in the form of gas or stars will keep a quasar shining brightly, such is the efficiency of the gravity engine at their centers.*

minutes. That's insanely fast, probably meaning the radiation is coming from the base of the jet, within spitting distance of the dark beast. To gather data that fast I needed a big telescope and a special instrument. I called a Chilean colleague who had a photometer that could be packed into three suitcases. Luckily, Santiago said yes immediately; he was always up for an adventure. We applied for time on the Russian 6-meter, one of the world's largest telescopes.

It's just after the fall of the Soviet Union, and Russia is a place full of chaos and euphoria. We fly into St. Petersburg and our hosts take us out for borscht and vodka. The city is epic and grandiose, but decay and banditry are never far from view. Our hosts feign optimism for the future but it doesn't fool us; it's tightly wrapped around a core of dark,

Slavic despair. If they have kids they're trying to get them educated overseas. Among scientists, the brain drain is a torrent. Santiago is prepping the instrument and I'm strategizing for the observations. In spare moments, I visit the library at the Pulkovo Observatory, maybe the world's finest, with original manuscripts by Copernicus, Kepler and Galileo. (Half the collection was lost to arson five years after my visit.)

It's a three-day train ride south to the observatory in the Crimea, through the vast grain baskets of Russia and the Ukraine. We sleep fitfully on rock-hard benches in the compartment, our instrument at our side. Soldiers patrol the train with Kalashnikovs at the ready, on the prowl for thieves who like to roam the trains at night. The soldiers scare me because they look like scared teenagers.

The telescope is in a remote and beautiful mountain area at 6000 feet elevation. We've brought a sturdy and low-tech instrument, which is a big advantage in a place where infrastructure is threadbare. Victor meets us at the airport and helps us set up. He's a bull of a man in ill-fitting clothes, with muscular shoulders, a wide face, and metal teeth strewn through his mouth.

It's early October and the temperature dips below freezing at night. Observing at the 6-meter is "old school." Our instrument is in the prime focus cage, a metal cylinder barely larger than a person where light collected by the telescope comes to a focus. The observer has to sit in the cage all night on a cold metal seat, suspended 80 feet above the mirror, pivoting from side to side as the telescope points toward different objects. Our first night is completely clouded out. I affect a Russian stoicism. In any case, the light from OJ 287 has to travel for 3 billion years to get to the Crimea; waiting another night is no big deal.

The next day, the dry-ice machine at the observatory breaks. Without coolant our instrument won't work. There's no dry ice to be found anywhere nearby, but Victor says there's an ice cream factory 50 miles away, just across the border in Georgia. I get in the back seat of the Lada and two goons with guns pile in on either side, like in a B-grade movie. What do we need them for, I ask? No reason, Victor flashes a metallic grin. Our trip is uneventful and we do a deal for a large brick of frozen carbon dioxide.

Finally, our luck turns. The weather clears, the instrument works, and I'm able to forget my frozen feet as we set our sights on OJ 287. The blazar doesn't disappoint us; the flux is varying minute by minute and it's 20 percent polarized. That's the signature of plasma shock-heated in the acceleration zone close to the black hole. We have an impressive beast by the tail and he's shaking us as hard as he can.

At the end of a long night, I'm cold and exhausted, but Victor and the night crew want to stay up and eat poor man's vegetable caviar and do shots of vodka; it will be rude not to join them. I'm hung over at lunch the next day, so I clear my head with a walk along a path high on the mountainside flanking a deep and green valley. The scenery is spectacular. That evening I tell Victor where I went walking. He arches an eyebrow. You should be careful, he says, Georgian gun-runners use that valley and they don't take kindly to strangers.

Are astronomers fooled by the bestiary, just as the blind men were fooled by the elephant? Are all the creatures just described really the same beast? Maybe. At the center of an active galaxy is a spinning black hole and an accretion disk, and they won't look the same from every direction. Starting in the 1970s, various "unification" schemes were proposed, where sources with different apparent properties might be intrinsically the same, but seen from different angles.[17] OJ 287, the blazar of my attentions, is seen pole-on, but edge-on the jets would be laid out on the plane of the sky and we'd see double radio lobes. Seyfert galaxies and quasars have broad emission lines, but they represent a large set of mildly active galaxies with narrow emission lines, where the broad lines are hidden from view by a donut-shaped cloud of gas and dust. The unification idea proposes that a common central engine inhabits all active galaxies, but obscuring material can alter our view of it, or even remove it from our samples entirely.[18]

FEEDING THE BEAST

The brightest active galaxies are extraordinarily energetic. If M31 had a quasar at its center, the spiral galaxy would still be barely visible to the naked eye, but the nucleus would be brighter than any star in the night sky. If the Milky Way had a quasar at its center, it would appear as a second sun in the daytime sky.[19] How do quasars form, how long do they live, and is there something special about a galaxy that becomes one?

Imagine you're a substitute teacher, assigned to teach one lesson to a third-grade class in an unfamiliar school. At the front of the classroom is a large bowl of Tootsie Pops. Evidently these kids are allowed to eat candy in class. Even though there's enough for everyone, only 4 out of the roughly 40 kids are sucking on a Tootsie Pop.

You're intrigued. Why are only four kids eating the Tootsie Pops? Do the rest of the kids not like them? If you stayed in the classroom and taught this class through its entire day, would all the kids eat Tootsie Pops at some time or other? Would the kids who are eating them now eat them throughout the day, or only some of the time? Would you be able to figure out the answer by counting how many Tootsie Pops were eaten in a day?

In this analogy, kids and Tootsie Pops are stand-ins for galaxies and quasars. We see that a small subset of galaxies host quasars—most kids aren't eating Tootsie Pops—but we don't know if it's something special about those galaxies or if at some time or other every galaxy hosts a quasar. Counting the total number of quasars doesn't help—how many Tootsie Pops are removed from the bowl—because of the ambiguity between a few galaxies being active all of the time and all galaxies being active some of the time. We're in a similar situation to the forest dweller mentioned earlier, trying to infer the life cycle of trees from a single snapshot of the forest. All we can say for sure is that quasars discovered in the early 1960s are still "there" and shining bright, so they live longer than 50 years.

We can learn more if we know how the beast gets fed. A luminous quasar, extracting mass-energy with 10 percent efficiency, can be powered by eating just 3 solar masses of material per year. If the black hole prefers snack food to three square meals (though stars are really spherical), it would get the same nourishment from scarfing a million earths a year. In practice, black holes rarely eat planets and stars whole, their normal diet is hot gas siphoned from the inner edge of the accretion disk.

If a medium-large black hole 100 million times the mass of the Sun grows by eating 3 solar masses a year, then it would take only 30 million years to grow the black hole from scratch, a small fraction of the age of the universe. That's if the black hole always eats at the rate we observe now. But if it takes long breaks between meals, the time for it to grow to the present size might be much longer.

Another argument comes from the available supply of fuel. Spiral galaxies have several billion solar masses of gas in their disks, but most of it is too far to have any way of getting to the center. Within the central 10 light-years, there are about 30 million solar masses of gas in the larder. Eating it at a rate of 3 solar masses per year, it would last about 10 million years. If a quasar is starved of fuel after 10 million years, then its active phase is less than 0.1 percent of the age of the universe. It then has to wait a much longer time for new fuel to dribble in from intergalactic space.

Research in the past decade has led to an intriguing picture of activity in galaxies. We've talked about accretion as the engine for quasars as "lightbulbs," but quasars don't just accrete matter, they also eject it. A luminous quasar has a hot wind coming off the accretion disk that can drive out gas from the nuclear regions, shutting off the supply of fuel.[20] The quasar is quenched and has to wait for gas to accumulate in the center of the galaxy before it can brighten again. So 10 million years of brightness are followed by an even longer period of darkness. Another strand of evidence comes from black holes in normal galaxies. The Milky Way has a modest black hole and other nearby galaxies also have black holes, but most of them are inactive. It may take a special set of conditions, and a very large mass, for a quasar to form.

Activity in galaxies is democratic. The biggest galaxies have the most fun, but every galaxy gets a taste. Around the universe, quasars are flaring up and sliding back into darkness all the time. Everyone gets a Tootsie Pop—you just can't suck on it all the time. Or to use an earlier analogy, every sheep has a bit of the wolf in them. Just as we all have a bit of the beast in us.

—

I've been so engaged in my connection to the ancient Earth that I've lost track of my surroundings. The large galaxy that had shrunk from view as I was ejected from the jet is growing again. A trillion star's worth of gravity is pulling me in.

I fall on an arcing trajectory into the central star cluster. Ahead is a white-hot gas disk; I swirl toward it like a piece of debris moving into a drain. Suddenly, I'm inside the accretion disk and surrounded by blinding light and heat. I circulate for a while; the steadily increasing heat and the light shading from white to blue tell me I'm spiraling inward. Then I'm free of the disk, on a final spiral into the beast. The dark object is enormous. Where the black hole left behind when a star dies would be the size of a city, this black hole is the size of a solar system.

There's beauty and tranquility in this intertidal zone between the radiance of the disk and the fathomless darkness of the black hole. I enter a luminal state of consciousness; all the events of my life are in clear view but ahead is a place where time will have no meaning. As I approach the event horizon, the light cone shrinks and the rays gain energy until the universe has shrunk to a circle of blue light.

I am not afraid. Inside the event horizon of this billion-solar-mass black hole, the density is less than the air in my living room (a living room that is now a quadrillion miles and billions of years away). Tidal forces in such a black hole are mild; I feel no more than a slight tug as I slide across the event horizon. There's enough gravity here to power the dreams of every madman in the universe.

9

THE GROWTH OF GALAXIES

THE SCENE IS EPIC. *I'm embedded in a vast, three-dimensional building site. But this construction is of galaxies and supermassive black holes, not houses and offices. At my previous stops a quiet stillness and a feeling of space and grandeur flowed over me. Here, I can't see actual movement but the sense of motion is palpable. The galaxies are small and close together. The feeling is claustrophobic, vertiginous.*

It's hard to see the forest for the trees, but the galaxies don't seem to be randomly placed around me. If I look carefully, I can see small blobs in the act of merging with larger blobs, and a gathering of young galaxies that looks like an incipient cluster. I search in vain for the familiarity of a grand design spiral like the Milky Way. But these stellar systems are ragged and blobby and have little coherent structure. The few large galaxies look like ellipticals but they're hotter and bluer than those I'm familiar with.

As I look longer I notice something more subtle: a hot web of gas that connects the galaxies to each other, like glowing lace. It's warm. I'm bathed in ultraviolet radiation from young stars and the accretion disks that girdle black holes. At the centers of many galaxies, black holes grow quickly as they feast on abundant gas. Some are shrouded in dust and hidden from view. Others have blown away the surrounding gas and triumphantly sent jets punching through the murk into space, the only linear features in the filamentary chaos. In this cosmic tanning salon, I put on my imaginary shades against the glare.

—

THE FASTEST THING THERE IS

To tell the story of galaxies, from their formation to the present day, we first have to look at light, and how it travels and is gathered in an expanding universe.

Imagine the inconvenience if light were sluggish. Suppose it moved at 1 meter per second, a fast walking clip, as opposed to its measured speed, which is a blistering 300,000 kilometers per second. Everything in this hypothetical world is the same except the speed of light.

You enter a darkened house and switch on the overhead light. You patiently wait a few seconds to see the light. Then you watch as the nearby walls—and a while later the far wall—become visible. The speed of sound is still 343 meters per second, much faster than light, which leads to some interesting consequences. Face-to-face conversations tend to happen at a fixed distance of a few feet; everyone is used to lip motions slightly lagging sounds, but it gets annoying if it lags by more than a few seconds. TVs have a delay button on the remote so you can adjust the sound to match the several seconds it takes the picture to reach you across the living room.

Walking to the supermarket you wave to a neighbor across the street and half a minute later he waves back; not because he's rude, but that's how long it took for him to see your wave and for his wave to reach you. You spot a friend you haven't seen for a while. He's at the end of your aisle inspecting some salsa. By the time you reach him, he's long moved on and is nowhere to be found. A new checkout line opens but people standing closer saw the light go on seconds before you and are all in line as you get there.

Driving is extremely dangerous. The road ahead of you contains older and older situations. People you see walking on the sidewalk 10 yards ahead might not be there any more since you're seeing them as they were 10 seconds ago. Approaching cars are already upon you as you see them. Traffic 100 yards ahead might be completely different when you get there, since your information is almost two minutes out of date. That's during the day. Don't even think about driving at night. All time lags double because light has to stroll from your headlights to an object ahead and then stroll back to your eye. If you drive faster than a meter per second you'll outrace the light from your headlights, rendering them useless.[1] In fact, just running down the street makes events play out backward, which can be quite amusing.

At the end of another long, disconcerting day, you console yourself with the sight of a beautiful sunset. Remembering, of course, that the hilly horizon is 20 miles away so the Sun set 10 hours ago and is almost ready to rise.

OK, I agree, it's more than inconvenient. It's downright disturbing. The "slow light" world is crazy because everyday events are morphed out of recognition. Light is our main information carrier. If the time it takes information to reach us is similar to the time it takes for events to occur, causality appears to be scrambled. About the only familiar object for which slow light makes no difference is the Sun. If light traveled 300 million times slower, sunlight would take 4500 years to reach us. That seems long, but the Sun is (thankfully) unerring in its light output, so a delay of a few millennia would make no difference to our lives.

Light isn't slow, it's incredibly speedy. How do we know? The Greek scientists debated the issue; Empedocles thought light from the Sun must take some time to reach us, but Aristotle thought that it travels instantaneously. The Greeks did no experiments to resolve the issue. Galileo was the first to try, standing on a hilltop and holding a lantern while a colleague stood a few miles away and held a similar lantern. Galileo uncovered the shutter and his friend uncovered his shutter as soon as he saw light from Galileo's lantern. Galileo tried to use his pulse to record the round-trip travel time of the light. No lag was detectable but Galileo was able to argue that light was at least 10 times as fast as sound.

The first true measurement of the speed of light was made in 1676. Ole Römer had been observing periodic eclipses of Io by Jupiter when he noticed that the timing of the eclipses sped up and slowed down as the years passed. He attributed it to variation in the distance between the Earth and Jupiter. When the Earth is farthest from Jupiter, the light from Jupiter must travel an extra distance: the diameter of the Earth's orbit of the Sun. As a result the eclipse occurs later than when Jupiter and the Earth are closest. Römer's elegant but crude method gave 125,000 miles per second, or three-quarters of the correct value.[2] Modern measurements nail the speed very precisely at 299,792,548 meters per second.[3]

It's not just pretty zippy; light speed is absolute and an absolute limit. So said Albert Einstein, who as a teenager wondered what it would be like to "ride on a beam of light."

Einstein's intuition was influenced by the result of a famous experiment in 1887 that showed that light traveled at the same speed regardless of the direction of motion of Earth in its orbit of the Sun.[4] His "special" theory of relativity postulated that there is no preferred reference frame for propagation of light. All the delightful weirdness of relativity follows from the constancy of the speed of light. The clocks of rapidly moving objects run slower. Objects shrink in the direction of motion. Objects gain mass as they approach the speed of light.[5] This last effect is the key to light being the fastest thing there is. Attempting to accelerate any particle or object to the speed of light makes it heavier; instead of going faster the object approaches the speed of light but never reaches it, while continuing to get more massive as energy is converted into mass by the relation $E = mc^2$.

It's just as well, since faster than light travel would lead to impossible situations. In the "slow light" world we imagined, by outrunning light you could make time run backward. Physicists saw that information traveling faster than light would shred common sense. Suppose you had a hypothetical gizmo called a tachyonic antitelephone that could communicate at twice the speed of light.[6] If you were traveling on a spaceship away from Earth at 90 percent of the speed of light while talking to your mother, you'd get the reply from her before you'd sent your first message.

Aristotle was essentially right about light in the everyday world. For practical purposes, it travels instantaneously. When I flick on a light switch, I don't have to stand and wait; light illuminates the room in a ten-billionth of a second. Galileo's hilltop experiment was doomed since light streaks between lanterns in less than a hundred-thousandth of a second. Those who watched the Apollo Moon shots will recall the awkward hiccups in conversation caused by the two-plus seconds it took for radio communication to go to the Moon and back. Moving out into the universe causes an increase in the time it takes for light to reach us.

The vignettes that open and close each chapter of this book visualize the increasing dislocation of a traveler who must experience his home in the past rather than the present. Light takes eight minutes to reach us from the Sun, so we see it as it was eight minutes ago. Light takes four hours to reach us from Neptune, four years to reach us from the nearest star, and 2.5 million years to reach us from M31.

It makes no sense to ask what the objects look like *now*; the speed of light imposes a limit on information-gathering. Imagine that the speed of light was infinite. We'd have a simultaneous view of everything in the universe. Everything happening anywhere would be visible now. The entire history of the universe would appear simultaneously and instantly. If the universe had an edge we'd see it, and if the universe had no edge we'd see an infinite distance in all directions.[7] That's like information overload! While 300,000 kilometers per second may seem incredibly fast, in the chasms of space it can take a substantial time to get here from somewhere interesting.

The corollary of the finite speed of light is the fact that we see distant objects as they were, not as they are. Distant light is old light. This phenomenon of lookback time is a central feature of cosmology because it adds the dimension of time to the science of the universe. Geologists study surface rocks to see how the planet behaves now, and they dig down through the strata or layers of rock to see how it behaved in the past. Similarly, astronomers study nearby galaxies for a current snapshot of the universe and distant galaxies to learn what the universe used to look like.

Hubble expansion means that the universe is getting larger and more diffuse as the distances between all galaxies increase. The universe is evolving but that need not be true of galaxies. After all, galaxies are like paper clips attached to a stretched rubber band or beads glued to an expanding balloon. They seem self-contained and separated from the underlying behavior of space-time. The only way to find out is to observe them at large distances and see whether galaxies then are different from galaxies now. For this purpose it's essential to define "now" and "then." The cosmological principle holds if we take a chunk of space 300 million light-years across, real estate that holds about 100 million galaxies. That fair sample of the universe contains so many galaxies that even the rarest types like luminous quasars will be represented. The lookback time of 300 million years is only 2 percent of the age of the universe, so it qualifies as now, or at least as recent. If the universe wasn't evolving there'd be no reason to look further. We'd just see more of the same.[8]

However, by the middle of the last century astronomers suspected that the expanding universe was very different in the distant past from how it is now. Not only was it smaller and denser, but it must have been hotter. So they were highly motivated to search for fainter galaxies at larger lookback times. To answer the question, how did it begin? we must look out in space.

BIG GLASS

In modern cosmology, telescopes are time machines, and astronomers are armchair time travelers, venturing back toward our origins.

To march on out through the universe and back in time, isn't it enough to gather more and more light? It's not that simple. If all galaxies were identical and had a well-regulated brightness like a standard lightbulb (try finding a 10^{38} watt lightbulb in a hardware store), the detection of galaxies at different distances would be straightforward. Suppose a galaxy like the Milky Way 100 million light-years away was marginally detected with a one-hour exposure on a 1-meter telescope. Let's crunch some numbers and see how to do better. Remember that light from any source goes down by the square of the distance. To see twice as far you'd need to gather four times the light. That could be done by collecting light for four hours, or by using a 2-meter telescope, where one hour of observation would gather the same amount of light. Light leaves any galaxy and diffuses in all directions through space, so the difficulty of detection increases with the square of the distance.

A big complication to this picture is the fact that galaxies aren't all the same. As with stars, there are many dwarfs for each large galaxy. The galaxies a lot more luminous than the Milky Way are very rare but for every Milky Way there are about three moderate galaxies with 10 percent of the Milky Way's brightness and about 10 puny dwarfs with 1 percent of the Milky Way's brightness.[9] Any survey that can detect a Milky Way–type galaxy at a particular distance can detect dwarf galaxies at a smaller distance, and although it searches a smaller volume for dwarfs than for giants that's offset by the larger number of dwarfs. The result is that any deep image of the sky will have a few luminous galaxies at large distances, but it will also be "littered" with many puny galaxies much closer. Sorting them out requires spectroscopy to determine redshifts, and that needs a big telescope.

The expanding universe giveth and it taketh away. Expansion bequeaths us with a large universe where by looking out we can look back in time and learn about the origin. But the expansion of space-time dims the light of a distant object by more than the amount predicted by an inverse square law. There are two extra effects.[10] The arrival rate of photons is reduced because each photon travels farther to reach us than the one before. Moreover, the energy of each photon is reduced by the stretching effect of redshift. Let's see how this plays out for galaxies at substantial redshifts and lookback times. Relative to a Milky Way type–galaxy at a distance of 300 million light-years, which is the edge of the "local" universe, it's two times harder to detect than the same galaxy 25 percent of the age of the universe ago, 10 times harder 50 percent of the age of the universe ago, and 60 times harder 75 percent of the age of the universe ago. That difficulty is the factor of additional light that must be gathered, which is why astronomers push to build larger and larger telescopes.

When last we left the history of telescopes, Edwin Hubble and Milton Humason were using the newly commissioned 200-inch at Palomar to survey the architecture of the cosmos. Nicknamed the "Big Eye," the Palomar reflector was the world's largest for nearly half a century, if

FIGURE 9.1. *Modern large telescopes use either a segmented design where light from the individual segments is combined, as in the Keck telescopes, or a single monolithic mirror, as in the 8.4-meter Large Synoptic Survey Telescope, shown here. The honeycomb molds create a mirror that is very light for its size.*

the initially flawed Russian 6-meter is discounted. Telescope building hit a cost "wall" because the mirror has to move on a mount and be protected by a dome. Over a century from Herschel to Hale, the cost of telescopes grew by nearly the cube of the mirror diameter. That's more than the growth in collecting area so it's no bargain!

Technical innovation was needed to break this ruinous cost curve. In 1993, the first Keck 10-meter telescope saw first light on Mauna Kea in Hawaii, followed three years later by its nearby twin. The Kitt Peak 4-meter telescope had been built in 1973 for a cost of $10.7 million. Predicting from the cost scaling and updating for inflation, the Keck telescope should have cost $400 million. In fact, it cost $100 million, quite a steal. By using lightweight hexagonal segments to mosaic the parabolic surface, the Keck telescopes were smaller and lighter than any big telescopes built previously. Each Keck scope has six times the collecting area of the Kitt Peak telescope but similar moving mass and dome size. Jerry Nelson at the University of California is the architect and pioneer of the mosaic approach (Figure 9.1).

Roger Angel took a parallel road to large telescopes at the University of Arizona. As a professor there, it furrows my brow to know that the football coach gets paid twice as much as the president and many times more than the faculty. As I point out to my students, the best thing that happens in the football stadium isn't on the turf but under the east stand, where Angel and his team make the largest and most accurate mirrors in the world.

Angel's approach is to fabricate a single large mirror, but keep it light and give it a tighter curvature so the telescope it's part of can be very compact. The genesis of his idea has a strange echo with the story of Phil Knight, the founder of Nike, who was a middle-distance runner at the University of Oregon and frustrated with the clunky running shoes of the time. Experimenting with his mother's waffle iron (and ruining it in the process), he realized that he could make shoe molds that were light and flexible. Angel also experimented with a waffle iron, but with plastics and Perspex, and found that a honeycomb structure gave the material strength, rigidity, and lightness. He probably trashed several waffle irons along the way, but at least he didn't piss off his mom.

So every year or so, under the football stadium, engineers carefully fasten 1680 hexagonal molds made of ceramic fiber into a cylindrical tub made of silicon carbide. These molds have 0.5-inch spaces on all sides. The tops of the molds trace the parabolic contour of the mirror surface; no two are the same. Then they load 52,000 pounds of glass in 10-pound chunks on top of the molds, after inspecting each one. The glass comes from a family-run foundry on the northern Japanese island of Hokkaido. It has a purity and uniformity that's unparalleled, and its creation in small batches on a remote, snowy island gives the project a Zen-like quality. The oven is sealed and it begins a carefully designed temperature cycle. As the glass passes through its melting point, it fills the gaps between the molds and leaves a 1-inch-thick layer across the top of the molds. The oven's spinning creates a parabolic surface. Heat-resistant cameras inside the oven monitor the process throughout.

I remember vividly the casting of an 8.4-meter mirror destined for a national project called the Large Synoptic Survey telescope. The oven seemed huge, even in the cavernous space under the football stadium. Forty feet in diameter and 20 feet high, the oven was covered with a tangle of electrical cables and conduits. It had already been spinning for three days and was about to reach its peak temperature of 1400°F. The noise as it rotated every 10 seconds was deafening, and with the flashing lights and pulsing waves of heat, it was like some maniacal merry-go-round from Hell. Angel and his crew were wearing white coats and goggles to protect against the heat and they looked exactly like mad scientists are supposed to look.

Next step: patience. Three months must pass before the team can crack the oven door and peek in to see how the "cake" is doing. That's how long it takes a large mirror to cool and anneal. Hopefully there are no tiny bubbles and the surface is pristine. (After one casting, the mirror was half a ton lighter than expected and nobody ever figured out where the missing molten glass went.) Then the mirror is gingerly lifted out of the oven and put on the polishing rig, one of several times when a single misstep would be disastrous. It takes six months to grind away at the surface using computer-controlled machinery and a substance like jeweler's rouge. When the polishing is done the parabolic mirror surface is accurate to a millionth of an inch. You could use it to read a newspaper 10 miles away. Scaled to the size of North America, the largest imperfections would be bumps and valleys 1 inch high.

Ground-based telescopes have gone through a renaissance. Eighteen telescopes of 6-meter or larger aperture have been built in the last

20 years. They all use either the laser-aligned mosaic approach of Jerry Nelson or the monolithic thin mirror approach of Roger Angel.[11] Behemoths ranging from 20 up to 40 meters are planned. About 40 percent of the time on the large telescopes is given through competitive peer review for the study of the universe beyond the Milky Way. "Big glass" has been essential in telling the story of galaxies.

Sometimes even 900 square feet of collecting area isn't enough. For the ultimate view of distant galaxies and their evolution, astronomers have turned to a telescope in space.

Astronomy in space is difficult and very expensive. It's possible to put a much larger telescope on a remote mountaintop than could ever be put in Earth orbit. Ground-based instruments can be tested, fixed, and upgraded much more easily than if astronauts are required. For a given collecting area, orbiting telescopes cost 30 to 50 times more than sedentary telescopes. Yet the deepest view of galaxies has come from the Hubble Space Telescope, a facility that wouldn't make it onto a list of the top 50 largest telescopes. How is this possible?

The vacuum of space liberates a telescope from image blurring caused by turbulent motions in the Earth's atmosphere. Until recently, image blurring has prevented any ground-based telescope larger than about 10 inches from realizing the image quality that its optics could deliver in principle (Figure 9.2). The best astronomers can do on the ground is locate high mountaintops far from city lights and civilization, above the inversion layer that contains most of the weather. However, most of the image motion and blurring happens high in the atmosphere. It's annoying—to say the least—when the galaxy light travels billions of light-years in straight lines only to be scrambled in the last 10 miles. Telescopes in space are immune from this problem.[12]

Recently, a technique called "adaptive optics" has been closing the gap between imaging from the ground and from space. A powerful sodium laser shines in the direction where the telescope is pointing. Laser light reflects back from the same upper atmospheric layer where

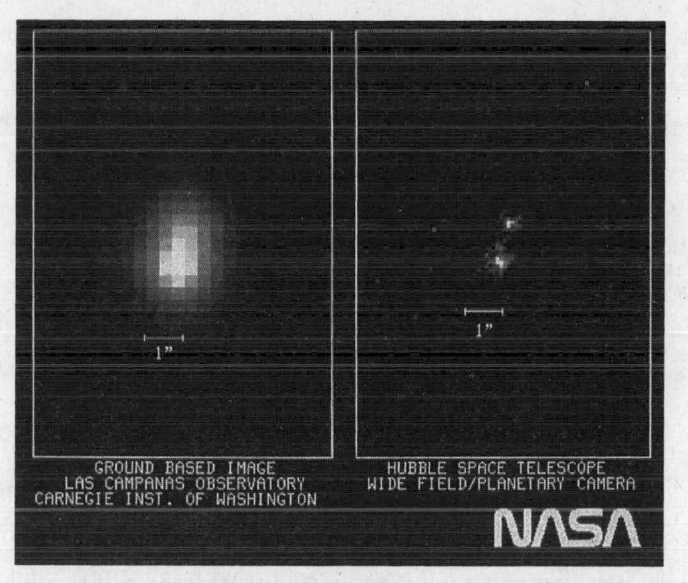

FIGURE 9.2. *A globular star cluster observed with the Hubble Space Telescope* (right) *and a similar-sized telescope on the ground* (left)*. The increased image sharpness also helps to reach a fainter limiting brightness. Much of the advantage of space observations in terms of resolution has been offset by the use of adaptive optics on telescopes on high mountaintops.*

the image blurring occurs, and deviations in the waves are recorded 50 times a second. The telescope has a special secondary mirror thin and flexible enough to be adjusted 50 times a second by actuators mounted on the back surface. The adjustment speed is set to take out most of the image blurring. In this way, the telescope can exactly compensate for what happens in the upper atmosphere and so restore the sharpest possible images. It's like looking at the complex and rapidly changing reflections on the surface of the ocean, and then putting on "magic" glasses that made it look as smooth as a millpond.

There's one way space imaging still has an unassailable edge over ground-based imaging: sky brightness. Light from stars and galaxies isn't detected against a black backdrop; diffuse emission fills the sky. Much of this emission is "pollution" from the lights of dwellings, cars, streetlights, factories, and advertising displays. Most people know the dramatic improvement in visibility of stars in the night sky when you go from a city or suburb to a remote or rural setting. But even if you are in the darkest place on Earth the sky isn't totally dark; the air is dimly lit by chemical reactions in the upper atmosphere. A telescope in space escapes this "airglow" so the Hubble Space Telescope sees a sky six times darker than the sky at the darkest mountaintop observatory. This advantage is especially important when looking at galaxies because galaxy light is spread over more area of sky than a star of the same brightness.

Space gives a darker sky, sharper images, and greater stability of the observing conditions. For these reasons, the modest 2.4-meter Hubble Space Telescope has broken records with the faintest and most distant galaxies ever observed.

BUILDING GALAXIES

Reaching back into cosmic time, we need a way to keep track of the changing properties of the expanding universe. Redshift is the main observed quantity. It's the percent increase in the wavelength of light of observed radiation relative to a similar source of radiation nearby, usually quoted as a pure number. Galaxies in the Coma Cluster have redshifts of 0.023, indicating their light is redshifted by 2.3 percent, and 3C 273 has a redshift of 0.158, meaning its light is redshifted by 15.8 percent. For redshifts less than 1 (less than 100 percent increase in wavelength), the redshift approximately equals the recession velocity as a fraction of the speed of light. Converting the redshift to distance and lookback time requires a model for the expansion of the universe (Figure 9.3). The cosmological redshift plus one has a very simple interpretation: it's

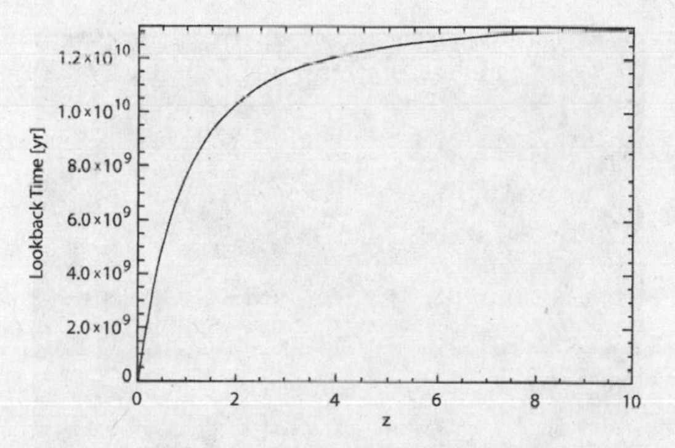

FIGURE 9.3. *Lookback time plotted against redshift. The relation assumes a cosmological model with an age of 13.7 billion years and a universe composed of roughly two-thirds dark energy, one-third dark matter, and a few percent normal matter. Redshift z is related to recession velocity through the cosmological mode. More fundamentally, 1 + z is the factor by which space has expanded since radiation was emitted by an object at redshift z.*

the ratio of the size of the universe now to the size of the universe when the light was emitted.[13]

Big trees from little acorns grow. The paradigm for the formation and evolution of galaxies is that the universe started smooth and became lumpy through the action of gravity—the bottom-up structure formation. Since dark matter dominates normal matter, the way galaxies form depends strongly on the properties of dark matter. Since we have no idea of the nature of that major component of the universe, cosmology is a bit uncertain! However, after a process of elimination of dark matter candidates, the last remaining option is a fundamental subatomic particle. It would have moved slower than light speed early in the history of the universe, so in physics parlance it's considered "cold." Structure is sculpted by cold dark matter.[14]

Invisibly, and under the cover of darkness, dark matter slowly turns a smooth universe into an undulating ocean of gravity. The gravity wells deepen over time and merge, and small mass "halos"

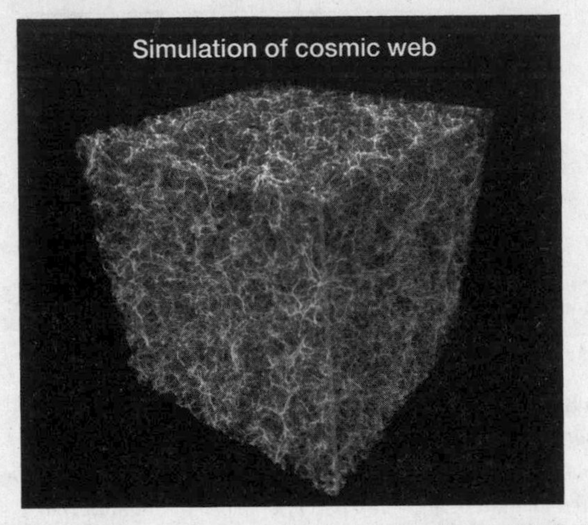

FIGURE 9.4. *A computer can simulate how a smooth gas distribution in the early universe evolves into objects like galaxies after 13 billion years. In this perspective view onto the simulation cube, whiter shades represent higher gas density. The region is 300 million light-years on a side, and the clustering of dark and visible matter creates millions of galaxies within this volume.*

gradually evolve into larger mass halos. Normal matter slides into dark matter gravity wells and collects there. Under the right conditions it condenses into a galaxy of stars. Our knowledge of galaxy formation is sketchy since the physics of star formation is not fully understood even for a nearby situation like the Orion Nebula, let alone for places where stars form billions of light-years away (Figure 9.4).

We see galaxies and can be fooled into thinking they're the story, but in fact they're like pretty marbles resting in the hollows of a thin black rubber sheet. The changing shape of the dark matter sheet dictates where to look for galaxies, stars, planets, and of course people.[15]

A bottom-up theory of structure formation makes simple predictions. At any time in the history of the universe there should be many more small dark matter halos than large ones. The small dark matter halos form first and gradually merge into larger ones. Since normal matter flows into dark matter halos, forming visible galaxies, theory predicts many more dwarf galaxies than large ones, and it predicts that dwarf galaxies should form first, with large galaxies forming later.[16]

How well does the theory do? It mostly checks out, and we've already seen that our galaxy has food stains on its chin that indicate it's grown and continues to grow by feasting on smaller galaxies. However there are some discrepancies that are the subject of active research.

Dark matter is aloof and not as concentrated as normal matter because it only feels gravity, while the stars and compact objects formed mostly by normal matter eject gas and interact with radiation and create a lot of complications for the formation and evolution of galaxies. Imagine a school. Dark matter is the school and its employees. The building is the framework and location for learning, and employees adhere to a hierarchy: substitutes and assistants report to teachers, who report to head teachers, who report to the principal. Add a bunch of kids to the mix and all hell breaks loose. They might or might not go to class or follow the rules, and they interact with each other and the teachers and the hierarchy in all sorts of interesting ways. In the analogy, the kids are normal matter, the raw material for making galaxies.

Here are some of the things that mess up the simple theory of how structure forms in the universe. There aren't nearly as many dwarf galaxies in the local universe as there should be, which is to say all large dark matter halos have galaxies in them, but many small dark matter halos are empty. The favored explanation is that these small halos did form galaxies early on, but the first burst of star formation drove out all the gas from the shallow gravity potential and later on the universe was too diffuse for new gas to fall in and form stars. So most small galaxies had a wild party early on and now the lights are dark as they sleep off the hangover.

Another problem is that astronomers observe massive galaxies that formed all their stars early in the history of the universe and smaller galaxies that continued to form stars until quite recently. That's the opposite of what the bottom-up theory would predict.

The solution to this puzzle is connected with the fact that every galaxy harbors a supermassive black hole, but they're only active a fraction of the time. Over the past decade, data from the Hubble Space Telescope have shown that black holes are ubiquitous in the center of galaxies and the black hole mass is proportional to the mass of old stars in the host galaxies.[17] A massive elliptical like M87 has a black hole several billion times the mass of the Sun, the Milky Way has a black hole 4 million times the mass of the Sun, and there are even examples of black holes tens of thousands of times the mass of the Sun in dwarf galaxies and globular clusters. Nature makes black holes ranging over a factor of a billion in mass!

Black holes grow and are fueled from the same drizzle of infalling gas and wholesale mergers that make galaxies grow. The story of galaxies and black holes must be parallel story lines of cosmic history. In fact, the connection is even more intimate. When galaxies become active, or "switch on" as quasars, gas can be driven out and star formation can be quenched. Supermassive black holes interfere with the simple progression of stellar buildup from small to large galaxies. Early in the universe, mergers rapidly created massive galaxies and massive black holes within them, and their active phases correspond to the luminous quasars. Galaxies like the Milky Way grow by mergers and steady infall of gas from intergalactic space. The buildup of stars in a middleweight galaxy is extended until recent times by outflows from the sporadically active black hole.[18]

We've grown a lot of trees from acorns, but it's easy to lose sight of the forest for all the trees, so let's step back and look at the broad sweep of galaxy evolution.

The present day universe is large, cold, and relatively quiet. Galaxies are so far apart that they rarely meet or merge, and there's not much supply of fresh gas for star formation in intergalactic space. Disks are kept at a moderate level of star formation by ingesting the occasional dwarf galaxy. Black holes have grown in proportion with their parent galaxies, but their glory days are behind them. Most are starved of fuel and have feeble nuclear activity. The most massive black holes can enter a quasar phase, but only for a small fraction of the time. This is your universe in its dull and plodding old age.

Rewind to the universe when it was half its present age. Light emitted from galaxies then has a redshift of 0.8, so the universe was half its current size and twice as hot. Things are pretty lively; there's lots of gas available to fuel star formation and nuclear activity, so both are 20 times their present-day levels. The most massive galaxies and black holes are fully formed but middleweight objects are being very actively assembled and fueled. This is your universe in its prime.

Now go back further to the universe when it was a quarter of its age, the situation of the opening vignette. The redshift is 2, so the universe is a third its current size and three times as hot. Acquisitions, mergers, and hostile takeovers are in full swing. Greed is good in the young universe. Most galaxies are not mature or fully formed but they are in the white heat of their growth phase; fuel for nuclear activity is abundant. The star formation and nuclear activity levels are 1000 times their levels today. Party time! This is your universe as a wild adolescent.

Before this time, information is fragmentary. Large telescopes on the ground and the Hubble Space Telescope work at their limits to go back any further in time. At the limits of observation a galaxy is 10,000 times dimmer than a nearby galaxy of similar size, and the bulk of the stellar energy has been stretched to infrared wavelengths, leaving an optical telescope grasping at straws of light.

SHADOWS AND LIGHT

The universe is made of darkness and light. It can be cyclic like the Chinese qi of yin and yang;[19] a star forms from dark gas and dust and becomes light, and at the end of its life it sends some material back into the darkness to be part of new stars and retains the rest in a dark core. When astronomers first became aware that the universe is expanding, they realized there was tension between light and dark. If the expansion had been much more rapid than we observe, the gas would have thinned out too rapidly for gravity to form stars and galaxies. On the other hand, if the expansion had been much slower than we observe, the gas would have congealed into a single massive object, or the universe might have collapsed too quickly back to its original state for stars and galaxies to form.

All the world's best telescopes, from the Keck Telescope to the Hubble Space Telescope, have attempted to trace the expansion history of the universe by looking out in space and back in time.

Galaxies aren't useful for this because, as we've seen, their properties vary widely and galaxy size or brightness isn't a reliable indicator of its distance. What about using Cepheid variable stars, as Hubble did in the 1920s? Cepheids are good standard "lightbulbs" but it becomes difficult to dig them out of a galaxy beyond 100 million light-years, which is still the local universe. In the 1990s, the favored tool for measuring the history of the expansion was the type of supernova that can occur in a binary star system. A massive evolved star siphons gas onto a white dwarf companion, and the companion detonates as a supernova. Since mass is "spooned" onto the white dwarf steadily and slowly, it explodes in a very predictable way and is a "standard bomb," like plutonium if you gradually added to it until the lump was just above the cricital mass. A supernova can rival the brightness of an entire galaxy, so it can be observed out to enormous

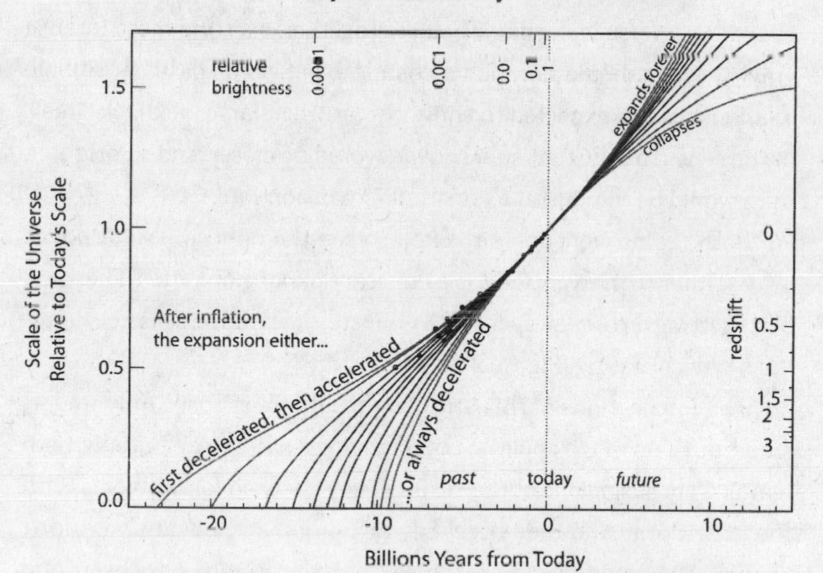

FIGURE 9.5. *The expansion history of the universe, measured using supernovae as standard lightbulbs or "bombs." Each black dot is a distant supernova, going back 7 billion years to when the universe was half of its present size. The data are described by a model with deceleration early on and acceleration in the past 5 billion years. This unexpected result has been used to infer the existence of dark energy and conclude that the universe is likely to expand forever.*

distances. Supernovae are excellent tools for measuring the far away and the long ago.

Two groups started working hard on this project in the 1990s, taking images of large areas of sky to find supernovae as they went off, and then using the peak brightness and the decay rate as the measure of absolute brightness. Absolute and apparent brightness combine to give distance using the fact that light diminishes by the inverse square of distance. Astronomy is a small community; researchers on both teams knew each other, and the rivalry was good-natured but intense. Both groups had a clear hope and expectation of what they would find. The younger universe had galaxies closer together, exerting more intense gravity, which should slow down cosmic expansion. The linear Hubble relationship was expected to show curvature at large lookback times, in the sense that distant supernovae would be closer, and so brighter, than would be the case at a constant expansion rate.

Both teams were gob-smacked to see the opposite: supernovae were fainter than expected for a constant expansion rate.[20] They had been getting farther away faster and faster. The universe is accelerating not decelerating (Figure 9.5).

The implications of this startling result are still not fully understood, but it upset the applecart of cosmology in a big way. Initially, both teams were suspicious of the result. They checked for problems with the supernovae and they checked each other, and the answer didn't change. The supernovae were fainter than they should have been, and the explanation was that acceleration since their light was emitted has carried them farther away than expected. In the last decade, the two teams have redoubled their efforts, trying to trace the expansion back even earlier. They've stepped up from 4-meter telescopes to 8- and 10-meter telescopes, and have had lots of time on the Hubble Space Telescope. This discovery was recognized with a Nobel Prize in 2011.

Here's what they see. Out to a redshift of 0.5, or a lookback time of 5 billion years, there's acceleration. However, from a redshift 0.5 to 1 or above, between 5 and 8 billion years ago, the supernovae aren't fainter than expected. They start getting brighter, the signature of deceleration.[21]

What causes the acceleration? That's the most profound mystery in cosmology. The evocative term given for the agent accelerating the universe is *dark energy*. But it's more of a placeholder for ignorance than a physical description. Dark energy acts in opposition to gravity, causing space-time to expand at an ever-increasing rate. Other than that, almost nothing is known about it. Physicists are intrigued and chagrined by dark energy because it represents new physics and is not part of the "standard model" of fundamental particles.[22]

The universe has a driver, and it's a bad and anxious driver; maybe it's never driven a universe before. Its feet are on the brake and the accelerator at the same time, but the braking foot is weaker and gets tired. Initially, the universe is barreling along with speed decreasing due to strong pressure on the brake. As the brake eases, the accelerator, which has always been pressed down, begins dominating. The universe accelerates because the brake is applied with diminishing force. Dark matter is the brake; dark energy is the accelerator. For the first 8 or 9 billion years of cosmic history, dark matter decelerated the expansion. About 5 billion years ago, dark energy eclipsed the diminishing tug of dark energy, and we've been accelerating ever since. Two entities that we don't understand control the universe, and the normal matter that includes us is just along for the ride.

Acceleration answers a long-standing question about the expanding universe: will it expand forever or one day collapse? Dark energy, as much as we know about it, is an implacable force making space expand faster and faster. With insufficient matter to counter it, victory is assured. The universe will expand forever.

Oh yes. Remember when I said that light was the fastest thing there is? I lied. Light speed as an absolute limit is a feature of special relativity. But when Einstein extended his reasoning to nonuniform motion, the framework changed to general relativity. General relativity places no limit on the expansion rate of the universe. It can expand as fast as it wants! Using an expansion model with the observed amounts of dark matter and dark energy leads to an interesting conclusion. When we observed ever more distant objects, that light was emitted when the expansion rate was much faster than it is now.[23] At a redshift of 1.5, which is 10 billion years ago, the universe was 2 1/2 times smaller than it is now. Any galaxy or quasar we observe at that epoch was moving away from us at light speed when the light we see was emitted. Any galaxy or quasar seen at a lookback time of more than 10 billion years—large telescopes have found more than 1000 of them—was receding from us faster than light when its light was emitted.

How can that be? Light from that distant galaxy starts off receding from us; even at 300,000 kilometers per second it can't make any headway against the rapid expansion of space-time. But as dark matter slows the expansion, the photons get traction and start to approach the place where our galaxy will be billions of years later. Light arrives with a speed of 300,000 kilometers per second, just as special relativity in our local situation would dictate. But it's had an amazing journey to get here.

The "yin" of dark matter and the "yang" of dark energy have been in near balance for most of cosmic time. But as dark energy dominates more and more, they will fall out of balance. Dark energy will carry all galaxies but our nearest neighbors away from us with ever-increasing speed. At some time in the distant future, galaxies will disappear from the reach of telescopes as expanding space whisks them away faster than light speed. We should enjoy the view while we still have it.

I could stay and watch the minuet of galaxies whirling and merging forever, but the tug on my heartstrings is the call of home. Even as I try to project myself back there, I have a sinking feeling that this time I've come too far, that perhaps I'm past the place where the road turns from a rutted and overgrown path into a trackless wilderness.

I turn my attention homeward, thinking that I never really left, that the spectacle around me is an entertainment, a confection of my imagination. I am home, but there's no home there. My worst fear is realized. At the place where I will one day be, but 10 billion years too early, I'm at the periphery of a small elliptical galaxy. Nothing looks familiar. And why should it? This modest galaxy will have to ingest dozens of smaller galaxies and suffer two major mergers with other galaxies before it can settle down to grow a disk. Billions of years will have to pass, and the central black hole will grow and flare up hundreds of times, and the disk will rotate dozens of times, before I will be at the place where a middleweight star will form from swirling gas and, around it, a small but fertile planet.

Instead of being there now and waiting, I could travel from where I am now, on the fleet feet of light. Einstein once imagined riding on a beam of light; I'll do the same. But the journey is daunting. As I leave, my home is moving away from me faster than light and it seems like a Sisyphean quest. Space is expanding beneath my feet. I surge forward but I'm still receding from my destination. For a moment I'm like Alice in the Red Queen's race, running as fast as I can just to stay even. But the unfolding of space beneath me slows and I begin to cross the void, approaching a galaxy that's taking familiar shape as I travel. I hope that I'll complete the journey, and that my path will converge on the right time and place, so I can find my planet and my home, and slip back into my skin in the here and now. But it might not happen. And I have miles to go before I sleep.

10

LIGHT AND LIFE

I'M EMBEDDED IN GAUZY PALE LIGHT. *Compared to the empty canyons of the later universe, this space is womblike and placental. I can sense motion, the outrush of space and the steady clenching of gravity, but nothing seems rushed. It's comfortable, like a warm bath.*

All around me stars are twinkling. They aren't distributed randomly but are in loose clumps and agglomerations, surrounded by denser gas and dust. But as I look closer, it's not like the twinkling of stars in my own night sky. Each spark of light is either the surging brightness of a star switching on for the first time or the paroxysm of a star dying as a supernova. Ultraviolet rays from these massive stars create spheres of glowing gas. These spheres overlap and run into each other like shiny soap bubbles. There's no sound—even at 20 times the density of the ancient universe I came from, it's an almost perfect vacuum.

I think I can make out the vague outlines of galaxies, but mostly this is a spectacle of stars. Sturm und Drang. Son et lumière—but without the sound. Gravity is biding its time; its great architectural project is still in the future. These stars are massive. They do not go gently into that good night. Most expire with an explosion, leaving behind a dark core. Into the distance, it's more of the same, like glitter in mist.

I open my mouth and breathe in. There's no grit. Just the lightness of hydrogen and helium. These first stars are formed from golem, a primordial and incomplete substance. There is no carbon. There are no planets. Biology lies in the future. I know I am truly alone.

—

FIRST LIGHT

To learn how galaxies formed, astronomers took their best facility and pushed it to the limit. The result was the deepest picture of the sky ever made.

The Hubble Ultra Deep Field grew from an earlier project called the Hubble Deep Field and a bold decision by the second director of the Space Telescope Science Institute, Bob Williams. In 1995, Williams devoted the 10 percent of the observing time that he had at his discretion as director to a very deep multicolor image of a single patch of sky. To see why this was bold, let's take a brief excursion into the culture and sociology of research astronomy.

For astronomers, the Hubble Space Telescope is the best game in town.[1] The Hubble has made more discoveries and generated more papers than any other research facility. It's a high-profile project with excellent name recognition among the general public—the gorgeous pictures have been downloaded by millions of people. Quality like this doesn't come cheap; over 20-plus years Hubble has cost over $7 billion. Every year, astronomers craft proposals for time and there are six or seven times more proposals than can get on the telescope. I've been on the review panel several times and there are hardly any weak proposals, so strong science gets rejected. I've dished out bad news and I've been on the receiving end too. In such a tough competition there's a natural tendency to spread the bets and keep more people happy by giving small proposals a little bit of time each. Director's discretionary time is used similarly, spread around small proposals to ease the oversubscription.

Bob Williams decided to put all of his eggs in one basket by giving 150 orbits—a huge allocation of time—for one deep image. Williams had been similarly bold in his career a few years earlier. As a tenured professor in my department, he'd grown weary of the politics and numbing bureaucracy that can afflict university life. He gave up a tenured job without any new job to jump to, intending to write and consider his options. After being unemployed for a year, he got a job as the director at the Cerro Tololo Observatory in Chile, and then he was recruited to direct the Space Telescope Science Institute, one of the top jobs in astronomy.

His decision to invest Hubble time in a single deep field changed the culture of astronomy. He let the research community decide where the telescope should be pointed and what color of filters should be used for the 140 hours, but he insisted that the data be processed and made public immediately for any astronomer to use. This tiny region in Ursa Major—1/28,000,000 of the sky—contained over 3000 galaxies, and the data paper for the Hubble Deep Field is one of the most oft-cited papers in astronomy.[2] The large investment of a scarce resource in a single target persuaded infrared and radio and X-ray astronomers to follow suit. Other telescopes invested lots of time into complementing the optical images with data across the electromagnetic spectrum and often that data were also made available quickly. An intriguing mix of competition and altruism spurred the research forward.

But what if the one field you pick isn't typical for some reason? The cosmological principle is based on isotropy, and there's been nothing

FIGURE 10.1. *The Hubble Ultra Deep Field, the deepest image of the sky ever made. This tiny region in Fornax contains 10,000 galaxies, most of which are 5 to 10 billion light-years away. The Hubble Space Telescope captured the images in four different filters to get color information for the galaxies, which helps to assign redshifts.*

to say that one direction in the universe is different from any other, but astronomers were nervous, so Williams committed Hubble time to a deep field in the southern sky in 2000. Since then, deep fields have sprouted like mushrooms. In 2002, after the fourth servicing mission installed a sensitive camera, the Space Telescope Institute director at the time, Steve Beckwith, upped the ante by putting 400 orbits, a million seconds of observing time distributed in four colors, into a tiny patch of sky in the direction of Fornax.[3] That's the Hubble Ultra Deep Field (Figure 10.1).

Let's try and get a sense of this incredible image. Hold a pin at arm's length; the head of the pin covers as much sky as the image produced by *Hubble*'s CCD camera. Astronomers harvested 10,000 galaxies from this miniscule region of sky. The faintest are 5 billion times fainter than the eye can see, and *Hubble* can only collect one photon per minute from them—think of seeing a firefly on the Moon. Covering the whole sky to this depth would take a million years of uninterrupted observing.

The numbers are staggering, and they can be used to derive some important information about the contents of the universe. Since the Ultra Deep Field covers 1/13,000,000 of the sky, the projected total number of galaxies in all directions is 130 billion. Each galaxy will on average contain 100 billion stars, so there are about 10^{22}, or 10,000 billion billion, stars in the visible universe. This number grows to 10^{23} if we include dim red dwarfs down to the fusion limit. That's a mind-bending number, made even more exciting by the implications for life. We've learned that planets are ubiquitous around Sun-like stars and we expect to learn soon about the abundance of habitable and Earth-like planets. The number of potential biology experiments in the universe may be similar to the number of stars. What odds would you put on our being alone?

The Ultra Deep Field is littered with faint galaxies, but to go beyond counting we need to place them within three-dimensional space. It's impossible to find distance indicators like supernovae for galaxies this faint, so we have to assume a cosmological model and use redshift to estimate age, distance, size, and luminosity. However, most of these galaxies are so faint it's hard to detect them, which means there's no chance of spreading the light into a spectrum to measure the redshift. Luckily, astronomers have a "quick and dirty" method to get redshifts based on the energy distribution. Stellar populations in a galaxy have colors that change predictably as the energy distribution is redshifted through the filters of the imaging data. So colors can be used to infer crude redshifts in most cases.[4]

With redshifts, the galaxies can be assigned a distance. The shape of the volume is much longer and skinnier than the survey of brighter galaxies that produces slices of the universe; it's deeper in the radial dimension, like a 3-yard-long soda straw. The radial dimension is look-back time as well as distance. There are signs that the heroic project to map out the evolution of galaxies is nearing completion. First, the Ultra Deep Field was three times more sensitive than the Deep Field but only added 30 percent more galaxies (per unit area). Next, the number of galaxies at increasing distance falls rapidly from a peak at redshift of 2 to a small tail at redshift of 4 and larger. Last, the very faintest galaxies are not massive galaxies at greater redshifts but less distant galaxies with 5 to 10 percent the luminosity of the Milky Way.

We've seen through most of the volume of the universe. A full census is unlikely to be much more than 130 billion.

Nonetheless, the rare, high-redshift galaxies are gold dust to people who work in this field. They want to find the epoch of "first light," the time when the first star galaxies were forged from hot gas. Let's see how far back they can reach with the Ultra Deep Field. As orientation, redshift of 2 is 10 billion years ago when the universe was three times smaller than it is now, redshift of 4 is 12 billion years ago when it was five times smaller, and redshift of 8 is 13 billion years ago when it was nine times smaller. The original Ultra Deep Field images were taken in 2003 and 2004, but the most spectacular results didn't occur until the fifth servicing mission in 2009 installed a camera that could take deep infrared images. The most distant galaxies emit little optical light as their energy is stretched by a factor of 10 to infrared wavelengths (Figure 10.2). This research field is highly competitive; in late 2009 five different groups submitted papers on very high redshift galaxies within a few months of each other.[5]

What are the earliest galaxies like? They're small, blobby, and irregular in shape. There are very few highly luminous galaxies, or spirals like the Milky Way, or ellipticals. These ragged galaxies are forming stars at a ferocious rate. The results are nicely consistent with the bottom-up model of galaxy formation, since we can watch the first

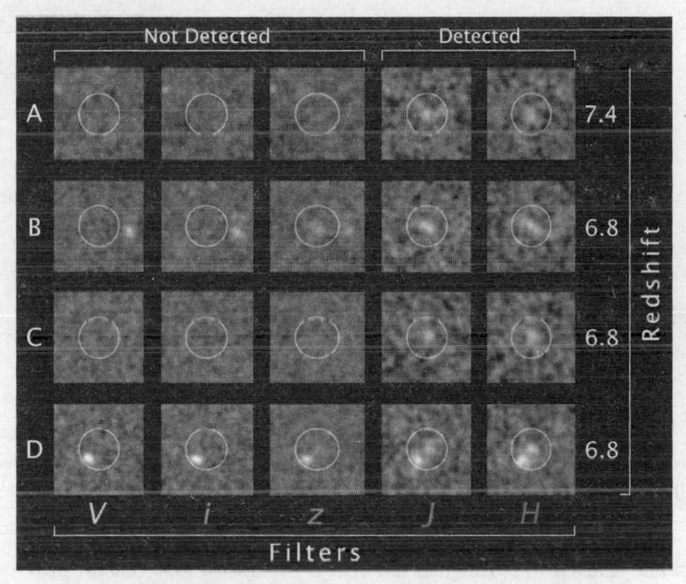

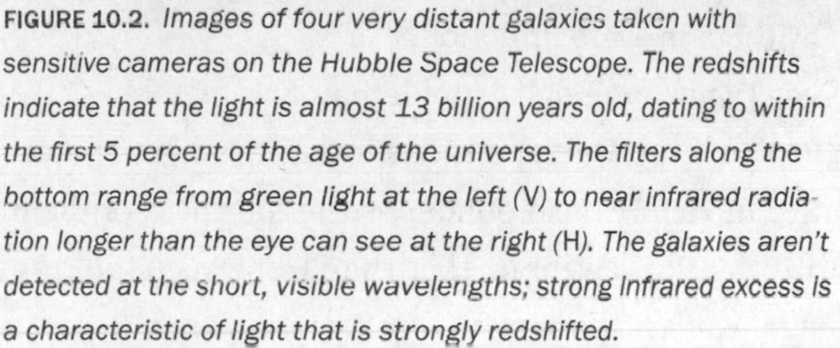

FIGURE 10.2. *Images of four very distant galaxies taken with sensitive cameras on the Hubble Space Telescope. The redshifts indicate that the light is almost 13 billion years old, dating to within the first 5 percent of the age of the universe. The filters along the bottom range from green light at the left (V) to near infrared radiation longer than the eye can see at the right (H). The galaxies aren't detected at the short, visible wavelengths; strong infrared excess is a characteristic of light that is strongly redshifted.*

generation of galaxies being built. There are over 100 galaxies with redshifts of 7 to 8, and more than a dozen likely to be at redshifts of 9 to 10.[6] That's the current frontier. We've seen galaxies at the dawn of time, looking back 95 percent of the age of the universe to when it was 10 times smaller and hotter, and 1000 times denser, than it is today.

Is this first light? Probably not. Some of these high-redshift galaxies have stars that are a few hundred million years old. Pity the Hubble Space Telescope. Astronomers are pushing it to its limits, flogging it hard, but it's only a 2.4-meter telescope, puny compared to the 10-meter behemoths on the ground. Progress will have to wait for the

Hubble's successor, NASA's James Webb Space Telescope, or JWST. It's an extremely ambitious project, due for launch in 2018. JWST is 6.5-meters in size and its hexagonal mirrors will deploy by unfolding like flower petals, all at a distance of a million miles from Earth, far from the reach of astronauts with handy tools. JWST was designed to work at infrared wavelengths, and one of its big science goals is to mop up what the Hubble left undone, and detect true first light.

LOST HORIZONS

Sometimes the simplest question has the most profound answer. Why is the sky dark at night? That's something a kid might wonder, looking at all the stars and thinking that if they went on forever, the sky would be lit up by them in every direction.

Thomas Digges, an English nobleman and member of Parliament, was the first to translate Copernicus into English and popularize the heliocentric model. He also pondered the darkness of the night sky. In his writings, Digges went further than Copernicus, and speculated that the stars weren't confined to a celestial sphere but were scattered through infinite space. There are two ways to deduce that the night sky should be bright in an infinite—and infinitely old—universe. First, every sightline through the universe will eventually terminate on a star, so all directions should be as bright as the surface of a star. Alternatively, think of the contribution of starlight from concentric shells moving out from the Earth. The light from each star goes down by the square of its distance. But the number of stars in each shell goes up as the surface area of the shell, proportional to the square of the distance (Figure 10.3). Each shell of stars therefore contributes an equal amount of light and an infinite number of shells adds up to infinite light.

Kepler worried about this problem, Halley did too, and then it was labeled "Olbers' paradox" after an eighteenth-century German ama-

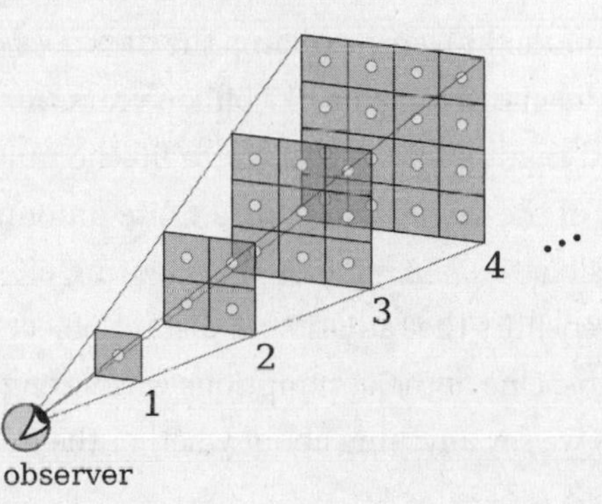

FIGURE 10.3. *This idealized diagram illustrates Olbers' paradox. Moving out from the Earth, the number of stars in any "shell" increases by the surface area, or square of the distance. However, the light from each star goes down by the square of the distance. Therefore, the light in each shell is the same, and the total of an infinite number of shells is infinite light.*

teur astronomer.[7] Kelvin provided a possible resolution of the paradox and, intriguingly, so did writer and poet Edgar Allen Poe, who wrote in 1848: "Were the succession of stars endless, the background of the sky would present us with a constant luminosity . . . The only mode, therefore, in which, under such a state of affairs, we could comprehend the voids which our telescopes find in innumerable directions, would be by supposing the distance of the invisible background so immense that no ray from it has yet been able to reach us at all."[8]

All of these people were working at a time when the Milky Way was thought to be the only set of stars in the universe. But the argument doesn't change if the stars are aggregated into galaxies; there still should be a galaxy at the end of every direction out from the Earth. That's not what we see. The Hubble Ultra Deep Field has 10,000 galaxies, but they're laid out like diamonds on velvet, with plenty of jet black sky among them.

Since the night sky isn't bright, one of the assumptions is wrong.

The finite age of the universe explains the darkness of the night sky. Light has only been traveling for 13.7 billion years through the regions we can see and that's not long enough to fill the universe with light. The finite age of the universe means a finite amount of star or galaxy light is visible to us. A smaller contributing effect is expanding space, which redshifts the radiation of distant objects away from visible wavelengths. One of the assumptions was wrong but what about the other? Can we say anything about whether the universe is infinite in extent or not?

The boundary between what we can and cannot see is a horizon. We're familiar with the idea of a horizon. On the curved surface of the Earth, a horizon marks the edge of what we can see, but not the edge of everything that exists. In thinking about the size of the universe—the part we can see and the part we can't—and the possibility that it's finite or has an edge, we define terms carefully. The familiar analogies from everyday life are useless. Crutches will be discarded. Walls, floor, and ceiling will fall away. Prepare for vertigo.

Here are the tethers we can hold on to as we start our journey. The universe is expanding, meaning that space itself is expanding and the galaxies are carried apart by expanding space, not that the galaxies are moving through space. The finite speed of light and the vast size of the universe mean we see distant objects as they were, not as they are. There's no speed limit to the expansion; the universe can and has expanded faster than light. The only true observable properties of distant objects are apparent brightness and redshift; distance, age, size, and luminosity must be derived from a model for the expansion based on general relativity. What follows will assume that the cosmic expansion is three-fourths governed by dark energy and one-fourth governed by dark matter.

The local universe has all its sidewalks and handrails in place. We see a linear and uniform expansion in all directions. As Hubble observed, more distant galaxies recede faster. The slope of the Hubble diagram is the local expansion rate. Recession velocities are much less than the speed of light. The region is also "local" in time since the light travel time is a small fraction of the age of the universe.

Let's leave the comfort zone. If we extrapolate the Hubble law to the distance where the recession velocity equals the speed of light, that defines a volume called the Hubble sphere. With the current value of the expansion rate, the distance to the edge of the Hubble sphere is about 14 billion light-years.[9] Galaxies beyond the edge of the Hubble sphere are receding faster than light. A galaxy that was moving away at the speed of light when its light was emitted is now moving away at less than the speed of light because of deceleration in the intervening time. The redshift at the edge of the Hubble sphere is 1.5, corresponding to radiation that has had its wavelength stretched by a factor of 2.5 due to expanding space. But the Hubble sphere isn't a horizon because we routinely observe even more distant galaxies and quasars.

How is that possible? Because the Hubble sphere has changed in size over the age of the universe. For most of the history, the expansion rate has been slowing. Therefore, the Hubble sphere has been getting larger; as time goes by we get to see more and more distant regions of space.

Imagine a galaxy so far away it's outside the Hubble sphere. Photons leaving it were initially in a region of space expanding away from us faster than light so they were receding. But as they travel and cosmic expansion slows, they travel into regions that are expanding slower than light speed. The photons eventually can reach us and so we see ancient light from the galaxy. However, that galaxy is and always has been moving away from us faster than light. Imagine riding a photon from that galaxy. You leave long ago and are dragged away from the Milky Way by rapidly expanding space. But the deceleration lets you claw your way to a place where for an instant you mark time like Alice in the Red Queen's game. Then you begin to approach the Milky Way, and eventually arrive, all the time moving at 300,000 kilometers per second. Tamara Davis and Charles Lineweaver have worked patiently to combat widespread misconceptions with this concept, which is subtle enough that even professional astronomers can get confused.[10]

What then is the size of the observable universe? The edge of the visible universe corresponds to the distance light has traveled since the universe began. It isn't 13.7 billion light-years—that would only be true in a static universe. The calculation incorporates the expansion history of deceleration followed by acceleration. Space has unfolded copiously and voluptuously over the past 13.7 billion years, so the distance to the edge of our vision is now about 46 billion light-years, three times more than the distance to the edge of the Hubble sphere. In cosmology, it's called the particle horizon, our information limit.

All good things come to those that wait. Before the discovery of the accelerating universe, it was expected that dark matter would make the expansion rate continue to slow. Light from ever more distant regions would come into view, even if the objects emitting the light continued to recede from us superluminally. The light would be really old—yellow newsprint from billions of years ago, not yesterday's tweets—but just by being patient we'd see more of it. The observable universe would get a bit bigger every day.

But the universe has been accelerating for 5 billion years, and the diminishing gravity of dark matter means it will probably accelerate forever. Acceleration rips objects from our view by moving them away from us so fast that not even their fleet-footed photons can reach us. It creates a boundary beyond which are events we'll never see—an event horizon. Like the event horizon of a black hole, it acts like an information barrier. The current distance to the cosmic event horizon is 16 billion light-years, well within our observable range. We can see galaxies that are receding across our event horizon now. They're the most distant objects from which we'll ever learn things that happened to them today. As the universe continues to accelerate, the observable universe will shrink as more galaxies are ripped from view. The event horizon will close down like the iris of a camera lens until all we have to look at is the Milky Way. The universe ends with us staring at our navel.

None of these weird properties of the expanding universe wreck the physics of everyday life. Light measured locally travels at 300,000 kilometers per second and nothing ever overtakes a photon. If you wanted, you could sit comfortably in your living room (which is not expanding) and ignore everything you've just read.

FIRST LIFE

The Methuselah planet has no right to exist. About 12.7 billion years ago, a Jovian planet formed around a Sun-like star on the outskirts of the globular cluster M4, 7200 light-years away in the constellation of Scorpius. For about 10 billion years, this planet system had an uneventful life. Then it headed toward the cluster core and encountered a neutron star, the dead remnant of a massive star. The neutron star was already mated with a white dwarf companion in a binary orbit, but a gravitational tug of war ensued, and the neutron star spurned its white dwarf, ejecting it and taking the Sun-like star and its accompanying planet as new mates.[11] Then the Sun-like star entered into a tight embrace with the neutron star, while the giant planet watched from a safe distance. The normal star aged, and just as the Sun will one day, it became a red giant. Swelling up, it spilled its gas onto the neutron star, energizing it and turning it into a pulsar that spins 100 times a second, which is 10 times faster than a hummingbird flaps its wings. Meanwhile, the red giant's fuel ran out and it turned into a dim and cool white dwarf. Today, the giant planet circumnavigates a pair of tightly orbiting stellar corpses.

The planet, about 2.5 times the mass of Jupiter, was quickly dubbed the "Methuselah" planet. There are several reasons it's very surprising. It has a tortured history.[12] After an uneventful youth and middle age, the planet survived passage through the dense interior of the globular cluster. It could easily have been ejected when its parent stars mated, and in a stroke of good fortune, mate-swapping ejected the system to the more gentle outer reaches of M4. It also survived the death of its parent star. Without the parent-swapping that put it into the orbit of a pulsar, we'd never know it existed!

But what blew astronomers away about this planet was its age. If you want to grow a pearl, you need grit. Conventional wisdom says that to make a giant planet you need a rocky core 5 to 10 times the mass of the Earth. That core then accretes a gas giant envelope fairly quickly. All the elements other than hydrogen and helium were made by multiple generations of stars. As cosmic time proceeds, stars forge the heavy elements in their cores and eject a fraction of them into space. So the universe gets more able to form planets, and more hospitable for life, as time goes by. There was a time, before "first light" and before the first stars formed, when there couldn't have been planets or biology.

In that context, a 12.7-billion-year-old planet is extraordinary. M4 has 5 percent of the abundance of heavy elements of the Solar System. Yet it was apparently enough "grit" to grow a giant planet "pearl."[13] If there was enough rocky material to make a super-Earth, there was enough to make an Earth-like planet in a more hospitable location. The giant planet formed in a nearly circular orbit two to eight times as far as the Earth is from the Sun, allowing room for a terrestrial planet to form in the habitable zone. There were 10 billion years of uneventful time for such a hypothetical planet to evolve life before the mayhem of passing through the cluster core and parent-swapping.

Think of it: a possible Earth-like planet that's three times older than the Earth. If life had formed as quickly on such a planet as life formed on Earth, it would have more than an 8-billion-year head start on us. That's a lot of time to play with. Once life moved from Earth's oceans onto land, it only took 400 million years to grow large brains and figure out space travel and computers.

As tempting as it is to try, we can't reach reliable scientific conclusions based on one object. So let's look at the broader question of when and how the first life in the universe might have started.

Answering that question means we have to know when first light was, because before there were stars, there were no elements other than hydrogen and helium. The origin of the universe, the time projecting the galaxies back to zero separation, is 13.7 billion years ago. We can track early time with redshift, which at these epochs equals the factor by which the universe is smaller than it is today. Galaxies are rapidly declining but there are plenty at a redshift of 6; this is a billion years after the universe began (the Methuselah planet is this old). Galaxies are reliably found at a redshift of 8, 700 million years after the origin, and more tentatively at a redshift of 10, just 500 million years after the origin. First light may have been glimpsed by the Spitzer Space Telescope (Figure 10.4).

In the bottom-up structure formation scenario, stars would precede galaxies, so the quest for first light involves looking for individual stars or their remnants. Amazingly, a few single stars have been detected at redshift 8.[14] They are gamma ray bursts, where a massive star dies in a titanic explosion, outshining the entire universe in gamma rays for a few seconds and leaving behind an afterglow that optical astronomers can study if they're nimble.[15] They're the most distant astrophysical objects known. Nature helps to detect them because the dimming caused by large distance is partly offset by time dilation, which stretches out the burst and makes it easier to catch. Current technology could see the bursts out to a redshift of 20, or 200 million years after the origin, and the WST should be able to detect them to a redshift of 30, a scant 100 million years after the origin. Cosmologists are confident that's far enough back in time to snag first light.

It's splitting hairs to worry exactly when the first stars sprang into life. What's more interesting is where, within the first 5 percent of the age of the universe, the first life might have sprung into action.

FIGURE 10.4. *Glimmers of first light. On the right, the "foreground" objects are stars and galaxies in optical light. On the left, they have been subtracted out in an infrared image, and the diffuse residual radiation may be light from the first generation of stars.*

The slow and steady buildup of carbon, nitrogen, oxygen, and other biogenic elements appears to make the early universe inhospitable for life. Exoplanet researchers have seen a strong dependence between the abundance of heavy elements in a star and the incidence rate of Jupiters: 30 percent of stars with three times the heavy elements of the Sun have giant planets but this drops to 5 percent for stars like the Sun and to 1 to 2 percent for stars with a third the heavy elements of the Sun (Figure 10.5).[16] By this reckoning the Methuselah planet shouldn't exist! Another idea is that there's a galactic habitable zone, or ring-shaped region of the disk, where conditions are just right for complex life. Inside that the porridge is ruined by too many supernovae and stellar encounters; outside that the porridge doesn't have enough grit to make planets. Terrestrial planets should be less than 8 billion years old.[17]

Countering these arguments, several billion stars in the bulge of the Milky Way (and all other spirals) are 8 to 12 billion years old and have 10 to 200 percent the heavy element abundance of the Sun. Even at low occurrence rates, that's enough to make trillions of ancient planets. Simulations suggest that while giant planets may be inhibited by low heavy-element levels, there's plenty of grit in the inner solar systems so terrestrial planet formation is *enhanced*.[18]

If I had to place a bet on where life's motor first turned over, it would be a high-redshift quasar, or rather in a galaxy that formed around one of the first supermassive black holes. The furious stellar activity that generates the black hole and fuels it creates plenty of car-

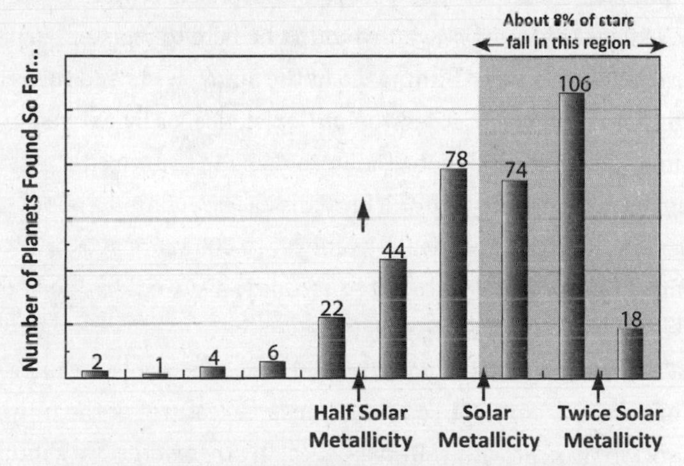

FIGURE 10.5. *There is a strong trend for the incidence rate of giant planets to be lower around stars with lower abundance of heavy elements. However, simulations show that terrestrial planets might actually be more prevalent around stars with less heavy elements.*

bon and other heavy elements—gas around distant quasars can have 10 times more than the Sun. It's good to be first to the feast, and life near a quasar would guarantee an entertaining night sky.

What would creatures be like who got to our level of intelligence and then went on to progress for millions or billions of years? We've no idea. The biological opportunities of space and time could be used to argue that we're neither alone nor first, but it's not a solid argument. We need to look. Our best bets for finding life are to detect microbial alteration of a nearby exoplanet atmosphere or to detect deliberate signals sent by some intelligent species farther afield. Meanwhile, *we* exist, *we* have used our brains to comprehend the vast universe that surrounds us, and some people consider that worthy of surprise.

WHY ARE WE HERE?

Let's start with logic. It's incorrect to be surprised by your existence, because if you didn't exist, you wouldn't be here to register surprise. It's also not legitimate to be amazed at the many twists and turns and contingent outcomes that led to your existence, or by extension led to human life on this planet. If mammals hadn't survived a gigantic impact 65 million years ago, or if humans hadn't endured a population bottleneck after a supervolcano went off 70,000 years ago, or if your parents hadn't crossed paths more recently, you wouldn't be here to ponder the meaning of these events.

We can extend this logic. If you'll excuse the triple negative, it's also incorrect to be surprised *not* to observe features of the universe that are *inconsistent* with your existence. In other words, we shouldn't be surprised that the Earth isn't overrun with flesh-eating bacteria, or that the Sun doesn't have a lifetime of 2 billion years, or that the universe doesn't have equal amounts of matter and antimatter. In each of those circumstances, we wouldn't exist. I'm hugely relieved and thankful I don't live in those hypothetical worlds, but so what?

Now it gets more interesting, and a lot trickier. It's possible to make a list, consisting of features of the natural world ranging from biological to cosmological, where if things were a little different we wouldn't be here. Examples include: if carbon were much rarer, if DNA replicated perfectly or didn't replicate, if the Earth didn't have a big Moon, if the Sun were twice as powerful, if the speed of light were twice as slow, if the amounts of dark energy and dark matter were much different, and if gravity were an even weaker force than it is. Out of an enormous range of physical possibilities, it has been claimed that carbon-based life can only exist within a narrow range of those possibilities. We're expected to be surprised that the universe has a set of properties that are both *unlikely* and *necessary* for our existence.

These arguments go under the guise of "fine-tuning" or the "anthropic principle" but they're descended from arguments from design, which have been used to argue for the existence of a Creator. As Bertrand Russell said in 1927, "Everything in the world is made just so that we can live in the world, and if the world were ever so little different, we could not manage to live in it. That is the argument from design."[19] A lot hinges on how we interpret the words "unlikely" and "necessary." These ideas have attracted some heavyweight thinkers from physics and astronomy and philosophy, but with their taint of teleology, and their proximity to the ongoing science-religion "culture wars," they're also very controversial.

You're here because of the chimichangas at El Rancho Grande. I'm looking my twenty-year-old son Ben straight in the eye and telling him his existence hinges on a deep-fried burrito served in a Mexican chain restaurant in Pasadena, California.

He stares back at me, blinking, uncomprehending.

Well, I explain, I was a postdoc at Caltech and there was a conference coming up in India. I wanted to do some adventure travel after it but I didn't know another astronomer who was game. There was a postdoc named Doug at another institute a few miles away, and someone told me he was a keen trekker and climber. I called him. He was very busy but I mentioned El Rancho Grande and their killer chimichangas and it was enough to lure him into dinner.

It turned out he'd planned a big trek in Nepal but his partner had just broken an ankle and had to withdraw. My timing was excellent. He had several replacements lined up but I persuaded him to take me instead. Of course, I had to tell a white lie about all my extensive climbing and high-altitude experience. The corners of Ben's mouth crease in a slight smile.

We went, I continue, and it was intense. Doug and I trekked for nine days and got to Kala Patar, a 19,000-foot peak across from Everest. We bonded through the physical challenges and by experiencing the awesome beauty of the roof of the world. Not long after we got back he set me up on a blind date with your mom. Now Ben is grinning. I remind him how when he was four he drew a picture for Doug with a touching note thanking him for being the reason he existed. With all the twists and turns of life, the opportunities missed and taken, it all came down to the chimichanga.

We've dipped our toes into the water of anthropic reasoning, and we'll return here later in our journey. Perhaps you've found it scalding hot, perhaps numbing cold, or perhaps the temperature is perfect, as if it had been prepared just for your liking. We've also ended the second part of our journey, through remote realms of space and time. We've reached a time and a place when there are no galaxies and stars, no planets and people. In what follows we'll explore an alien landscape. The body of water facing us—opaque and alluring—is called "Why?"

First light. I look for the first and most massive star to form, because it will be the first to die. That one—nearby—it may not be the very first, but it will do. It has just died, soundlessly, and the blast wave is approaching. It reaches me and I'm shaking like a rag doll. I open my mouth and taste... soot. Yes! I'm disembodied and homeless, but I need a narrative. I turn and surf the blast wave. I am a microcosm of the cosmos: carbon.

I float for who knows how long; I have lost track of time. Around me, structure forms like dew condensing on the web. I'm aloof for eons but then get drawn into a newly forming star. Buffeted in its furnace, I'm assailed by other atoms, but they cannot gain access to my heart so I am unaltered and get ejected on a gentle stellar wind. The roller coaster ride continues as I'm drawn into another star. This time I'm nearly trapped for eternity, only luck keeps me free. My neighbors are pulled into the crystalline white dwarf core, but the star uses its last guttering spasm of energy to blow a smoke ring into space. More drifting, more eons pass.

On the cool periphery of yet another young star, I become part of a morsel of rock. In a delicate dance, the morsel becomes a flake and then a boulder. After many jarring collisions, I'm in a dark place. It's completely quiet except for a low-frequency rumble, and the cold is only leavened by tepid heat from my heavier cousins. Perhaps this is my resting place.

Then I'm riding a viscous seam of magma, rising through rock. For the first time in a long time I see the night sky. Dissolved in water, I ride to the ocean. Bound with oxygen, I circulate, riding the breeze and the ocean currents.

Something magical: I land on a leaf and am pinned there by light like a butterfly to green baize. I've entered the world of living things. Combining with others, I'm industrious and purposeful. This is what I was meant for. I cycle through the biosphere many times. Each adventure is slightly different and all of them are completely engrossing. Then, as close as I will get to a true return, I am ingested by an individual of the dominant species.

They are just a cipher. I, after all, am just an atom. I have my job to do in a cell; I keep my nose to the grindstone. Then unaccountable stillness. It makes no sense, it's not their time. I'm overwhelmed by something; it can't be sadness, I can't parse sadness. It's absence. It's void. Then heat, intense heat, and I take flight on the wind again. It is the fate of carbon to be free.

PART III
ALIEN

BIG BANG

THE FOG AROUND ME GLOWS A DULL VERMILION. *It's hot enough to melt steel but the color makes the plasma seem cool. It's disorienting not to be able to tell the difference between up and down or in and out. I have the vague sense that the fog is steadily thinning, but if there's any motion it's almost imperceptible.*

Something remarkable is happening. The glow around me is sliding toward a deeper and richer color, something like dried blood. Meanwhile the texture of the light is changing. I sense diminishing opacity, and a shift from translucence to transparency. The effect is subtle; no objects appear out of the mist because there are no objects to appear. The light is sharper. Distances seem greater. I'm no longer inside a cocoon. It seems like I'm suspended inside the invisible lattice of an endless ruby.

This is the infant universe. Around me electrons have paired off with protons and shimmering plasma has turned into mundane gas. Light is no longer the servant of matter. It moves through space unimpeded, affected only by the unfolding of space. As the waves stretch beyond my ability to detect and comprehend them, I'm slightly sad. The universe is fading to black. Before the color slides off the end of the rainbow and disappears I want to tell someone about it, but there's no one to tell.

I wonder: If nobody is there to see it, does this amazing quality of redness have any meaning? And if the scene around me is a dream, is the dream shared by anyone else?

—

FIREBALL

This story of beginnings began with the proximate universe, the realm of voyages. For now we can only send our robots to the planets. One day we may venture out and embrace the stars. Next we explored the remote universe, the realm of history. The messenger is light and we reel in the photons and the years until we encounter darkness. Finally, we reach back so far in time that it's hard to call it a story; after all, a story needs a teller and a listener and in the early universe there was no person, no object for them to describe, no *thing*. Yet not nothing.

In a universe so alien, we will reach for a metaphor, recalling Robert Frost's comment that "all metaphors are imperfect; that is the beauty of them." The metaphor is the senses. To humanize what is inhuman we will imagine that the early universe can be experienced sensorily. The term "big bang" suggests we should start with sound.

To the Australian aborigines, Dreamtime is the time before time, a precursor state of the cosmos. The Creation is called the "Dreaming" and every person is eternal in the Dreaming, with an individual's life just an interregnum in their endless existence.[1] Songlines are paths traveled by the creator-beings during the Dreaming. For example, the Yolngu tribe of the Northern Territories tells of a creator-being who is associated with the planet Venus and who led the first humans into Australia and flew from east to west, naming and creating plants and animals and features of the land. In modern times, as they travel the land, aborigines "sing" it into creation and so pay homage to their totemic ancestors.[2] As we will see, modern cosmology also tells of a "song" by which the universe is created.

Given that the universe began with unspeakable heat and will end with unimaginable cold, our place in it is almost perfect. Next time you sit outside on a sunny day, give thanks for the fact that your porridge is just right. We might piss and moan about the weather but most extremes can be handled by piling on or peeling off clothes. But 50 miles below our feet is heat so intense we'd expire in a few minutes, and 50 miles above our heads is the lung-busting and frigid vacuum of space. In simple terms, the universe is made of stars and the space among stars. We huddle in a slender Goldilocks zone near the Sun where life is possible, but the fires are sparsely scattered and most space is so cold that air would freeze.

The universe wasn't always so cold and empty. Hubble was strangely reticent about investigating the implications of his discovery. He was slow to embrace the idea of expanding space, and largely uninformed about general relativity. He showed no interest in speculating about the history of the expansion.

George Lemaître had no such qualms. Lemaître was the first scientist to present a realistic model for the expanding universe using general relativity. As he put it, "We must have a fireworks theory of evolution. The fireworks are over and just the smoke is left. Cosmology must try to picture the splendor of the fireworks."[3] Lemaître knew that "running the clock backwards" implied a universe that was smaller, denser, and hotter in the past, and he speculated about an original state that he called the "primeval atom" and a "day without a yesterday."

Lemaître was a Jesuit priest. Born in Belgium, his university studies in civil engineering were interrupted by World War I. He served in the army as an artillery officer and was decorated with the Military Cross. After the war, he changed tack and studied physics and mathematics, becoming excited by Einstein's new theoretical work on gravity. After getting his doctorate in mathematics, he was ordained in the Catholic Church. He got a second doctorate in science and learned astronomy from the experts: Eddington at Cambridge and Shapley at Harvard. In 1927, he published a paper in a little-read Belgian journal that presented his new idea of the expanding universe. That year he talked with Einstein in Brussels, but the great man seemed unimpressed, telling him, "Your math is correct, but your physics is abominable."[4]

Two years later, Hubble showed that the universe was expanding. Eddington became a champion of Lemaître's work and helped get his paper translated into English in 1930.[5] At a meeting in London on the relationship between the physical universe and spirituality, Lemaître proposed an origin to time and space as a point that had contained all matter and energy. Many scientists were uncomfortable with this idea; Eddington praised Lemaître for his "brilliant solution" to the problems of cosmology, but he thought the idea of a beginning to the universe was "repugnant."

Einstein eventually came around. When the two lectured together in California in 1933, he stood and applauded after Lemaître spoke and said it was "the most beautiful and satisfactory explanation of creation I've ever heard."[6]

Big bang. That's the slightly cavalier term used by cosmologists for the birth of the universe, the instantaneous creation of all space and time with enough mass and energy to form 100 billion galaxies and disperse them across a volume of a million billion billion billion cubic light-years. With self-conscious understatement, astronomers write it lower case: big bang. Not a huge deal. Don't sweat it.

In fact, the term "big bang" was coined in a 1949 radio interview by Fred Hoyle, who had the previous year proposed a rival theory, called the "steady state," along with colleagues Thomas Gold and Hermann

Bondi. Hoyle averred later that it was never meant to be pejorative and claimed the term was just striking imagery intended to contrast the two theories. The three men were a study in contrasts. Hoyle was a blunt Yorkshireman, intuitive and versatile but undisciplined. Bondi was precise and mathematical. Gold was gifted with a daring physical imagination. They'd worked on radar research in World War II, and in 1947 they all saw the celebrated English horror movie "Dead of Night." The movie had lots of twists and turns but it ended the way it started and this made the three men imagine a universe that was unchanging yet dynamic. As Hoyle put it: "One thinks of unchanging situations as necessarily being static. What the ghost story did for all three of us is remove this wrong notion. One can have unchanging situations that are dynamic, as for instance a smoothly flowing river" (Figure 11.1).[7]

Steady state theory has an expanding universe but adds the feature that matter is being created at a modest rate in the spaces among galaxies. This spontaneously created matter—one hydrogen atom per cubic meter every billion years—is enough to form galaxies and keep

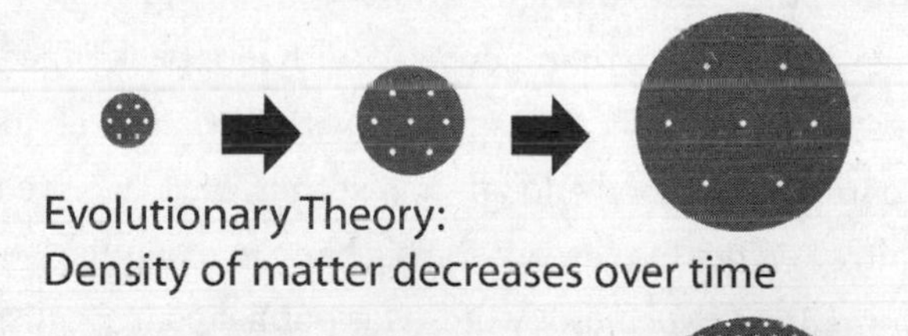

FIGURE 11.1. *The big bang theory is an evolutionary theory, where the mean density of the universe changes with time as space expands. In contrast, the steady state theory has new matter being created to fill the gaps in expanding space, so the mean density of the universe is unchanging.*

the universe appearing the same everywhere in space and time. The theorists called this the "perfect cosmological principle." Hoyle, Bondi, and Gold thought it more elegant to apply the cosmological principle over time and space, and a modest level of matter creation from the vacuum seemed a small price to pay. New stars and galaxies could thus fill the void left behind as the older ones separated. Hoyle was sure that it was more plausible to create a dribble of matter steadily from the vacuum of space than to instantly create the stuff to make 100 billion galaxies from nothing. The big bang seemed like an extreme version of a magician pulling a rabbit out of a hat.

Through the 1950s, there was no observation that could decisively discriminate between the two theories. For a while, the steady state theory was ascendant. It seemed impossible to talk meaningfully of the universe as a singularity, as required by the big bang, and it was unnerving to cosmologists that the laws of physics might be different in the very different conditions early on. Also, Hubble measured the age of the universe to be less than the age of the Solar System! This age problem abated by the 1960s, but was immediately solved in a steady state universe that had no origin.[8] Moreover, the big bang model predicted an expansion rate that increased with lookback time and older galaxies seen at higher redshifts, whereas the steady state model did not. The data weren't good enough to test these last two predictions.

Lemaître laid out his case for the big bang in a popular book called *The Primeval Atom Hypothesis*, which was published in English in 1950. To many nonscientists, it was incongruous that a Jesuit priest had made a major contribution to cosmology. Lemaître was unperturbed. He was generally unperturbed. His vows of poverty, chastity, and obedience didn't mean he was denied all Earthly pleasures. Friends found him to be sociable and cheerful. He loved to play the piano even though his neighbors weren't enamored of the noise and they encouraged him to move. Many walks from his apartment ended at a nearby pastry shop. His favorite tipple was whiskey on the rocks. As one of his biographers noted, "He liked the good things God had put at our disposal. He didn't scorn a good cake, a good bottle, a tasty dinner; everything within the limits of reason."[9]

My own insights into the concordance between science, religion, and hedonism came while floating in the pool at Kino House in Tucson. Kino House is the part-time residence of half a dozen scientist-priests who work at the Vatican Observatory. They split their time between the papal summer palace in Castelgondolfo, just outside Rome, and southern Arizona, where they work at my university and operate a telescope in the Pinaleño Mountains. When I first visited Tucson in 1981, I stayed with the Jesuits at Kino House and marveled at their well-stocked fridge and liquor cabinet. One Sunday, I was relaxing in the heated pool after a night of observing when a mass started in the outdoor cabana. Margarita in hand, I pondered the etiquette of the situation: was it better to finish my drink and wait out the service or make my excuses and walk dripping through the congregation in my Speedo?

Over the years I got to know many of the priests and count them as valued friends and colleagues. Several times I've lectured at their summer schools in Castelgondolfo, where class breaks for coffee and also lunch on a patio high in the ramparts of the papal palace, with a panoramic view of the nearby lake formed in a volcanic crater. Wine is always served, made from grapes grown in the papal gardens, and the atmosphere is relaxed and convivial. So convivial that I've weakened slightly in my nonbeliefs. But not so much I could sign on and agree to the vows, especially that one about celibacy.

The staff members of the Vatican Observatory are the true heirs to Lemaître. They all have PhDs in astronomy and they're all practicing priests. One likes to race his bike in snazzy team colors through the narrow streets of Lazio, another has been spotted wearing fake angel wings to entertain his students, a third is an occasional and avuncular BBC radio show host. These are not dry, colorless clerics. I've taught cosmology with another member of that small clan, whose research specialty is general relativity. I tease him that he believes in physical theory as far back as the unification of the four forces. So his "God of the gaps" has been squeezed into a tiny fraction of a second after the big bang and scientists get to lord over the rest. He smiles benignly at my jibe.

MICROWAVES FROM CREATION

Clouds are mundane yet magical. Sometime, go outside and lie on the grass and recall when you were a child and the clouds entertained you with a cavalcade of exotic creatures, shifting and undulating across the sky. Remember how they could mirror or influence your mood, from a drizzling slab of stratus to feathery wisps of pure white cirrus. Think of all the times you've been diverted by the warm palette of colors when clouds refract and reflect the rising or setting Sun. And who hasn't felt a sliver of amazement when their plane passes through a crisp cumulus boundary without a bump and the cloud turns out to be less substantial than uncooked meringue?

Light can't travel far inside a cloud, yet the difference between inside and out is slight. A cloud defines the region where the density of tiny water vapor droplets is higher than average, high enough that photons bounce off the droplets quite frequently. As a result, the cloud is opaque. At the edge, there's less water vapor and photons will on average not hit a droplet traveling outward, so we see a sharp edge. The seeming solidity of a cloud is an illusion.

It's the same story with the Sun. We've never done the experiment, but an astronaut entering the Sun in a hypothetical spacecraft that could withstand the high temperature wouldn't feel a bump.[10] Going through the region we see as the Sun's "edge," they'd observe smooth increases in temperature and density; there would be no discontinuity.

One moment they'd be heading toward a surface of clearly delineated light and dark patterns, and the next they'd enter undifferentiated fog. The visible edge of the Sun or any other star is called the photosphere. It marks the boundary between opaque and transparent regions of the star. The surface of the Sun is an illusion, no more concrete than the boundary of a cloud.

The early universe may seem alien from the comfort of a cloud or the warmth of the Sun, but the physics is the same. If we turn back the clock on the expanding universe, the big bang model predicts a hotter and smaller universe looking back in time. Redshift measures how light and space are stretched by the expansion, and at large look-back times redshift is essentially the factor by which the universe was smaller and hotter when the light was emitted.[11] When the universe was 5 percent of its present age, at a redshift of 10, it was 10 times smaller and hotter than it is now. When the first stars formed, 13.6 billion years ago, or at a redshift of 30, the universe was 30 times smaller and hotter than it is now. Before this time—100 million years after the big bang—the universe had lots of mass and energy so it wasn't empty, but it contained no "things."

The infant universe is in many ways easier to understand than the ancient universe we see around us.

That's counterintuitive so let's explore the reason. The universe is interesting because it contains people, planets, stars, and galaxies. Structure extends over a huge range of distance scales and even though it's the result of the well-understood forces of gravity and electromagnetism, the details aren't predictable. No simulation can begin with a completely smooth distribution of matter and energy in an expanding universe and generate *our* galaxy or *our* star or *our* planet, and it certainly couldn't predict *us*. All theory and computer simulations can do is predict the general nature of structures and their statistical properties.

Complex structure is gratifying but it comes with a price in terms of understanding. When a galaxy forms and stars and planets congeal from the condensing gas, the process is chaotic, turbulent and highly nonlinear.[12] Remember that you're a billion billion times denser than the universe that gave rise to you; it's very difficult to reliably predict evolution over so many orders of magnitude in density. Astronomers have only a rudimentary understanding of the formation of galaxies, stars, and planets, and they sensibly leave the formation of people to biologists.

However, for its entire early history, the universe behaved like a gas and gas is described by a simple relationship between temperature, density, and pressure. Before the universe creates things by the force of gravity it can be treated as one thing. If the early universe was a gas, then it must have emitted radiation of a wavelength appropriate to its temperature. The existence of radiation "left over" from the big bang was predicted by the clever and quirky physicist George Gamow.

Gamow was born in Odessa, then a part of the Soviet Union. He had an early interest in science. He used a telescope his parents gave him to gaze at the stars, and when his father gave him a microscope, he used it to see if the communion bread from the local church was any different from normal bread. (It wasn't, and Gamow was disappointed in the lack of magical properties.) Gamow made major contributions to the theory of radioactivity and was one of the youngest people elected to the Russian Academy of Sciences. He routinely met with three other colleagues to discuss frontier issues in physics; Gamow had a waggish wit and he called them the Three Musketeers. However, his brilliance was overshadowed by the increasing repression of Stalin's regime. One of the musketeers was executed in a pogrom. When Gamow and his wife saw the danger they were in they tried to flee. Twice they tried to kayak across the Black Sea to Turkey but each time bad weather forced them to turn back. Finally they were able to defect while attending a physics conference in Brussels.

In the late 1940s, Gamow started work with his student Ralph Alpher on predictions that would distinguish the big bang from its rival theory, the steady state. They wrote a paper showing how the universe could have produced the observed abundance of helium and other light elements when it was as hot as the center of a star. Gamow added Cornell physicist Hans Bethe's name to the paper, even though he'd done no work on it, so the author order Alpher, Bethe, Gamow would be a riff on the first three letters of the Greek alphabet. A separate paper included a prediction that the afterglow of the big bang should have cooled down after billions of years to a temperature of just 5 degrees above absolute zero.[13]

Gamow had a rich and varied career. He wrote a series of whimsical children's books about physics, illustrated with his own drawings. He played a seminal role in molecular biology, being the first to suggest that triplet combinations of the four DNA bases form the amino acids that code for proteins. Unfortunately, he loved alcohol almost as much as he loved physics, and he died from complications associated with alcoholism at the age of sixty-four.

The prediction of relic radiation from the big bang languished for over a decade. Thermal radiation with a temperature of just a few degrees above absolute zero has a wavelength of a few millimeters and even if somebody had wanted to look for the radiation, the technology didn't exist to detect it.

THE FOG LIFTS

The curtain-raising of the universe happened 380,000 years after the big bang. That's when the expanding gas cooled enough for electrons to join with protons and form stable atoms. With atoms preoccupied, light could travel unimpeded without careening off energetic electrons.

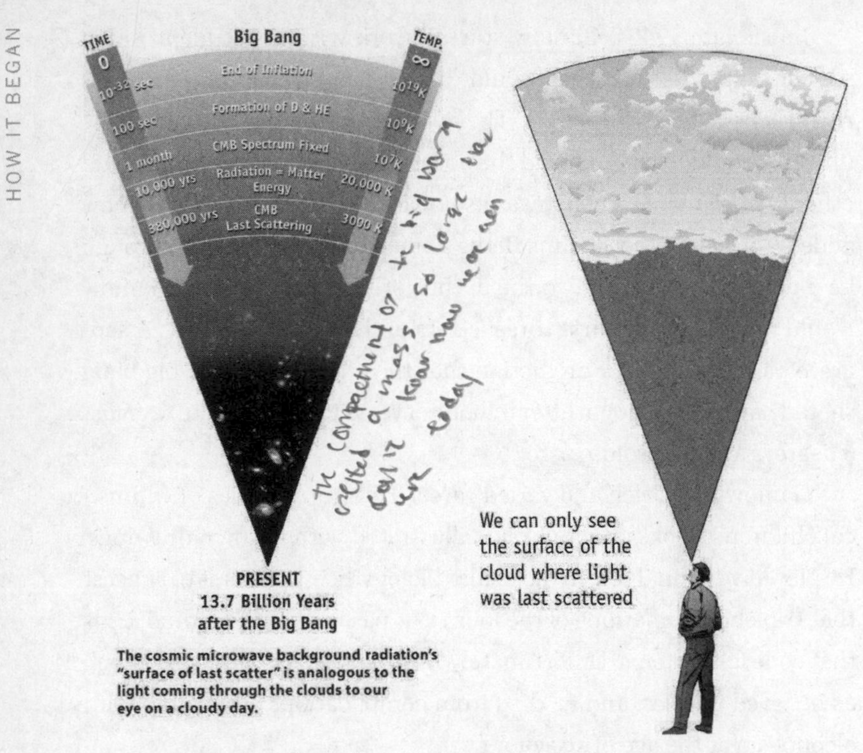

FIGURE 11.2. *The limit of our vision in the universe corresponds to a time 380,000 years after the big bang, when the density became low enough that photons traveled freely. Events before that are hidden from view. Photons from that time have been redshifted by expanding space and are all around us. Similarly, the edge of a cloud is the place where photons can travel freely; inside the cloud, light bounces around and the cloud is opaque.*

The universe went from being opaque to being transparent. When the fog lifted, the universe was 1000 times hotter and smaller than it is now (Figure 11.2).

We're reaching back to the first thousandths of a percent of time in the life of the universe. If the universe now is a forty-year-old in his or her prime, the universe then was a mewling baby on the day it was born, still reeling from its passage through the birth canal.

The universe is like a cool red star except that its boundary is in time rather than space. A star with a surface temperature of 3000 Kelvin seems to have an edge but there's no sharp transition in any of its properties. At a radius corresponding to the apparent surface, the gas density drops to the point where photons no longer interact. As they travel freely we see that radius as a surface; inside is opaque. In the expanding universe, less than 380,000 years after the big bang the density and temperature are so high that there are no atoms. All electrons and protons are flying around separately and light can't go far before being deflected. Before this time the universe was opaque. After this time, electrons attach to atoms to form neutral matter and light rarely interacts with atoms. The curtain rises and the universe becomes transparent.[14]

Back then, the radiation that filled the universe had a wavelength of 2 microns, in the infrared region of the electromagnetic spectrum, just beyond the range of vision. Since then the universe has grown 1000 times bigger and all the waves have been stretched by a factor of 1000 to a wavelength of 2 millimeters, which is the microwave region of the electromagnetic spectrum. The temperature has dropped by the same factor, from a toasty 3000 Kelvin to a frosty 3 Kelvin. In the big bang model, the microwaves from creation should suffuse and permeate every cubic meter of space.

Cue one of the most remarkable, accidental discoveries in the history of science.

The birth of modern cosmology depended on two unlikely ingredients: noise and shit. In 1964, radio astronomers Arno Penzias and Robert Wilson were working for Bell Labs in New Jersey. The phone company at that time encouraged fundamental research, since it might lead to technology with commercial value. They got their hands on a 20-foot horn antenna—resembling a huge ear trumpet—that had been used to test early satellite communication systems, and they carefully prepped it to observe faint radio sources. Radio astronomers have to deal with interference of all kinds and Penzias and Wilson had taken great pains to suppress or remove the effects of radar and radio broadcasting and cool their detector with liquid helium to reduce its background noise. So they were surprised and frustrated to detect a persistent level of noise or "radio hiss" in their data. The mysterious noise didn't vary in strength, was evenly spread on the sky, and persisted day and night (Figure 11.3).

Penzias and Wilson were systematic and thorough experimenters, and they noticed that pigeons used the horn for shelter during the rugged New Jersey winter. As they delicately described the situation in a Bell Labs technical memorandum, a "thin, white dielectric film" might be acting as a source of noise. But after they cleaned out the

FIGURE 11.3. *Arno Penzias and Robert Wilson in front of the 20-foot horn antenna with which they discovered relic radiation from the big bang in 1965. They worked at Bell Labs and found a weak microwave signal of very low temperature that didn't originate from any known terrestrial or celestial source.*

pigeon shit, the excess noise was still present. They concluded it was coming from beyond the Milky Way, though the emission couldn't be identified with any known source of radio waves or microwaves. Penzias and Wilson had stumbled on "smoking gun" evidence for the big bang.

Meanwhile, just 40 miles up the road, a research group in Princeton was in hot pursuit of microwaves from the early hot universe, but they weren't quick enough to make the first discovery. Robert Dicke worked on microwave detection in the 1940s as part of the effort to put radar in Allied fighter planes during World War II. With better equipment he might have made the discovery in 1946. Gamow's work languished in the research literature, unrecognized by Dicke and his group. Then in 1964, two Russian theorists drew new attention to the prediction of radiation from the hot big bang and suggested it might be detectable. Dicke was spurred by this paper to construct a small antenna on the roof of the Physics Department at Princeton University. But before he could gather data, Penzias and Wilson made their discovery and were told of its significance. Dicke gathered his team and told them, "Boys, we've been scooped."

The Princeton group and the Bell Labs team published back-to-back papers on the observation and interpretation of microwaves from the big bang, but accolades often go to discoveries, so it was Penzias and Wilson alone who won the Nobel Prize in Physics in 1978.[15] Their boss at Bell Labs, Ivan Kaminow, summed it up nicely in recalling the luck of the young researchers with a laugh, "They looked for shit but found gold, which is just opposite of the experience of most of us."

Where's the big bang? The big bang is all around us. There are tens of thousands of microwaves from creation in every breath you take. This sounds dangerous, but their energy is feeble—the radiant intensity of this relic of the big bang is 10^{-5} watts, a ten-millionth of a lightbulb's worth of radio power. This observation was a strong affirmation of Lemaître's "primeval atom."

PRECISION COSMOLOGY

Astronomers have spent nearly four decades trying to make better pictures of the microwave sky. As we've seen, optical telescopes find galaxies at large lookback times, but the search runs out of steam a few hundred million years after the big bang, because the universe was a trackless wilderness with no sources of light. Microwaves give astronomers a snapshot of the universe when it was just 0.003 percent of the present age. We've learned more from these baby pictures than from any other type of evidence in cosmology.

The cosmic microwave background radiation, or CMB radiation as it's called, thrusts us deep into the weirdness of the expanding universe. Sunlight is simple. The Sun is there, we're here, and sunlight travels from there to here. Microwaves from the big bang are all around us and they travel in all directions. Radio telescopes gather them as we might pluck motes of dust from the air around us. But they haven't traveled neatly from A to B like sunlight. They've traveled on many different paths through tens of billions of light-years of space while being stretched like taffy to 1000-times-longer wavelengths. Radio telescopes gather the microwaves and make an image of the last time the radiation interacted with matter. Even though the result is a "map" of the sky in all directions, we're neither the center nor the target of this ancient radiation. Hypothetical astronomers looking out from a distant galaxy would see a similar microwave sky.

It's very strange to think of us as bathed in low-level radiation that comes from every direction, where each photon has traveled without interruption for over 13 billion years. But this situation allows us to form a picture of the entire universe at a much earlier time, just after the big bang, when its physical state was very different.

Penzias and Wilson established the basic attributes of the microwave radiation in their discovery observations. That fact that radiation is detectable at all means there are a lot of photons left over from the big bang. A thousand in each chunk of space the size of a sugar cube, to be precise. How many photons are there in the universe? OK, let's do the math—just shout it out if you get there first—the volume of the universe is 10^{33} cubic light-years, and 10^{53} sugar cubes fit into a cubic light-year—making for a very sweet universe—times 1000—now I'm sweating a bit—to get the answer, 10^{89} photons. That's the largest pure number in science and it's even impressive to astronomers, who are cavalier in tossing around their billions and trillions.

Penzias and Wilson knew the microwave signal was unusual because it didn't come from any particular source, either terrestrial or celestial. The smoothness of the radiation is one of its most important features. How smooth? The 1965 experiment was only good enough to say that it didn't vary by more than a few percent. It seemed featureless. To do any better, astronomers needed a satellite to escape the cacophony of radio and radar and TV transmissions that plague anyone who tries to detect microwaves from the ground. In 1975, NASA commissioned a science team to design a satellite optimized to measure microwaves with high precision. The Cosmic Background Explorer satellite (COBE) was launched in November 1989.

COBE was a stunning success, surpassing all its design goals. One of its three instruments measured the spectrum of the radiation for the first time and showed it was perfectly thermal and consistent with one temperature for the whole universe.[16] That ruled out any possibility of the radiation being the sum of spectra of many discrete objects like stars since stars don't have perfectly thermal spectra and they'd all have different temperatures. COBE measured the temperature to be 2.725 Kelvin with an uncertainty of only 1/1000 of a degree (Figure 11.4). The coldest thing in the universe is the universe!

With much greater sensitivity, another of COBE's instruments looked for variations in the intensity of the microwaves in different

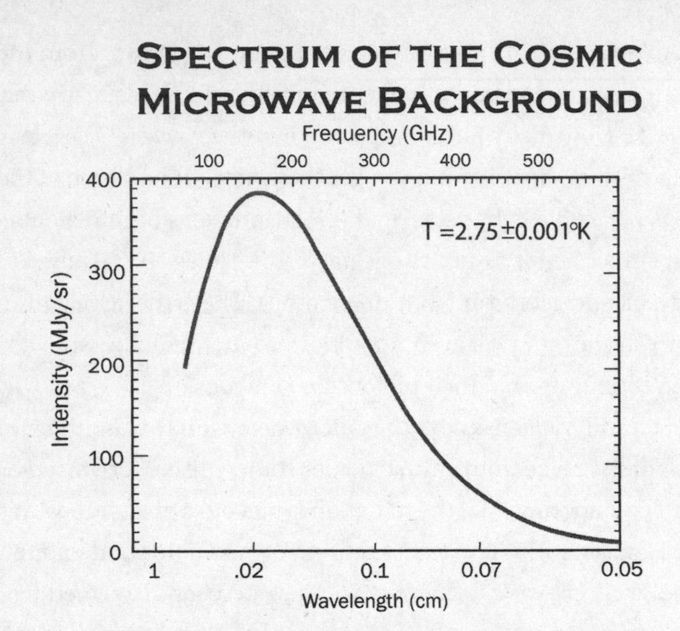

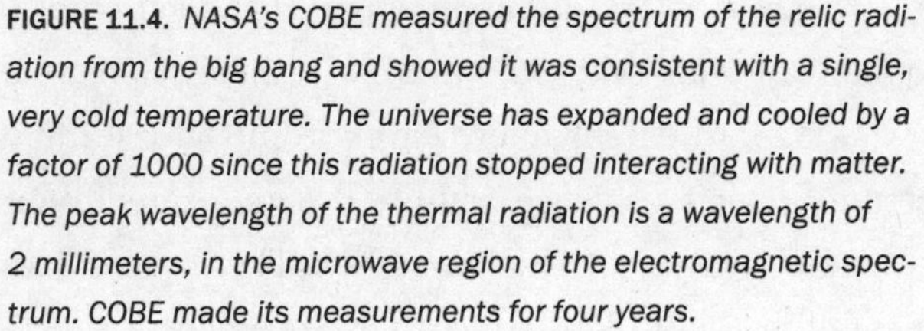

FIGURE 11.4. *NASA's COBE measured the spectrum of the relic radiation from the big bang and showed it was consistent with a single, very cold temperature. The universe has expanded and cooled by a factor of 1000 since this radiation stopped interacting with matter. The peak wavelength of the thermal radiation is a wavelength of 2 millimeters, in the microwave region of the electromagnetic spectrum. COBE made its measurements for four years.*

directions. The map showed a smooth gradient over the sky, with a temperature 0.0034 Kelvin hotter in the direction of Leo and a temperature 0.0034 Kelvin cooler in the direction of Aquarius.[17]

Wait. How can the universe be hotter in one direction than in another? The interpretation is that the Earth isn't stationary with respect to the universe as a whole. We orbit the Sun. The Sun orbits within the Milky Way. The Milky Way moves within the Local Group, which is falling into the Virgo Cluster, and the Milky Way and all the galaxies in Virgo are being tugged toward an even more massive agglomeration of matter 100 million light-years away. The sum of these nested "Matryoshka" motions is 370 kilometers per second, or nearly 800,000 mph. That Doppler shift gives the microwaves shorter wavelengths (and makes them slightly hotter) in the direction of motion and longer wavelengths (and makes them slightly cooler) in the opposite direction. We assume in this argument that the universe isn't moving. (After all, where would it go?)

COBE scientists gathered four years of data before the satellite ran out of its liquid helium coolant. When they modeled and subtracted out the variations resulting from the systematic motion just described, and subtracted out a band of emission from cool dust in the plane of the Milky Way, low-level "speckles" were left over. These subtle variations had no obvious pattern and they were 0.00003 Kelvin above and the same amount below the average temperature of 2.725 Kelvin. Try to imagine the microwave sky as the surface of a pond 100 meters across; the biggest ripples are 1 centimeter high (Figure 11.5).

The picture of the infant universe now had features. It wasn't very detailed, just good enough to make out a head, hands and feet, and large splotches on the skin. Astronomers were excited because they knew these speckles must be the seeds for galaxy formation. Hold down the plunger of a bicycle pump with the valve covered up and the cylinder of air will become warm to the touch. The temperature of a gas rises up when its density rises. Gas in the universe behaves the same way; it was hotter at earlier times when it was denser. The same is true of the variations across space—slightly hotter regions are slightly denser. Great oaks from little acorns grow. The variation in gravity from a very slightly denser to a very slightly less dense region was only a tiny fraction of a percent 380,000 years after the big bang but over tens of millions of years the variations in density grew steadily. Then, about 100 million years ago, the gravitational collapse process accelerated as the first stars and galaxies formed.[18]

The popular media enthusiastically picked up the story and there were breathless quotes about finding the "fingerprints of God." Pro-

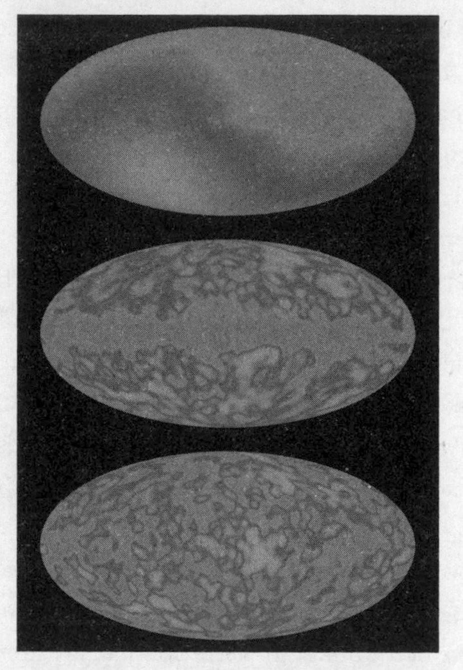

FIGURE 11.5. *These all-sky maps show the main results of the Cosmic Background Explorer, a NASA satellite that made the first accurate measurements of the microwave background radiation in the late 1980s. The top frame shows a low-level gradient in temperature from one side of the sky to the other, indicating the motion of our galaxy and others relative to the universe as a whole. The middle frame shows emission from gas in the Milky Way running across the middle and when this is subtracted, the lower frame shows a very low level of temperature variations that are the seeds for later galaxy formation.*

fessional accolades followed and, in 1996, the lead investigators of the mapping and spectroscopy experiments, George Smoot and John Mather, won the Nobel Prize in Physics for their work. (Smoot has another claim to fame as the million-dollar-prize winner on the Fox TV show "Are You Smarter Than a Fifth Grader?" It's reassuring that he in fact is.) The Nobel Prize committee said the pair of scientists had ushered in the era of "cosmology as a precision science."[19]

COBE was a stunning success, and for a mere $100 million, it was a steal. However, the speckles were at the limit of sensitivity and the angular resolution of the maps was 10 degrees, or 20 times the Moon's diameter, making for a very crude view of the early structure. There were several successful balloon missions to detect microwave background radiation in the 1990s, all launched from Antarctica, but astronomers had their sights set on a successor to COBE. Work began on the Wilkinson Microwave Anisotropy Probe (WMAP) in 1995, and it was launched by NASA in 2001 to a destination 1.5 million kilometers from Earth. Its mission was supposed to last two years, but the data were gorgeous and the satellite worked flawlessly, so NASA extended the mission several times and it took its last science data in August 2010. WMAP improved on COBE as much as COBE improved on the experiments that preceded it.

WMAP has made maps of the microwave sky with 50 times better angular resolution than COBE. In pictures of the newborn universe, that lets us count fingers and toes and start to see the shapes of the mouth and ears (Figure 11.6). WMAP enabled the profound measurement of the shape of the universe!

How can the relic radiation from the big bang reveal the shape of the universe? It works like this. The speckles or ripples in the microwaves have a characteristic angular scale of about 1 degree; this property is consistent with the amount of dark matter and dark energy that we observed in the more proximate universe. The 1-degree features appear across billions of light-years of expanding space as photons travel for 13.7 billion years to reach us. The universe is like a gigantic optics experiment. In general relativity, the shape of space is related to the density of mass and energy. If the universe is closed, it has a positive curvature like the surface of a balloon (but in three rather

FIGURE 11.6. *WMAP made images of the microwave sky with 50 times better angular resolution than COBE. This all-sky image, the result of seven years of data, shows the tiny variations, a few parts on 100,000 of temperature across the sky, that indicate the seeds for galaxy formation. It would take more than 100 million years for these slight density enhancements to turn into galaxies.*

than two dimensions). In that case, parallel light paths converge, the universe acts like a magnifying lens, and features in the microwave sky are larger than 1 degree in size. On the other hand, if the universe is open, it has a negative curvature like the surface of a saddle. In that case, parallel light paths diverge, the universe demagnifies, and the features in the microwave sky are smaller than 1 degree across.

WMAP showed that space is flat and Euclidean to within 1 percent. In a way, that's a bit disappointing. General relativity provides a framework for space to undulate and curve and turn inside out if it wanted. Instead, we live in a "straight back and sides" conventional universe. Given a choice of a tempting array of flavors, it chose vanilla. The door is still open on more exotic possibilities since there's more space-time out there than we can observe, so the topology of the universe could be complex and interesting beyond our horizon.

The ultrasharp image of the ancient universe pins down many of the parameters of the big bang with unprecedented accuracy. The age of the universe is 13.73 billion years with an accuracy of 1 percent, attested to by the Guinness Book of World Records. WMAP has also measured the proportions of normal matter, dark matter, and dark energy with an accuracy of 1 percent. To do any better than this might even be overkill; in the case of dark energy and dark matter it doesn't seem helpful to measure very accurate proportions of things so poorly understood.

With microwave eyes, we could see the big bang. As we'll see in the next chapter, if we had massive ears we could hear it too. The universe is a feast for the hyper-senses. All this information comes from a tiny early fraction of the universe's history. If I live to a ripe old age, it's like the picture that tells the story of the day I was born.

The universe doesn't have good or bad days; it's imperturbable and implacable. No such luck for sentient creatures that happen to live in it. We and other entities like us ride a roller coaster of emotion as we navigate our mortal coils. My day job involves figuring out how the universe works.

Not too long ago I had a "bad cosmology" day. I got into work at the university and my e-mail inbox included a bland message from the National Science Foundation telling me that my grant proposal had been rejected. With an oversubscription of applications of six to one, that's no great ignominy, but it represented weeks of work down the drain. Later in the morning I got a negative referee's report on a paper I'd written. The referee used the cloak of anonymity to be snarky and I was sure the next time I encountered them at a meeting, they'd smile at me innocuously. I nursed my half-cold cup of Student Union coffee and retreated into shuffling admin paper for safety.

In the middle of the afternoon, one of my graduate students stopped for our weekly meeting and sheepishly told me one of the data tapes from a recent observing run in Chile was corrupted. We were without vital calibration data for those three nights of observation. Dozens of spectra of high-redshift quasars were now indecipherable. Their light had traveled 10 billion years to be captured by the telescope, and we then dropped the ball. I sighed and consoled my student—we learn by making mistakes and I've made plenty of my own.

The afternoon wore on. I graded some quizzes from my introductory astronomy class. The students were freshmen, most of whom were taking the class as a General Education requirement. They were a snapshot of the nation's youth: hip and Internet-savvy, but diffident about science. My mistake was giving a quiz requiring short-answer questions. Multiple choice tests are broad brush stroke and statistical. But in a short written answer, errors and ignorance are fully revealed. I winced at the number of students who, impervious to my instruction, thought that a galaxy was smaller than a star or the universe was a few million years old. It was a day when the universe didn't have to work very hard to guard its secrets.

Enough.

I left the building weighing more than my usual 155 pounds. I'd been happy to snag a prime parking spot under the shade of a tree, but it was obviously the hangout of a gang of starlings, and they'd left their calling cards over the front of my car. I drove home squinting through the white splatter on my windshield. A pile of unpaid bills was staring at me dolefully on the kitchen table, but I ignored it, prepared some comfort food, and sat down to veg out and watch something on cable. No such luck. The cable was out and the TV screen was a blizzard of flickering speckles.

In life, you can either cry or laugh. My mind cast back to the delicious moment when we learned that everything around us, all that we hold dear, was once part of a cauldron of creation. To the time when two engineers working for the phone company stumbled on a feeble signal from that fantastic event. The bird shit on my car seemed an ironic reminder of their discovery. A small percentage of the speckles in front of me were interactions of the phosphor on the screen with photons of the cosmic microwave background. So I popped the top off a cold beer and watched the big bang that evening, confident it was more entertaining than anything I might watch in a normal evening of TV.

—

The show's over. I missed the fireworks and as the gas flies away from me in every direction and the red light fades, I'm melancholy. As if I arrived too late for an outdoor festival to find everyone had dispersed and darkness was falling. It's an undifferentiated wilderness. I've no idea where I am.

Patience. A phantom in this alien space, I yearn for the familiarity of my breath, my body, my home, my world. There's nothing to do but wait. So I take my cue from the universe itself, for it seems not to be mindful of time. If there's a pattern in the mingling of atoms, I can't discern it. If there's a purpose in the outrushing plasma, it's not obvious. In this austere setting, I can only hope that I'm built into the equations.

I stay put and wait for events to unfold. If I moved slightly to the left or right, or slightly forward or back, or slightly up or down, the density of matter would be infinitesimally lower and gravity's power to gather and sculpt would be compromised. A single sideways step would cast me into barrenness and doom me to eternal darkness.

I've made a good choice.

It gets much colder. Photons stretch like rubber bands and slide off into the infrared. Atoms thin out into an ever more perfect vacuum. But the tiny enhancement of my starting position leads in a different direction. The excess gravity corresponds to a meager interest rate, but by compounding over time the gas around me starts to grow thicker, as if the arrow of time has been reversed.

Without warning, a light goes on nearby, then another, and a third, until I'm part of a web of light. These first stars don't last long, but they're markers of brooding clouds of dark matter into which gas is now falling. I watch as the millions of years flick by. Legions of stars ignite in small, ragged galaxies all around me. I'm drawn to the periphery of two larger galaxies as they swoon into a tight embrace and merge. I watch a spiral pattern develop and I gently fall onto the disk, noticing that I'm part of a gentle rain of gas falling in from the depths of space. Doing nothing but follow the dictates of the expansion and gravity, I've ended up riding the slow merry-go-round of a spiral galaxy. For 8 billion years and 30 orbits I watch with mild interest until a middleweight star ignites nearby and a moist rocky cinder catches my eye.

Earth.

12

WHITE HEAT

BLINDING LIGHT. *Searing heat. Violent expansion. I've no reference or experience to help me make sense of what's happening around me. Luckily, I'm disembodied, or the radiation would cook me, the heat would vaporize me, and the outrushing gas would tear me asunder in an instant.*

I stop trying to figure out why I'm here, or even what "here" means, since the universe is an undifferentiated cauldron of particles and the view is the same in all directions. It's far denser than lead, yet it's a supple gas, burgeoning outward in all directions, with no center and no edge to the expansion. Radiation and particles are moving at light speed, but space is unreeling even faster. Nearby regions are being whisked away at impossible speeds, and I realize they'll never be seen again, by me or anyone else. My own pocket of space-time is a cozy quadrillion miles across; it's amazing to think that eventually it will blossom into a billion-light-year void painted with stars and galaxies.

Interesting. In the firestorm of particles and photons, it seemed as if they were just bouncing off each other like rubber balls. But now I can see that's not true. A fraction of the larger particles are sticking together and continuing to ricochet around in the plasma. A small fraction has grown to four times their original mass. These little clusters of neutrons and protons are the first hints of a vast construction project.

Something unfathomable has happened just moments earlier. I can't comprehend what could unleash such titanic forces or how this place—if everywhere and nowhere can be called a place*—could be any hotter or denser than it is now.*

—

PIPER AT THE GATES OF DAWN

Cosmology has its root in the ancient Greek concept of "cosmos," meaning an orderly and harmonious system. In the Greek view, the antithetical concept of "chaos" referred to the primeval state of the universe, which was total darkness or an abyss. Supposedly, order emerged from disorder when the universe was born.

The term *cosmos* was first used by the philosopher and mathematician Pythagoras, born on the rugged island of Samos in the Aegean Sea, in the sixth century BC. Pythagoras is also thought to have come up with the notion that the universe was based on mathematics and number, although in truth so little is known about Pythagoras and his followers that direct attribution of these ideas is impossible.[1] Humans had used counting systems for thousands of years but Pythagoras was the first to invent a layer of abstraction such that numbers could underlie and explain physical objects. He's credited with "harmony of the spheres," the mystical and mathematical idea that simple numerical relationships or harmonics were manifested by celestial bodies, with a result having commonality with musical harmony. Pythagoras developed the rules of musical harmony by experimenting with whole numbers to fractionally divide a plucked string. He and his acolytes didn't really think that the music of the spheres was audible, but it was an example of their belief in unity through mathematics.

Two thousand years later,[2] Kepler applied Pythagoras's ideas to the orbits in the Solar System. Kepler's life was so difficult and chaotic we can imagine why he sought harmony in the celestial realm. He was sickly, myopic, and covered in boils. His father abandoned the family when Kepler was a teen, and his mother dabbled in the occult and was later put on trial as a witch. Greek geometers had discovered that only five solids can be constructed from regular geometric shapes: these perfect "Platonic" solids have 4, 6, 8, 12, or 20 sides. Kepler realized that these solids nested would give the relative spacing of the six planets known at the time.[3] He was even more excited when he found that the ratios of maximum and minimum angular velocities of the planets corresponded to musical intervals. By combining pairs of planets he was able to derive the intervals of a complete scale. Kepler thought the music of the celestial realm manifested spiritual perfection that humans could only aspire to.

The resonance between mathematics and music was embodied more recently by Einstein, who was quoted as saying, "Mozart's music is so pure and beautiful that I see it as a reflection of the inner beauty of the universe."[4] A competent and passionate violinist, Einstein liked to improvise late at night while he ruminated on physics problems. We can imagine an unbroken connection in space-time from Pythagoras and his plucked string, through Kepler via Plato and Ptolemy, to the violin of Einstein.

We also hear echoes of the tradition of the Dreamtime, the aborigine creation story where the universe is sung into existence. Also in the modern tradition of cosmology, a series of harmonies brings forth the material world, providing the seeds for growing stars and galaxies.

There was a piper playing at the gates of dawn.

The cosmic microwave background radiation is decisive evidence that the big bang occurred, but the recent maps are so detailed we can use them to learn a lot about the early universe. We can easily rule out some possibilities. For example, if all the seeds that formed galaxies had the same physical scale, the speckles in the radiation map would all be of similar size and the pattern would repeat like a printed fabric or wallpaper. Or the universe might behave like a fractal, with equal numbers of tiny, medium, and large speckles. And another intriguing possibility is that the universe is a "hall of mirrors," where locally flat space-time is embedded in a globally curved space-time, so that light wraps around and travels through the observable universe multiple times. The signature of this would be patterns of speckles mirrored across the sky two or more times.

WMAP's data support none of these signatures. However, even to the eye, there does seem to be a typical size to the speckles, suggesting the situation intermediate between a regular pattern and randomness. Astronomers subject the WMAP data to something they call power spectrum analysis, measuring the fraction of the variations on different angular scales. In simple terms, *l* is the angular frequency of the variations. For example, an *l* of 2 corresponds to 2 cycles over the sky, or variations over 100 degrees, which shows the motion of the Milky Way relative to more distant galaxies. The 7-degree angular resolution of COBE corresponded to *l* of 30, and the much better 0.3 degree resolution of WMAP reaches almost to *l* of 1000. The shape of the angular power spectrum can be compared to predictions from the big bang models. In the musical analogy, each value of *l* is a different "harmonic" of variation of the radiation.[5]

The physics of the early universe is certainly esoteric, but the analogy with sound helps as long as we don't stretch it to the breaking point. Less than 380,000 years after the big bang, radiation was coupled to matter, and electrons and photons behaved like particles in a gas with photons ricocheting off electrons like bullets. As in any gas, density disturbances moved at the speed of sound as waves, or a series of compressions and rarefactions. The compressions heated the gas and the rarefactions cooled it so the sound waves manifested as a shifting series of temperature fluctuations. After 380,000 years, the electrons combined with the protons to become neutral atoms, and light from slightly hotter and cooler regions traveled unimpeded for 13 billion more years. The temperature variations we see now are the "frozen" record of fluctuations from that time.

What song was the piper playing at the gates of dawn? The power spectrum, or distribution of temperature fluctuations on different angular scales, shows that our visual impression of the map of microwaves is correct. There's a strong enhancement of speckles with angular size of 1 degree, about twice the diameter of the full Moon. There's a second, weaker peak on scales of a third of a degree, and a third peak around a quarter of a degree (Figure 12.1). After that, WMAP isn't capable of seeing finer structure.[6]

Let's imagine you're playing a flute. The fundamental tone is a wave with maximum compression where your mouth is, and with minimum compression at the open end. But the flute has a series of overtones, excited by the suitable placement of holes and the strength of the breath, with wavelengths that are integer fractions of the fundamental tone. The wavelengths of the first, second, and third overtones are one-half, one-third, and one-quarter as long as the fundamental tone. Overtones add richness to the sound.

Now let's turn to the cosmic piper, whose instrument is the universe. In the early universe, the sound waves are similar, with the important distinction that the waves are oscillating over time in expanding space as opposed to oscillating in fixed space. Also, sound travels incredibly fast in these unusual conditions, a million times faster than on Earth, or over half the speed of light! Imagine the waves originate at the big bang and end when the gas becomes transparent about 380,000 years later. The fundamental tone is a wave with maximum compression

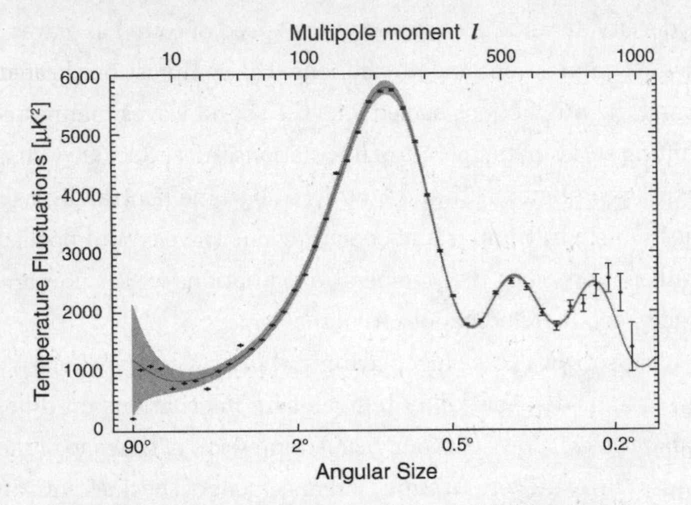

FIGURE 12.1. *The angular spectrum of fluctuations in the microwave background captures information on resonances and harmonics in the universe 380,000 years after the big bang, as particles and light waves oscillate in the primeval plasma. The strong peak at an angular scale of a degree is the fundamental tone, and two harmonics are seen at smaller angular scales.*

(or maximum temperature, since they vary together) at the big bang that has oscillated to minimum compression (minimum temperature) when the gas becomes transparent. The overtones oscillate two, three, four, or more times faster and so they cause successively smaller regions of space to reach maximum displacement after 380,000 years.

We now have all that we need to interpret the graph of temperature variation power versus angular frequency as measured by WMAP. The strongest peak is the fundamental tone, which compressed and rarified the plasma to the maximum extent during the first 380,000 years. The size of the region corresponding to 1 degree at a redshift of 1000 is about 1 million light-years. So the frozen seeds of galaxy formation were galaxy sized back then, but space has expanded by a factor of 1000 in the 13 billion years since, so microwave fluctuations that we observe now are a billion light-years across. As predicted, the first and second overtones have angular scale one-half and one-third of the fundamental, and the third overtone is on a scale too small for WMAP to measure (refer back to Figure 12.1).

There's another subtlety that explains the weaker fluctuations at small angular scales. Sound is carried by the collisions between particles and when the wavelength is shorter than the typical distance traveled by particles between collisions, the wave dissipates. In air, this distance is only 10^{-5} centimeters. But in the nearly perfect vacuum of the universe before it became transparent, photons could travel 10,000 light-years before colliding. So the high harmonics are reduced or damped out. After the thousandfold expansion, those scales are now 10 million light-years. So we wouldn't expect to see significant structure in the local universe on scales much more than 10 times that size. Clustering of galaxies is indeed weak on scales larger than 100 million light-years, so chalk up another success for the big bang model.

The microwave harmonics yield another vital piece of information. If the sound waves had been produced through the period from the big bang to the era of transparency 380,000 years later, the disturbances wouldn't be synchronized and the harmonics would be smoothed out and erased. Think of the muddy sound from a flute with many holes at irregular or random intervals. Only if the sound originated close to the origin of the universe would the sound waves be synchronized in a musical way.

What then was the sound of the big bang? Is it like music or more like a primal shriek? Mark Whittle at the University of Virginia has worked it out.[7] If anyone had been there to hear, it would have had a volume of 110 decibels, like a loud rock concert. Because the infant universe was large compared to a flute or an organ pipe, the sound frequency was incredibly low and inaudible, 50 octaves lower than middle C on the piano. Speeded up by a factor of a trillion and shifted up in frequency by 50 octaves, it would have sounded somewhat like the final crashing chord and subsequent reverb in the classic Beatles song "A Day in the Life."

We hear the piper play in microwaves from the dawn of time,[8] but is any echo of the music still audible? Yes, though not in the conventional way we would hear sound waves traveling in air. The preferred scale of variations in the microwave background is 1 degree; we can think of those slightly hotter and denser patches as the centers of the sound waves traveling out into the young universe. Each of these patches is a seed where a galaxy will one day form. But after 13.7 billion years, the waves have traveled 500 million light-years, so there's a very slight density enhancement 500 million light-years from every galaxy and a slightly larger chance that galaxies in the present universe will have this separation.

To extend the wave analogy, each galaxy is a pebble in a pond with a ripple spreading outward, while the universe is a very large number of pebbles tossed into the pond, whose ripples have spread outward and formed a complex pattern of overlapping ripples. The echo of the fundamental tone is found by counting pairs of galaxies and searching for a slight excess of pairs with a separation of 500 million light-years, but not 400 million or 600 million or any other number. This signal is very subtle—you'd never notice it by staring at a map of the galaxies in the sky, but it was found with statistical techniques in 2005.[9] Using the physical scale of this "acoustic peak" as a meterstick, astronomers are trying to map it across time since the big bang, since its evolution is governed by a tug-of-war between dark matter and dark energy, and this is one of the few ways of getting a handle on both.

The piper played very soon after the big bang, but even now we can hear an ancient echo of the music of the spheres.

TESTING THE BIG BANG

How confident are astronomers that the big bang actually happened? It seems to describe a hypothetical event that took place 13.7 billion years ago. Projecting the expansion back to time zero implies infinite temperature and density, and that's physically impossible. We have to be careful what the theory says and doesn't say, and avoid common misconceptions. It's unfortunate that "big bang" is widely interpreted as an explosion or a time when all matter was concentrated in a point. Neither notion is correct. The theory doesn't explain the origin of the universe; it describes its evolution from a denser and hotter condition. Testing the big bang means confronting the theory with as much data as possible, as early as possible in cosmic time.

Is the big bang theory just a clever idea that can't be proven? Theories are never proven, so in a sense, yes. But in science, certain hypotheses can be checked and verified so thoroughly that they are considered proven beyond a reasonable doubt. A good theory makes predictions that are unique, specific, and testable. You might propose that birds navigate by infinitely elastic, invisible threads that connect them to their nests (this was in fact suggested in the eighteenth century). It's a very hard idea to test. The modern view is that birds navigate long distances by a combination of visual cues, magnetic fields, and the orientation of the Sun and stars. Each part of the explanation makes predictions that can and have been tested. That doesn't mean there are no arguments—the research literature on bird navigation is very rich. But progress is made *because* the hypothesis is testable.

The big bang theory or hypothesis is supported by a web of evidence, and the rival steady state theory couldn't explain the evidence, so it's not considered credible anymore. Why hasn't a new challenger risen up to take potshots at the champ? Because the bar is set high (to mix sporting metaphors) for any new theory. An alternative to the big bang would have to naturally account for the same evidence as the big bang while doing better in new arenas where the big bang fell on its face.[10] That hasn't happened in the last 35 years.

Would I bet my life that the big bang theory is correct? No. Would I bet my dog's life? No, and it's odious to imply that my dog (assuming that I had one) is worth less than I am. Would I bet a body part? That depends. Probably not a major limb but maybe something emblematic but minor, like the last joint of my pinkie. If I were proved wrong, I'd wave the stump as a badge of honor to show that I'm a really serious scientist.

The big bang sits on a sturdy stool with four legs. The first leg is the Hubble expansion. The linear relation between galaxy distance and redshift shows that the expansion of the universe projects to a time about 14 billion years ago when all galaxies had zero separation. If we're not at the center of the universe—and no observation indicates such a privileged position for us—there must be a uniform expansion everywhere. The data didn't have to point this way. For example, with a mixture of redshifts and blueshifts, there might be some regions expanding and some contracting, or alternating phases of expansion and contraction. The similar pattern of expansion in all directions points to a common origin for galaxies.

A second leg is the evolution of galaxies and quasars, discussed two chapters ago. Assuming that redshift indicates distance and lookback time, the big bang predicts that high-redshift galaxies will be younger than low-redshift galaxies. Hubble Space Telescope observations back this up because high-redshift galaxies are smaller, bluer, and less well formed than low-redshift galaxies. They also have younger stars than massive red galaxies in the nearby universe. All these attributes are consistent with evolution by mergers in an expanding universe. The epoch when the first galaxies formed from nearly smooth gas is close to being observed. These observations don't exclusively indicate a big bang but they're inconsistent with the steady state theory.

Next, and most crucial, is the cosmic microwave background radiation. This evidence is compelling, because there's no natural explanation for a uniform bath of cold radiation in the steady state theory or any other plausible rival to the big bang.[11]

The fourth leg is the cosmic abundance of light elements, particularly helium. You might not think it if you've ever driven around looking for helium to fill balloons for a kid's birthday party, but the universe has an embarrassingly large amount of helium. Overall, roughly 90 percent of all atoms are hydrogen atoms and 10 percent are helium atoms, and since helium atoms are more massive than hydrogen atoms, helium is a quarter of the universe by mass. No problem, you think, there are plenty of stars like the Sun making helium. True, but not enough. When stars like the Sun fuse hydrogen into helium, their light is energy "leaked" into space according to $E = mc^2$. If stars had turned a quarter of their mass into helium by fusion, the night sky would be ablaze. Another sign that stars didn't make the helium we put in kid's balloons is the fact that the oldest galaxies have almost the same amount of helium as the youngest galaxies. If the stars in galaxies were churning out helium, we should see it ramp up in abundance in younger galaxies but it doesn't. Evidently most of the helium was already there when the first galaxies formed.

Rewind the expansion to about 100 seconds after the big bang. From the simple relationship among size, density, and temperature, we can infer that the temperature was a billion degrees. Soon after, the universe mimicked the Sun by fusing the available protons and neutrons into deuterium, then tritium, then helium. By five minutes after the big bang—less time than it takes to boil an egg—much of the hydrogen had been "cooked" into helium,[12] and the temperature had dropped too low for any more fusion to happen. The density had dropped to a mere 100 times that of water.

This firestorm of fusion abated because beryllium is radioactive and its nucleus falls apart before another neutron or proton can be added. In stars, the "bottleneck" caused by beryllium decay is overcome by triple collisions of helium nuclei to form carbon, but that process is too slow for carbon to be made in the big bang. Also, creation of much heavier elements would require high temperatures to overcome the electrical repulsion of the nuclei and, due to the expansion, the temperature was dropping. Therefore, the chemical composition of the universe is fixed at that time and doesn't change until "first light," roughly 100 million years later.

Imagine the universe as a crime scene. As the detective working the case, bear in mind the following. Helium is a footprint. Deuterium is a stray hair. Lithium is a bank account balance. It's easy to measure a footprint size and it fits the big bang exactly. A stray hair at a crime scene is a rare find, but its composition also matches the big bang and it's hard to produce that evidence in any other way. The bank balance evidence is more subtle. Everyone's bank balance fluctuates, but if the victim's bank balance goes down just before a suspect's bank balance goes up by the same amount, you don't have to be Sherlock Holmes to see that as incriminating evidence.

We've seen that helium is too abundant in the universe to have been produced by stars, and that abundance is already at a level of 1 in 10 atoms when the first galaxies formed. Deuterium is an isotope of hydrogen containing one proton and two neutrons, with an abundance of 1 part in 10^4 relative to hydrogen. In the big bang theory, deuterium abundance is much more sensitive to physical conditions than helium, just as a matching stray hair is more telling than a matching footprint. Deuterium is a fusion stepping stone to helium in stars, but the cosmic measurement of deuterium is made in diffuse space among galaxies where stellar processes can't reach. Lithium is even rarer, at 1 part in 10^9 relative to hydrogen. Lithium is tricky to use as evidence, since it's created and destroyed in stars. So astronomers look at the pattern of lithium abundance and—like telltale deposits in a bank account—the pattern matches the big bang theory and not the predictions of stellar evolution.

At this cosmic crime scene, the combination of pieces of evidence is compelling. Helium, deuterium, and lithium have completely different abundances, are measured with different methods, and exist in quite different regimes in the universe.[13] Yet all three measurements

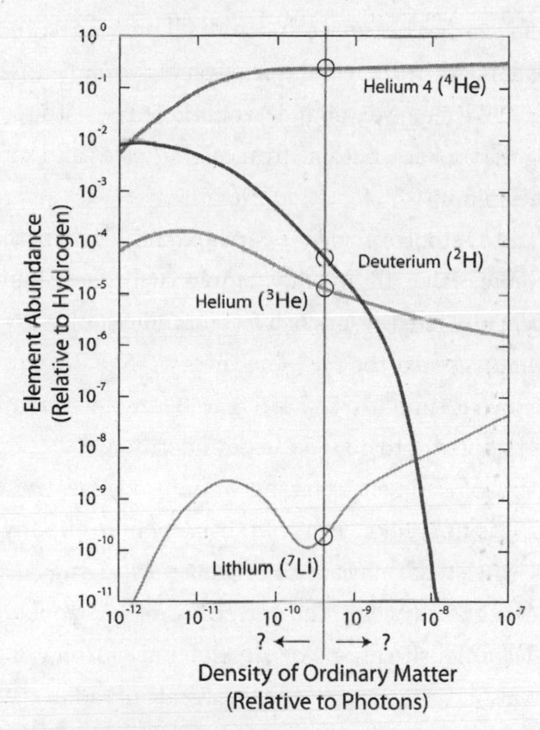

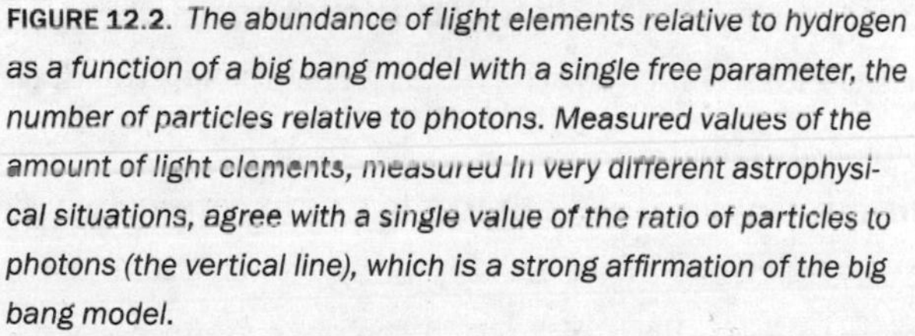

FIGURE 12.2. *The abundance of light elements relative to hydrogen as a function of a big bang model with a single free parameter, the number of particles relative to photons. Measured values of the amount of light elements, measured in very different astrophysical situations, agree with a single value of the ratio of particles to photons (the vertical line), which is a strong affirmation of the big bang model.*

agree with a big bang model that has only one variable: the ratio of photons to protons and neutrons (Figure 12.2). This is a stunning success and it didn't have to turn out that way. As the detective, imagine that you've discovered a footprint, a strand of hair, and a pattern of bank account deposits, all tying the crime to a single suspect. You don't need to get a confession. Your work is done. Put your feet up and wait for your promotion and raise.

Maybe my earlier bet was too timid. I'll offer up my entire pinky that the big bang is the correct description of the early universe. But if I'm wrong, I'd like a nice extendible, robotic pinky, with realistic latex skin, letting me pull down items from high shelves and scratch those hard-to-reach places.

Such a sturdy stool of evidence can hold the weight of the universe, but there's more. Other arguments combine with these four to make a web of evidence, where the loss of a few strands wouldn't compromise the whole. In this sense, the big bang theory is like Darwin's theory of evolution, which is supported by a disparate set of observations that it would be very difficult to dismantle or undermine.[14]

An important cross-check on the big bang involves the age of stars. If the universe is 13.7 billion years old there shouldn't be any stars older than that. Using stellar physics to estimate a star's age is completely independent of the physics of the early universe. Astronomers have fit stellar models to observations of old globular clusters in the halo of the Milky Way and they've used the decay rate of radioactive elements to estimate the age of the Milky Way itself. The results are 13 and 14 billion years, in agreement with the uncertainties in the data and the models.

There are also clever ways to test for the expansion of space-time, which is an underpinning of the big bang theory since high redshift corresponds to a smaller universe in the past. In an extended source of light like a galaxy, surface brightness, or intensity over a particular area, goes down as the fourth power of the redshift. Carrying out the test is hard, because galaxies evolve over cosmic time and they don't have sharp edges, but the results are consistent with redshift caused by expansion and they're inconsistent with "tired light" theories for redshift. Another neat test is to look at the rise and fall of the light from the distant supernovae that are used to trace cosmic expansion. If the redshift is caused by expanding space, then as the wavelength increases the frequency of light variations should decrease and distant supernovae should have a slower rise and fall in their light curves. The expanding universe passes this test with flying colors.

Other tests involve the behavior of the photons from the microwave background radiation. Astronomers have observed distortions of the radiation spectrum in the direction of clusters of galaxies, caused by the photons interacting with hot gas in the clusters (a subtle effect since only 1 percent of them interact, the same fraction that interact with a blank TV screen). They've also seen tiny variations in the temperature of the photons as they "fall into" and "climb out of" gravitational wells of large-scale structure along the way to the Earth. These signatures prove that the microwave radiation isn't a local radiation bath in the nearby universe.[15]

Finally, the temperature of the big bang photons at different redshifts has been measured by observing the excitation of carbon in diffuse regions of intergalactic space. The radiation really was hotter at large lookback times and at high redshifts, as the big bang predicts. This is gratifying, since observation of excited atoms in space should have led to the discovery of the background radiation back in 1940, but at that time nobody was able to connect theory and observation.

It's pretty darned convincing. Betting my pinkie was too timorous. I'm close to putting a limb or my first born on the line, but I would expect impressive recompense if the big bang is proved correct.

COSMIC INGREDIENTS

As Carl Sagan once said in his TV show *Cosmos*, "In order to make an apple pie from scratch, you must first create the universe."[16]

Apples and pie crust contain lots of carbon and oxygen atoms, so we can't make any apple pies until generations of stars have been born and died and have ejected heavy elements into the regions where the planets gather them. The formation of apples and people to make pies from them takes billions of years and is a complicated story that we'll ignore for the moment. Stars are stepping stones to apple pies.[17]

However, the hydrogen in an apple pie is primordial and was present from the first moment of cosmic expansion. When the universe was still as smooth and pure as a baby's bottom, a minute after the big bang, its behavior was governed by the ferocious energy of gamma ray photons. There are roughly a billion photons for every particle in the universe so to a good approximation the universe is composed of photons. In the early expansion, the photons outnumber the particles and bully them mercilessly so the universe is "radiation dominated." No structure can form. It doesn't seem that expanding space would change that situation since photons and particles thin out at an equal rate. But photons are subject to an extra effect: redshift. This lowers the energy of each photon and after about 10,000 years they are feeble enough that the universe is "matter dominated." Gravity can then steadily gather matter and form structures and make the universe a more interesting place.

After billions of years of redshifting, the radiation from the big bang is now a feeble microwave signal. For men who might be worried, the leakage from your microwave oven is much more likely to make you sterile than the microwave background radiation. Apart from radiation, what's the universe made of?

If we look at the cosmic "pie chart" (this graphical form is just a tool and no apples or other fruit are involved) 380,000 years after the big bang, when radiation and matter went their separate ways, we see 12 percent hydrogen and helium atoms, 10 percent neutrinos, 15 percent photons, and 63 percent dark matter. Now, the situation is quite different. The relative proportions of visible and dark matter have remained at about six to one, but neutrinos have diminished in importance due to their weak interactions with other forms of matter, while the dark energy has increased in importance since it seems to

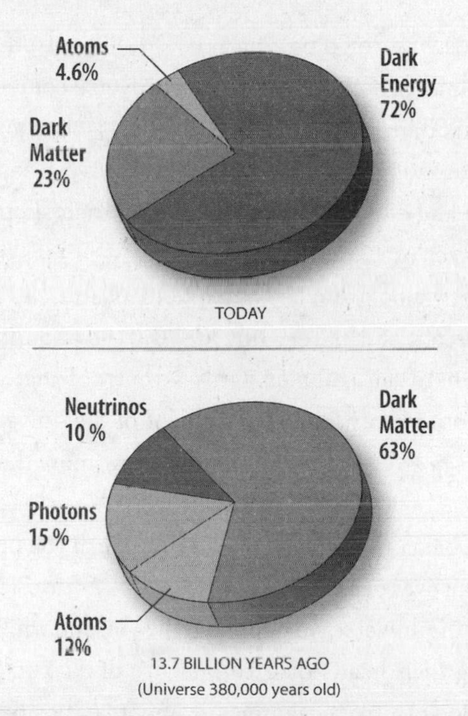

FIGURE 12.3. *Ingredients of the universe, now and 380,000 years after the big bang, when the universe became transparent. The universe now is governed by dark energy and dark matter, with normal matter as a small component. Closer to the big bang, the relative influence of dark energy was less, and photons had not been redshifted so their energy and influence were greater.*

be a constant attribute of the vacuum, even as the vacuum expands! The current breakdown of the cosmic pie is roughly 5 percent normal matter, 23 percent dark matter, and 72 percent dark energy. Neutrinos comprise 0.3 percent and all heavy elements, without which humans and apple pies can't be made, are only 0.03 percent, or 3 parts in 10,000, of the cosmic pie (Figure 12.3).

Imagine you had 20 kids. That's an implausible family these days but not unimaginable hundreds of years ago. If you only knew where one of your kids was, you'd be a bad parent. But astronomers are in a similar situation. We believe we live in a universe composed largely of dark energy and dark matter, but we can only isolate and understand the normal matter, one-twentieth of the universe. The twin enigmas of dark energy and dark matter mock us with their slippery ubiquity.

It gets worse. The 5 percent of the universe that's normal matter remained mostly aloof from the cacophony of star and galaxy formation. So 90 percent of it is very hot and diffuse and dispersed in the near-vacuum between galaxies. Only a tenth of that gas feeds star formation. In other words, just 0.5 percent of the universe ends up as stars. Yet that's been enough to make 10^{23} stars, more than the number of grains of sand on all Earth's beaches.

Is this like bald men fighting over a comb or, if you prefer, toothless women fighting over peanut brittle?

No, this is the universe we happen to live in, and while astronomers are scratching their heads over the nature of dark matter and dark energy, they don't think these measurements of cosmic composition are wrong. The percentages are rooted in observations of the cosmic microwave background. We've seen that the characteristic size of the speckles means that space-time is flat. The height of the peaks in the spectrum is a measure of the normal matter content of the universe; more mass increases the strength of the oscillations. Other aspects of the spectrum fix the relative amounts of dark matter and dark energy. Each of these quantities is also diagnosed by observations in the low redshift, or "nearby" universe, so the attributes of the universe aren't dependent on the measurements of the microwave background only. The agreement of different methods is so good that the universe with dark energy and dark matter running the show is widely accepted and there are no viable alternatives at the moment.

RE-CREATING THE UNIVERSE

Simon White is an unlikely sorcerer's apprentice. It's hard to imagine him letting any experiment get out of control. He speaks in measured, sonorous phrases, thoughtful and reflective. Tall and slender, his hair seems to have been pushed out in curls by a large and active brain.

But a closer look reveals a mischievous twinkle around his eyes, and he's quick to smile or laugh. He looks nothing like Mickey Mouse, but then the original sorcerer's apprentice hailed not from Disney, but from Goethe, and his poem of the same name written in 1797. White is the director of the Max Planck Institute for Astrophysics in Garching, near Munich. In the German scientific pantheon, that's a status nearly godlike. Max Planck directors have their jobs for life and are beholden only to the Minister of Science. It's easy to imagine two stout brass pipes traveling from central Berlin to his office, one carrying a steady flow of euros and the other carrying beer adhering to the German purity law of 1516.

White was my colleague for five years in the 1980s. I remember him inviting me to a gathering that he described as "a few friends playing folk music." I was slightly taken aback to enter a room full of grown men wearing colorful britches with bells attached to their ankles and knees. They were dancing in circles and waving handkerchiefs and carrying sticks that they banged together above their heads. It was Morris dancing, a European folk tradition even more ancient than the German purity law.

Simon White is a sorcerer of dark matter, and the black art that he's mastered is conjuring light and structure out of darkness. He's known primarily as a theorist who made seminal contributions to the theory of galaxy formation when he was fresh out of grad school. Theorists simplify the universe as much as possible to make it submit to their equations, and while people of White's caliber can make great progress, the complexity of the universe taxes their underlying assumptions. Galaxies aren't spheres and neither are their dark matter halos. It might be appealing to think of them as big meatballs coming together by the action of gravity in dark matter gravy. However, the better description of large-scale structure is the chaos of albumen in egg drop soup.

In the 1980s, White and others realized they could use the growing power of computers to simulate the universe. The idea is simple: put the ingredients of the known universe inside a computer, let it expand and let gravity act, and see what forms. Everything is virtual—there's no real space and no real ingredients, nothing really gets bigger, and the only gravity is what keeps the computer from floating away. This is a lab with no test tubes and no oscilloscopes and nobody in white coats. Everything is done with the protean power of transistors and integrated circuits working at blinding speed.

In Goethe's poem, the apprentice can't control an enchanted broom, so he splits it in two with an axe and the two pieces each start fetching water at twice the speed. For astrophysicists wanting to simulate the universe, the rapid doubling of computer speed resulting from Moore's law is a blessing, not a curse. It has increased the power and speed of simulations by a factor of a billion since 1970.

As Max Planck Institute director, White commands impressive computing power.[18] In the basement of the Leibniz Computing Center in Garching, 10,000 parallel processors keep track of gravity in the evolving universe, and the room has to be kept so cold you'd need a winter coat to be there for more than a few minutes. They do nearly 100 trillion floating-point operations per second, use 40,000 gigabytes of memory, and store the data on a million gigabytes of hard disk. The building that houses the computer uses more power than the sizable town of Garching.

That's the hardware; what about the software? Simulation software is designed to efficiently calculate the force of Newtonian gravity between a lot of "particles," move them all within the "space" of the simulation according to that force, and then redo the calculation in the following time step, ad infinitum.[19] Newton's law is simple but there's a problem. The figure of merit of a simulation is the number of particles it follows. Since the gravity of each particle acting on every other particle must be calculated, the number of calculations for every time step goes up as the square of the number of particles. So 1000 particles require a million calculations, and a million particles require a trillion calculations, and so on. A lot of algorithmic ingenuity has gone into mitigating this nasty scaling and as much of the gain in the past few decades has come from clever algorithms as from speedy processors.[20]

How many particles are contained in a state-of-the-art simulation? The Millennium Simulation from White and his group had 10 billion particles, it followed the evolution of a cube of the universe 2 billion light-years on a side, and it consumed 350,000 processor-hours. With parallel processors at work, nobody had to get old and gray waiting for the results; 13.7 billion years of cosmic history were compressed into a month of real time.[21] In simulations, there's a fundamental trade-off between the size of the chunk of the universe that's modeled and the mass represented by each particle. In the Millennium Simulation, the "particles" were about a billion times the mass of the Sun. Each one could stand in for a dwarf galaxy and a Milky Way–sized galaxy would be made of a few hundred such particles. At one extreme, it's possible to apportion one particle per large galaxy and so simulate 10 percent of the universe. At the other, the box could be made small enough to just hold the Milky Way and each particle would be standing in for about 10 stars. Simulators use both strategies, but not at the same time.

There are other limitations. Most of the particles in the simulations, as in the real universe, are made of dark matter. Dark matter is subject to gravity and doesn't interact with normal matter, so its behavior is simple. Normal atoms are fairly simple, but as we've seen, when they gather together to make stars and galaxies the physics is complex and messy. The simulators still have trouble making realistic

FIGURE 12.4. *A simulation of the formation of structure in the universe, starting from the nearly smooth conditions at the time the microwave background radiation was created. The simulation box is 320 million light-years on a side, and it expands with redshift, an effect that has been removed in this image.* Left to right: *the universe at 1 billion years after the big bang (when the first galaxies formed); 5 billion years after the big bang; and now, 13.7 billion years after the big bang.*

galaxies. Like clumsy bakers, their cakes come out lopsided or small and hard, or too sweet to eat. Also, just as a cake is no better than the ingredients that went into it, a simulation is only as good as the physical assumptions it was based on. Our rudimentary understanding of dark matter and dark energy is a small dark cloud hanging over the results of trying to make a "universe in a computer" (Figure 12.4).

That said, the simulators have racked up some impressive successes. The best simulations beautifully reproduce the filigree of large-scale structure, and the basic properties of galaxies. By varying the initial conditions, it's possible to see how assumptions about the ingredients affect the outcome. A minor industry has sprung up to "observe" the simulations, where researchers cull galaxies from the computational output and study their properties.

That's certainly easier than dragging yourself up to a telescope and staying up all night for data, but pulling galaxies from computer code lacks a certain charm. I once had a conversation with a bright young simulator where he looked sadly at my noisy galaxy spectra and said, "The problem, Chris, is your data is crappier than our data." I looked at him, stunned, and reminded him that his computer simulations did not represent real data any more than imagined sex represents real intercourse. Well, that's what I thought of saying. Instead I settled for a nerd curse: "May the God of Epistemology strike you dead!"

Simulation is the "third way" to do cosmology, along with observation and theory, and it's become increasingly important in the past decade. White has long ago graduated from apprentice to guru. His last update of the Millennium Simulation upped the ante with 300 billion particles and a box size that spans the entire observable universe. If the "Sorcerer in Chief" doesn't watch out, the Wizard of Garching will be angling for his job.

Is it possible to run a simulation "backward" to learn about the very early universe? Unfortunately, no. Simulations usually begin with the conditions observed when the universe became transparent, 400,000 years after the big bang. It's possible to project back to temperatures of millions and even billions of degrees, but the limitation is no longer computers, but the uncertain physics that operated in the very early universe. Venturing closer to the big bang forces us to consider the fundamental nature of matter.

An atom is an atom is an atom. Each the same as any other, and no way to track any one as they careen around a hair's breadth away from the speed of light. Disembodied as I am, I'll have to use my imagination to track matter. I focus on two nuclei, one hydrogen and one the rarer helium, as for a fleeting instant they come close to me and each other and then veer off into the expanding plasma.

I'm reminded of a poem I memorized as a child, about two paths that diverged in a wood. By a famous poet—I think he loved metaphors. It's a comfort to remember this rhyme—so homespun—when around me is nothing but the surreal and austere flux of physics. Robert Frost also speculated, I recall, about fire and ice. Perhaps his name tilted his hand; in which case he got it right, the furnace around me will ultimately end in frozen desolation.

I'm luckier than the laureate; I can follow both paths.

The helium is aloof. Unwilling to bind to any other atom, it floats through space, not partaking in any interactions. The hydrogen quickly mates with a twin and forms a molecule. It rides the gentle chemical cycle within a molecular cloud and is drawn into a swirling eddy. By chance, the helium atom is drawn into the same eddy. They are entombed in the same rock in space. A place that will one day be my home.

What happens next is frankly incredible, the odds against it so long they're not worth calculating. The "me" as I will one day be is stressed out, driving around town preparing for my young son's birthday party. As an astronomer, I know enough to be frustrated by how hard it is to find a cylinder of helium; every tenth atom, after all, is helium. But I finally do succeed and am sitting at the back of the very noisy room, after the clown has left, nursing a headache. My son grips the string of his balloon tightly, as if his life depended on it.

But his attention is drawn by the cake as it is wheeled out, candles lit, wafting a cloying scent of frosting in our direction. His fingers relax, and the balloon wafts toward the ceiling. I know, beyond a shadow of a doubt, that the hydrogen atom from the dawn of time is bound in a large molecule near the surface of the pinkish skin of his hand. And the helium atom that was once its close neighbor is making a tiny but finite contribution to the buoyancy that causes the balloon to yearn for its origins in deep space.

13

SOMETHING RATHER THAN NOTHING

I'M IN THE REALM OF INDIVISIBLE UNITS. *The building blocks of matter move at near-light speed in a blizzard of motion all around me. Democritus guessed that color, taste, and smell were secondary properties. He suspected that fundamental particles were ciphers to the senses. I cannot sense mass, charge, and spin. I'm far beyond the scale of normal measurement, and the familiar metrics of temperature and density don't apply. The scene is suffused with radiation so energetic it's beyond white-hot, it's more ultra than ultraviolet; it occupies the realm beyond gamma rays.*

I blink, and a frozen instant of the infant universe is painted on the inside of my retina. With motion stilled, I see that in some places particles appear in pairs from pure radiation and elsewhere they disappear and are converted into radiation. The chaos isn't absolute; these miniature acts of creation and destruction imprint a subtle pattern of branching points onto the scene. Some of the branches reconvene to make an evanescent loop as matter momentarily bubbles out of pure energy. With my eyes open again, the pattern is changing so quickly it's imperceptible. Matter is ephemeral.

If I had sharper senses, I'd be aware of a hierarchy in the myriad comings and goings of particles. There are light and heavy particles, with lower and higher barriers to their creation. Some particles have electric charge that makes them veer and swerve, others are neutral and travel straight and true, and some act as if invisible to radiation.

These structureless, colorless, odorless, and flavorless particles have strangely evocative names: electron, neutrino, meson, quark. It's difficult to imagine how this anarchy could turn into pleasing regularity of stars and superclusters, planets and people.

—

WHAT'S THE MATTER?

Why is there something rather than nothing? The German philosopher Gottfried Leibniz wrestled with this question over 300 years ago. I know what you must be thinking: only a philosopher would worry about existence at this level.[1] Most people would accept the simple riposte: why not?

Leibniz argued that nothing was more natural than something because it was simpler; after all it takes work or effort to make something. But he went further, saying that because we live in the best of all possible worlds, it must contain something, because something is better than nothing. He then made the leap to the need for a benevolent Creator. Even if we set aside the value judgment that something is better than nothing and the deistic inference that follows, Leibniz does seem to be on to something in that the question is profound. Ironically deists and atheists are united in having to explain creation out of nothing.

Let's look a little closer at the "something" that we call the observable universe. It consists of about 10^{33} cubic light-years of expanding space or 10^{86} sugar cubes' worth. That's the universe as a container; space in relativity is an abstract, mathematical construct. Space is also supple since we've described a history where it grows by a huge factor, with no apparent cost. Astronomers observe this space to hold about 10^{89} photons and 10^{80} particles (ignoring for now the ineffable ingredients: dark matter and dark energy). That's rather a lot of something!

The something has two flavors: matter and radiation. In physics, they are called fermions and bosons after the two scientists who described their statistical properties. The interaction or dialog between matter and radiation leads to the rich phenomena of the natural world.

Fermions are aloof and individualistic. The word means more than "particles" because it also describes the statistical properties of an ensemble of particles: no two particles can share the same set of microscopic properties. This is true whether the fermions are in the plasma of the early universe or in a chair on the opposite side of the room. Fermions were named after Enrico Fermi, the Italian physicist who described their properties in the 1920s. Fermi came to the United States in 1938 and soon after demonstrated nuclear fission for the first time with a reactor built in a squash court at the University of Chicago. Fermi was equally adept in theory and observation; his colleagues saw his physical judgment as so infallible they nicknamed him "the Pope." He used simple math and logical reasoning to estimate the answers to problems, a method that's still taught to physics students today.[2] In a somber example, Fermi witnessed Trinity, the first atomic weapons test. When the blast wave hit, he let torn paper fall from his hand, and used the drift rate of the confetti to estimate the yield of the bomb.

Bosons are promiscuous and communal. The word means more than "radiation" because it includes the carriers of all the forces in nature. Photons are bosons and they can have the same energy and occupy the same space; there's no limit in principle to the energy density of radiation. Bosons were named after the Indian mathematician and physicist Satyendra Bose, who had a bumpier road to preeminence than Fermi. The journals rejected his landmark paper on the statistics of photons and it was only after Einstein interceded on his behalf that it was published in 1924 and the physics community gave him his due. Bose traveled to Berlin to work with Einstein and eventually became a senior professor, even though he didn't have a doctorate.

The universe has far more photons than particles, more bosons than fermions, more ephemera than grit, and this begs for an explanation. The tiny fraction of stuff is enough to make 100 billion galaxies and countless stars and planets so it's nothing to be sneezed at. It's noteworthy for another reason: the near total absence of antistuff.

To see why we should care about antistuff, or antimatter, we turn to another important episode in 1920s physics. Paul Dirac was an English physicist who worked out crucial parts of a new theory called quantum mechanics. The quantum theory described the bizarre and sometimes counterintuitive world of atoms and subatomic particles, where atoms had wavelike properties and were fuzzy and not hard-edged. In this world, behavior was described with probabilities and knowledge was limited by a veil of uncertainty that no measurement, no matter how clever, could circumvent. Dirac solved the fundamental equation that described the behavior of the electron and found two solutions, one that was a positive square root and the other that was a negative square root. Instead of discarding the solution with the negative number as nonsensical, since it was like having negative energy, he decided based on the mathematics it was equally valid and he left it in his paper (Figure 13.1).

Dirac was one of the titans of twentieth century physics. Famously modest and socially awkward, he essentially worked all the time except for taking long Sunday walks, and climbing trees wearing a three-piece suit.[3] He spoke best with his elegant papers, saying little enough to his colleagues that at Cambridge a "dirac" was defined as a rate of speech

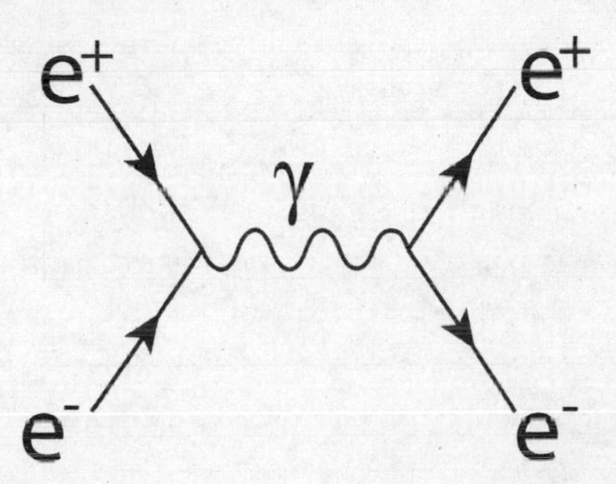

FIGURE 13.1. *Each particle has an antiparticle counterpart, a form of "shadow" matter mirrored in space and with the opposite charge. The antiparticle of an electron is a positron. Electron-positron pairs can be created from pure energy, and if they meet they annihilate into pure energy.*

of a word per hour. In Graham Farmelo's biography, Dirac's colleague Freeman Dyson is quoted: "His discoveries were like exquisitely carved stones falling from the sky, one after another. He seemed to be able to conjure laws of nature from pure thought." Dirac and his father were probably both mildly autistic, and in the history of science there have been few more baffling geniuses.

Four years later, Carl Anderson was watching cosmic rays interact in his cloud chamber when he saw something peculiar. Inside a cloud chamber, a supersaturated liquid like water or alcohol registers the passage of an energetic particle as a vapor trail. A magnetic field is applied so that a charged particle will make a curved trail. Anderson was familiar with the trails left by electrons but he noticed a few that curved in the opposite direction. Apparently, there were particles with the same mass as the electron but opposite charge. Anderson called them positrons for "positive electron." In the context of Dirac's work they're antielectrons.[4] This was the discovery of antimatter.

By the 1950s, scientists had succeeded in creating antiprotons and

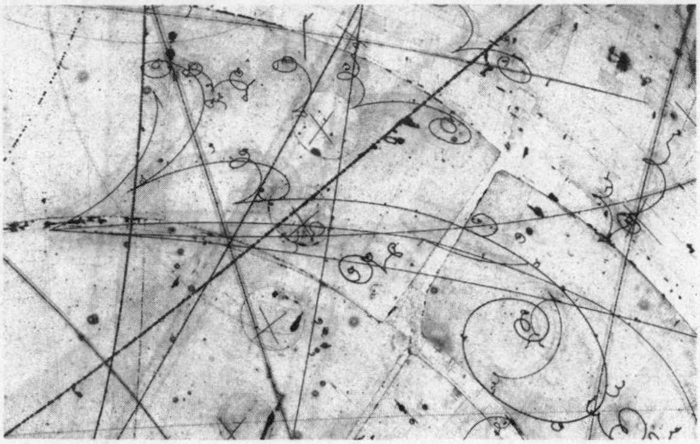

FIGURE 13.2. *Antimatter was first detected in cloud chambers, where a particle interacts with the supersaturated vapor in the detector and energy is converted into a particle-antiparticle pair. With an external magnetic field applied, particles and antiparticles curve in opposite directions because they have opposite electric charges. This image is produced from a more modern piece of apparatus called a bubble chamber, where the tracks are left by a stream of tiny bubbles in a compressed liquid.*

antineutrons as well, so all normal components of matter can exist in this "shadow" form. Antimatter has some of its quantum properties reversed but is otherwise indistinguishable from normal matter. When a particle and its antimatter analog come into contact, they disappear and turn into radiation or energy. Similarly, pure radiation can create particle and antiparticle pairs, but not one or the other. Inside a cloud chamber, the particle-antiparticle pairs appeared as oppositely curved vapor trails diverging from the creation point, like horns (Figure 13.2).

How do we know the universe is bereft of antistuff?

The violent and instantaneous demise of matter when it encounters antimatter gives us very strong limits. There are no anticans of antipeaches lurking on the supermarket shelf, no anticlouds floating in the sky, and no antibees buzzing in your garden. If you had an evil alien antitwin, you need not fear the encounter and the fatal handshake; they'd have disappeared in a blaze of gamma rays the first time they arrived on Earth and their antifeet touched the ground.

On Earth, antimatter is created and detected in huge accelerators. Bringing it into the realm of the senses might be possible. Imagine a microscopic dollop of antimatter dropped onto your tongue. Its anti-atoms would annihilate instantly with the atoms in your taste buds, releasing a flurry of gamma rays. This slight radiation burn would be the distinctive "taste" of antimatter. No billion-dollar hardware needed!

Further afield, the Apollo astronauts had no nasty surprise when they set foot on the Moon, proving it's made neither of green cheese nor antimatter. The probes we've landed on Mars and Venus and Titan didn't vaporize on contact. The Sun bathes us in a continuous stream of high-energy particles called the solar wind and we can confirm when they hit the atmosphere that they're almost all particles, not anti-particles.[5] In principle, antimatter could remain undetected if it's well separated from matter by almost pure vacuum. But the space among stars isn't truly empty and stars interact with their surroundings when they age and die, so there are strong limits on antistars in our galaxy. Beyond the Milky Way, the same reasoning applies. Intergalactic space isn't a perfect vacuum and galaxies do occasionally interact and even merge. Astronomers have seen no sign of telltale gamma rays from galaxies meeting antigalaxies.

The largest structures are superclusters several hundred million light-years across. Even if matter and antimatter are separated in domains this large, annihilation at the boundaries would produce a gamma ray signal large enough to be detected by current satellites. So symmetry between matter and antimatter is ruled out on scales up to the size of the observable universe.[6] We see lots of stuff and almost no antistuff. The question of whether domains of antimatter could exist beyond our cosmological horizon is interesting and we'll return to it later.

The extreme rarity of antimatter is reassuring since nobody likes to be annihilated into pure energy. But it's surprising because fundamental physics does not discriminate between matter and antimatter. To see why, let's venture into the world of particle physics.

THE STANDARD MODEL

To come to grips with the Standard Model of particle physics, we first must review how a Lagrangian is constructed from relativistic quantum field theory, subject to the global Poincaré symmetry . . .

Just kidding. This is an area of science where the weak and the frail fall away, and even grown-ups who aced college calculus can be brought to tears. It's very unfortunate that the quest to understand the fundamental nature of matter has led to a formalism that makes the field opaque to all but an elite cadre of theoretical physicists. To them, the theory contains math that's achingly beautiful and elegant. To others, it's inscrutable and the practitioners are high priests from a strange cult who speak in tongues.

Let's start with the simple question that any child might ask: what's stuff really made of?

Until the turn of the twentieth century the answer would have been atoms. Atoms were then thought to be hard, indivisible units of matter that combined to make all the materials of the physical world. But elegant experiments by Ernest Rutherford and others showed that atoms were mostly empty space—a diffuse cloud of electrons surrounding a dense nucleus of protons and neutrons. In the 1960s atom smashers showed that protons and neutrons were not fundamental; they were composed of bizarre fractionally charged particles called quarks. There was also a weightless particle called a neutrino that could pass through thousands of miles of solid iron without flinching. However the answer to the child's question was still fairly simple: two types of quark,[7] the electron, and the neutrino. Quarks are fantastically small. If an atom were the size of the Earth, a proton would be the size of a football stadium and quarks within the proton would be the size of tennis balls.

Through the sixties and seventies, results from particle accelerators showed that the microscopic world was considerably more complex. Smashing atoms into each other at near light speed created many new particles, more than the number of elements in the periodic table. But they all seemed reducible to 12 particles, or fermions, arranged in three levels of increasing mass. The lowest level is conventional matter: "up" and "down" quarks, the electron, and the neutrino. The second level includes "strange" and "charm" quarks, the muon, and its associated neutrino. The third level includes "top" and "bottom" quarks,[8] the tau particle, and its associated neutrino. The more massive particles in the second and third levels tend to be unstable and are only produced fleetingly in particle accelerators. Each of the 12 particles has a corresponding antiparticle, which can be created with a sufficiently energetic collision but which then quickly disappears when it meets a particle.

As a counterpart to the 12 fermions, there are four fundamental forces that together describe all interactions in the universe. Each force has carrier particles, or bosons. We're familiar with the photon, which carries the electromagnetic force. Less familiar are the W and Z bosons, which carry the weak nuclear force that causes radioactivity, and the gluons that carry the strong force and ensure that quarks are tightly confined inside protons and neutrons.[9] Completing the picture is a hypothesized particle to carry the force of gravity, the graviton. The graviton has never been observed. The theoretical framework to describe 12 fermions and four bosons is called the Standard Model (Figure 13.3).[10]

The Standard Model is like a glass half full and half empty. On the one hand, it chalked up an impressive series of successes over the past

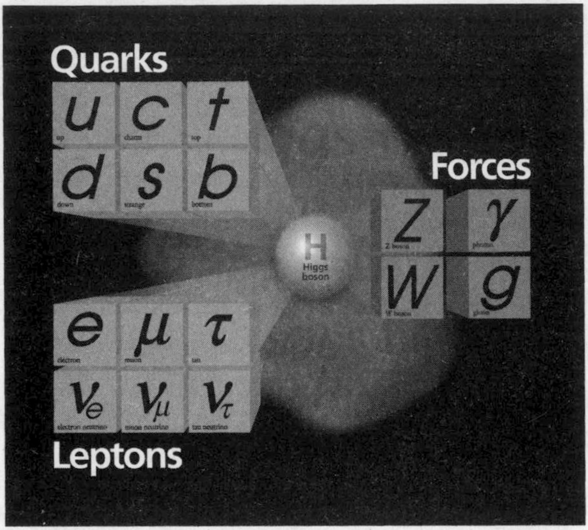

FIGURE 13.3. *The Standard Model of particle physics includes 12 fundamental constituents of matter and four fundamental forces. In normal matter, there are only three particles: up quarks, down quarks, and electrons. The second and third generations of the particles have much higher energies and are unstable. There are four force-carrying particles with widely differing strengths in the everyday world, ranging from gravity to the strong force that binds the atomic nucleus.*

30 years. The four most massive quarks were detected as predicted and the theory beautifully explains the myriad interactions seen in particle accelerators. It also predicts that at sufficiently high energies, the fundamental forces, which widely vary in strength, will be unified. The first stage of this unification, where electromagnetic and weak forces merge, was demonstrated in the 1970s.

On the other hand, the Standard Model is a bit like an aging movie star whose best work is decades old and whose flaws once seemed slight but are now becoming glaring. It gives no explanation for why there are three levels of quarks and light particles. It leaves open a possibility that there's a deeper level of structure. It can't predict the masses of all the particles. This omission is big enough that physicists have hypothesized a particle called the Higgs boson that exists to give all other particles mass and are currently using the Large Hadron Collider at CERN and the Tevatron at Fermilab to find it.[11] So far, $20 billion of hardware has failed to find this tiny particle that makes everything in the universe have heft. If the Higgs is not found in a plausible energy range, the Standard Model will suffer a major failure. Forget about bit parts; it will never work in this town again.

The Standard Model is incomplete because it doesn't include gravity, the weakest of the four fundamental forces. It predicts that neutrinos should be massless so it's flummoxed by the recent discovery that they have mass.[12] Then it has the glaring omission of failing to predict, or even account for, dark matter and dark energy, which form more than 95 percent of the universe. To complete the embarrassment, it doesn't explain why the universe contains an abundance of normal matter and virtually no antimatter.

At the microscopic level, matter and antimatter have equal status. Particles and antiparticles are created in strictly equal proportions from energy. Quantum theory predicts that every particle will have a corresponding antiparticle, and these have been observed in physics experiments.[13] In 2002, physicists at CERN combined antiprotons and antielectrons to make antihydrogen. This is a fiendishly difficult trick since the antiparticles are produced at nearly the speed of light; they have to be gathered and cooled and kept from meeting normal atoms. It's a very inefficient and expensive way to make antistuff.[14] About 100 antiatoms are created per second and they survive less than a minute. The premise of Dan Brown's *Angels and Demons*—where the bad guys steal half a gram of antimatter from CERN and transport it to Rome to make a bomb—is outlandish. With current production rates, it would take 10 million years to make a gram of antimatter, and so far it has cost a billion dollars to make a billionth of a gram. Antimatter is by far the most expensive thing (or rather, antithing) there is.

In the Standard Model, symmetry between matter and antimatter is a foundational principle. More broadly, all successful theories of physics embody symmetry in the combination of space, time, and charge. You can think of symmetry as a "mirror" in which the property is reflected or reversed. For space, this would be like a reflection in a real mirror, where the image of a right-handed glove is a left-handed glove, and the image of a particle spinning to the left is a particle spinning to the right (although the reversal must be done in three spatial dimensions). For time, this is a reversal of the arrow that connects the past and the future. For electric charge, it's reversal of the sign of the charge, which is equivalent to replacing every particle with its antiparticle.

What does this symmetry principle mean? It means that any particle interaction that's observed in the lab should also work if the particles are replaced by their antiparticle twins; they're mirrored in space and time runs backward, which means that the initial ingredients and final products of the interaction are reversed. It's verified. Familiar normal matter interactions can be re-created using antimatter and a reverse time sequence. Nature at the microscopic level is "unaware" that time has any particular direction.

In theory, a mirror-imaged universe is possible, made of antimatter, with time running backward, yet obeying the same physical laws as our universe!

So far, the combined symmetry of space, time, and charge has proved to be perfect. But there are cracks in the individual mirrors. In 1957, a type of radioactive decay was observed where there was a difference between the occurrence of left-handed and right-handed outcomes. It was the first physics experiment to distinguish between an object and its mirror image.

The physicist Richard Feynman told a story to illustrate this discovery. Suppose you had established two-way communication with aliens where you used light signals to convey information. You could convey your height in terms of the wavelength of light, and your age in terms of its frequency. But suppose you described how you greet someone by holding out your right hand. "Wait a minute," replies the alien, "What do you mean by 'right?'" Since 1957 we can answer that question. We could tell them how to set up a physics experiment that can distinguish left from right. Feynman added a mischievous coda to the story. Suppose you finally travel out in space to meet the alien. He extends his left hand. Watch out! That means he's made of antimatter because the same physics experiment using antimatter would give the opposite result.

A second crack in the mirror appeared in 1964, when the decay of a particle called a K meson was found to have a preference for matter over antimatter.[15] Is this the key to the dominance of matter in the current universe? No. The effect is too weak and could only result in an excess of matter sufficient to make one galaxy, rather than the 100 billion we observe. The Standard Model doesn't contain the answer to why there is something rather than nothing.

CREATING MATTER

Picture the scene: It's a microsecond after the big bang. The visible universe is not much bigger than the Solar System and it has a density not that much less than the air we're breathing. This sounds quite cozy and familiar, but it's also 100,000 times hotter than the core of the Sun and filled with a firestorm of gamma rays. Particles and antiparticles are appearing and disappearing, always in pairs, at a bewildering rate.

Flash forward to 10 microseconds after the big bang. Expansion has cooled the universe enough that quarks and antiquarks can no longer be created from radiation. The quarks and antiquarks existing at that moment are "frozen in" and they mutually annihilate. An iota of time later an analogous process happens with the lighter particles.

Electrons and positrons can no longer be created from radiation so those around at the time annihilate. It's a balmy 10 million degrees. By 100 microseconds—a twentieth of the time it takes a honeybee to flap its wings—it's all over. The particles and antiparticles have paired up, like guys and gals at a dance, with nobody left over. The universe contains only radiation. It continues to expand and photons stretch lazily into the yawning void. Without matter, the universe is devoid of structure, and devoid of the possibility of *us*.

That's the history, with one crucial difference. The symmetry between matter and antimatter wasn't perfect. Nature was very slightly skewed so that there was an excess of matter over antimatter. For each billion antiquarks there were a billion and one quarks, and for every billion positrons there were a billion and one electrons. When antimatter and matter annihilated for the last time, the result was a billion photons for every particle and no antiparticles.[16] The photons continued to stretch with the expansion to become the tepid radiation bath detected as the microwave background, while the particles gathered under the force of gravity to turn into 100 billion galaxies, one of which includes a middle-aged star orbited by a watery planet that we're quite fond of.

The asymmetry that led to us was tiny, one part in a billon. Analogies may help to visualize it. Imagine that someone built a steel ring 1000 kilometers across in the desert. Hovering above it in a helicopter, you can't detect a departure from perfection because the circle is distorted by just 1 millimeter. Or imagine someone has laid out pennies over an area 3 miles on a side, and they are all showing heads except one. Or imagine you have to find the single person in India who wears a particular type of shirt. I'm thankful for such a slight departure from perfection, for without it, you wouldn't be reading this book as neither of us would be here.

The physical requirements to create matter where none had previously existed were summarized succinctly in a three-paged paper by Andrei Sakharov in 1967. The paper wasn't taken seriously and didn't get a citation from any other paper for over a decade, but it has now been cited by a thousand other papers and is considered seminal. Sakharov was one of the first people to link cosmology and high-energy physics.

Sakharov's life represents an amazing journey from architect of war to proponent of peace. As a young man, after World War II, he moved to the secret Russian city of Sarov to work on the hydrogen bomb. He was the lead designer for a series of increasingly powerful devices that culminated in the Tsar Bomba, the largest nuclear weapon ever detonated. The Tsar Bomba was tested above Siberia in 1953, and it had 10 times the destructive power of all the munitions used in World War II. Sakharov was rewarded with the Stalin Prize, three Hero of Socialist Labor prizes, and a luxurious dacha outside Moscow.

He was the ultimate Soviet insider, but Sakharov became increasingly concerned about the moral and political implications of his research. Back in 1950, he'd come up with an idea to harness fusion power for peaceful purposes and his tokomak design is still used today. In 1963, he played a major role in the Partial Test Ban Treaty. Five years later his influential essay, "Reflections on Progress, Peaceful Coexistence, and Intellectual Freedom," became an underground classic within the dissident movement. He married human rights activist Yelena Bonner in 1972, and in 1975 he was awarded the Nobel Peace Prize but was prevented from leaving the Soviet Union to collect it. He spent most of the 1980s under house arrest in Gorky and he wasn't "rehabilitated" until the glasnost policy of Mikhail Gorbachev. Since his death in 1989 several major prizes for human rights have been set up in his name.

The origin of matter is a frontier topic in physics and cosmology. We still don't know how it happened and the only plausible ideas involve untested extensions to the Standard Model. However, we know it must have happened very early in the history of the universe because any process to create matter would have affected the radiation left behind. The microwave radiation we see that dates from 380,000 years after the big bang represents a single temperature perfectly, which pushes any matter-creating process back into a tiny fraction of a second.

It's possible that the imbalance leading to the dominance of matter was built in from the beginning. But this is logically suspect, because the universe contains everything, and so nothing should exist outside it or before it. In fact, as we'll see in the next chapter, there probably were processes in the very early universe that would have wiped out or diluted any initial asymmetry. Which leaves us speculating about physics in the first iota of time after the big bang.

Sakharov gave three conditions for there to be something, rather than nothing except radiation, in the universe. First, the normal rule where particle number is conserved in interactions has to be violated. Next, the perfect symmetry with respect to mirroring space and replacing all particles with antiparticles must be broken. Last, the universe has to be changing so quickly or violently that the matter excess isn't erased or washed out. Theorists have used great ingenuity to imagine ways to create matter in the very early universe. All the ideas are clever; none has been tested.

Is it possible to learn what happened in the early universe from a lab experiment? Yes. There's no way that the extraordinary conditions of the expanding early universe can be replicated on Earth. But by accelerating particles to within a fraction of a percent of the speed of light and making them collide, accelerators can briefly re-create temperatures not seen since the first instants after the big bang.

The Relativistic Heavy Ion Collider at the Brookhaven National Lab on Long Island was the first machine to do so. It smashes gold ions together. The current champ, though, is the Large Hadron Collider at CERN, which straddles the French-Swiss border near Geneva. It uses iron nuclei. Both machines create fireballs for a tiny fraction of a second with temperatures of trillions of degrees and pressures of 10^{30} times

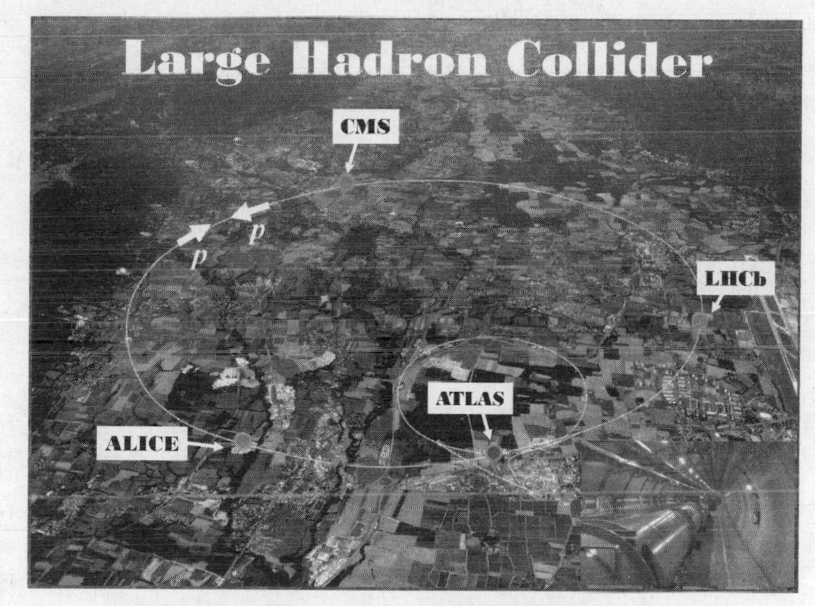

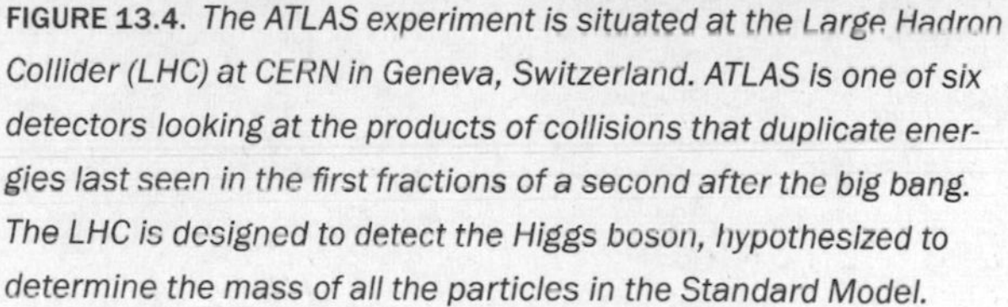
FIGURE 13.4. *The ATLAS experiment is situated at the Large Hadron Collider (LHC) at CERN in Geneva, Switzerland. ATLAS is one of six detectors looking at the products of collisions that duplicate energies last seen in the first fractions of a second after the big bang. The LHC is designed to detect the Higgs boson, hypothesized to determine the mass of all the particles in the Standard Model.*

the air we breathe. In those conditions, quarks freely interact with each other and the gluons that normally keep them tightly confined within the atomic nucleus. The only other place such conditions ever existed was the universe less than a nanosecond after the big bang. The key to the asymmetry between matter and antimatter hasn't yet emerged from these experiments, but in the next few years the Large Hadron Collider will test many of the possibilities (Figure 13.4).

We've reached the limits of conventional story-telling. In cosmology, telescopes are time machines, enabling us to peer back through the eons toward the big bang. The trail of evidence is unbroken back to a few minutes after the origin, when light elements were created in the cosmic fireball. Speculating back into the first fraction of a second of history, the clues come as much from physics as from astronomical observations.

This austere world is a realm of the senses, if we stretch the metaphors and use our imaginations. We've seen that with supersensitive hearing we could detect the last interactions of radiation and matter before they parted company 380,000 years after the big bang. With supersensitive sight we could observe the fireball as it shifts through the visible spectrum after 10,000 years. Smell is based on molecules with particular shapes fitting into receptors in our noses. Perhaps with supersensitive sniffing we could detect the shapes of nuclei fused in the first three minutes. A microsecond after the big bang, we'd need supersensitive tastebuds to enjoy a microscopic dollop of antimatter. Despite the slight radiation burn, the taste of the very early universe would be, well, cosmic.

HINTS OF FINE-TUNING

The existence of something rather than nothing is gratifying but not surprising. If there had been perfect symmetry between matter and antimatter, the universe would have evolved like a huge lightbulb on a dimmer switch, fading from white-hot to cold and black without the complicating froth of superclusters, stars, planets, and people. A few chapters ago we learned we shouldn't be surprised that the universe has features compatible with our existence, and lacks features incompatible with our existence.[17]

But it's intriguing to ask the question, What if? Without assuming that we live in the best of all possible worlds, we can ask what the universe would have been like if its core attributes had been different.

Asymmetry of 1 part in 10^9 gave rise to us and all the galaxies in the universe. If the asymmetry had been much smaller—1 part in 10^{11}

or less—there wouldn't have been enough matter for galaxies to form. If the asymmetry had been much larger, the abundant matter would have congealed without forming stars and galaxies. Another dimensionless number from cosmology is the level of fluctuations in the microwave background, 1 part in 10^5, which is the "graininess" of the early universe. If it had been much smaller—1 part in 10^6 or less—stars would not have formed. If it had been 1 part in 10^3 or more, gravity would have formed massive black holes but not normal stars. Apparently the graininess of the early universe was "just so" to generate normal stars.

The cosmic expansion poses other puzzles. The geometry of space is very close to flat. As we saw when discussing the microwave radiation left over from the big bang, the characteristic "speckles" haven't been magnified or demagnified in their long travel through space, and that means space is flat to 1 percent. Flatness results from a particular combination of the matter density and the dark energy density, which are two seemingly unrelated components of the universe. Very soon after the big bang, when space was curved and expanding insanely fast, there must have been exquisite finetuning toward this eventual situation. Much less dark matter and expansion would have been too fast for structure to form; much more and the universe would have recollapsed before stellar evolution could get established. Dark energy plays a role in this delicate balance. It's been dominating the expansion in the last 5 billion years; if it had been much stronger, the rapid acceleration would have squelched all structure formation.

Hypothetical outcomes are as intriguing in the microworld as they are in the macroworld. For example, if the strong force that binds quarks in atomic nuclei were a few percent stronger, quarks wouldn't form protons, and if it were 5 percent weaker, stars couldn't make heavy elements past hydrogen. If the weak force was much stronger, the big bang would have cooked atoms all the way up to iron, and if it were weaker the stars would have converted all the mass into helium. A stronger gravity force would have led to stars as feeble red dwarfs; a weaker force would have led to fast-burning blue giants. Either way, normal main sequence stars like the Sun would be rare or absent. There are other mass relationships between elementary particles where matter would be unstable if the values or ratios were different.[18] Roger Penrose has estimated that the combined probability of all the physical constants having their measured values is 10 to the power of 10 to the power of 123, a phenomenally unlikely outcome.[19]

These counterfactual universes are physically sensible, even if they operate under altered laws of physics. However, in almost every case, matter collapses, or heavy elements can't form, or stars can't function, or the universe is very short- or long-lived. Many scientists and philosophers have taken this situation to imply that the universe is somehow "finely tuned" or exhibits coincidences that have no natural explanation. They then go one step further and say that if the universe were slightly different it would be a sterile wasteland, devoid of life. Stephen Hawking puts it this way: "these numbers seem to have been very finely adjusted to make possible the development of life."[20] Another set of philosophers and theologians have argued that the supposed fine-tuning points to the work of a "Cosmic Designer." In a bizarre turn of events, physics and cosmology can be coopted in support of a theistic agenda.

The evocative analogy of a firing squad has been used to understand these issues. You've been sentenced to death and are facing a squad of 100 highly trained sharpshooters. There's a deafening volley and a cloud of smoke.[21] Amazingly, you survive. How are you to react to this outcome?

Clearly, you shouldn't be surprised that you're alive, since if you had been killed, that would be the end of the story. But you *are* alive and have survived in the face of overwhelming odds. It seems more likely that you survived for a definite reason than by sheer chance and luck. The low odds of your survival are an analogy to the fine-tuning of the universe and the improbability of life if the conditions had been slightly different. In the argument from design, the fact that you were spared is evidence for God.

Should we be seduced by these arguments or maintain a stout and healthy skepticism? Let's be skeptical. As Hippocrates once said: "Men think epilepsy divine, merely because they do not understand it. But if they called everything divine which they do not understand, there would be no end of divine things." It's likely there are natural explanations for many anthropic coincidences, although some will only emerge from frontier theories that aren't yet well-tested. Also, some instances of fine-tuning are not as fine as has been claimed,[22] and they're only remarkable in the context of a vast number of hypothetical physical realities. Most importantly, anthropic arguments depend on knowing the full range of conditions for life and intelligent observers to evolve. Without a general theory of biology we can only guess. If life elsewhere is stranger than we imagine, then a much wider range of hypothetical universes could support life and our surprise should be correspondingly reduced.

I'm skeptical. But I'm also grateful. Why? Because 13.7 billion years ago a microscopic bubble of space-time unfolded that would one day contain me. Because nature was skewed enough to make something rather than nothing. Because the universe expanded fast enough and for long enough to congeal stars. Because carbon nuclei had resonances that allowed them to form and not leave a sea of useless helium. Because asteroids brought water to Earth but didn't impact late enough to take out my ape forebears. And because my dad didn't go to the pub as planned with his best friend but went to the library, where he caught the attention of a slender girl with long brown hair and soulful eyes, with the eventual, improbable outcome that I've written this and you're reading it.

—

Something is wrong with the math. One plus one makes zero. Zero equals two. The dance of particles and photons seems to violate the parsimony of nature. Somehow, the radiation is not making like plus like, it's making like plus unlike, like plus its opposite, a form of shadow matter defined solely by being oppositional.

Antimatter.

To call it antimatter is nothing more than sleight of word, because the idea of antistuff makes no sense—it's antisense. But now the math works. A number plus its negative equals zero. And any number of numbers paired with their negatives can disappear. Just as zero can spawn any number of numbers along with their negatives. The infinite possibility of the vacuum is no more mysterious than mathematics. However, the zero sum rule must be obeyed; matter and antimatter are created in equal measure.

Contemplating this, the seething universe around me suddenly seems ominous. Particles and antiparticles are coming and going with no sense of permanence. Each morsel of matter has a doppelganger and when they meet the result is annihilation into gamma rays. The ubiquity of matter and antimatter isn't reassuring because they're so evanescent. Should a proton somehow survive long enough to make an atom, it will meet its antiatom doppelganger. Should an atom somehow survive long enough to become carbon, it will meet anticarbon and that will be that. And should matter somehow become part of a star that nurtures life, the jig will be up when it runs into an antistar.

Yet somehow I am here.

14

UNIFICATION AND INFLATION

FEVERISH. *It's hot beyond measure. Temperature implies temperate but conditions are so intemperate that no structure is possible. Crushing. The density is trillions of times that of lead, yet there's lightness because everything is in motion. Seething. The flux of interactions is epic. Particles and their antiparticle twins are created and destroyed continuously in a bath of blinding radiation. Inchoate. Space is warped on the scale of subatomic particle sizes. Time's arrow darts forward and backward with equal facility.*

There's pleasing unity here. The source of all things is one thing. It's anarchy but it's also democracy. Everything is possible.

Without warning, the scene changes. In a time as short compared to a femtosecond as a femtosecond is to the current age of the universe, a minute skewness in the cosmic überforce dumps energy into the neighborhood and drives a prodigious expansion. Space stretches out flat and smooth, like a wrinkled sheet snapped by an invisible hand. Almost everything that had been in view disappears. The zoo of relics and defects in space scatter to the four winds. The cacophony turns into a pristine void. But minute quantum fluctuations, unseen in the everyday world, have inflated along with the space-time.

In another instant, the energy that was borrowed from the vacuum is repaid. The infant universe is crowded with energy, matter, and antimatter. Expansion continues at a more sedate pace. The quantum fluctuations are subtle, barely discernable ripples in temperature. But they are the seeds to form galaxies, the grit that grows the pearls. And one day this space will hold 100 billion galaxies, including all that I hold dear.

—

BEYOND THE BIG BANG

We have a theory of the universe that spans most of its 13.7-billion-year existence. It's been tested and confirmed and it gives a robust explanation for how a small, hot, dense universe filled with radiation and matter turned into the vast arena of space-time we see today, containing feeble radiation and scattered clumps of matter.

Cosmologists should be happy. But they don't seem satisfied. They are apparently not constructed to be happy. They always want to push a successful theory past the breaking point. As Richard Feynman said, "We're trying to prove ourselves wrong as fast as possible, because that's how we make progress." And they're not immune from the existential angst induced by cogitations about space and time, as illustrated by this dour quote from the physicist Steven Weinberg: "The effort to understand the universe is one of the very few things that lifts human life a little above the level of farce and gives it some of the grace of tragedy."[1] Steven, dude, lighten up.

Here are some problems and limitations of the big bang model: It doesn't explain why the universe is flat. In the initial state, gravity was so strong that space could have substantial curvature. Flat space implies fine-tuning of the initial conditions for the universe to be so nearly flat this long after the big bang. This is the flatness problem.

It doesn't explain why the universe is smooth. Microwaves on opposite sides of the sky have almost exactly the same temperature. That can only happen physically if those regions have been in thermal contact. Yet at the time the microwaves were released the regions on opposite sides of the sky were separating at 50 times the speed of light and were beyond each other's horizons. If those regions had had different temperature at the beginning, there's no way they could have reached the same temperature later.[2] This is the smoothness problem.

It doesn't explain why the universe is lumpy. As we've seen, radiation from the big bang has deviations of 1 part in 100,000 that acted as the seeds for forming stars and galaxies. If the seeds had been much smaller, the universe would have stayed smooth and we wouldn't be here. If they'd been much larger, the universe would have congealed into dense structures early on. The big bang model has no explanation for such fine tuning. The porridge isn't thin like gruel and it isn't a big indigestible wad—it has a pleasing texture throughout. But there's no explanation for this happy outcome. This is the structure problem.

It doesn't explain the absence of exotic particles in the universe. At the very high temperatures of the early universe, theories that unify the forces of nature predict copious numbers of exotic and massive particles, in particular knots of magnetic field called monopoles. The universe doesn't have these entities; for example, monopoles have never been detected in a lab or accelerator.[3] This is the relic problem.

It doesn't explain why the universe is expanding. Distant galaxies are redshifted. Some force has caused the observable universe to grow to a prodigious 92 billion light-years across, despite the gravity from the normal and dark matter in 100 billion galaxies acting to counter the expansion of space-time. The big bang is mute on the cause. It would be nice if the theory explained the expansion from first principles.

If you were an artist and had a canvas that was flat and smooth, and you had started a painting with just the right amount of structure, and had formed your own artistic vision without relics of previous schools of art, and had steadily expanded your time for painting—you wouldn't complain. You'd be as tranquil as a still life. Or if you were a sailor on an ocean that was flat and smooth, with structure added by a steady wind, relics of yesterday's storm left behind, and your sails filled and expanded—you wouldn't fuss. You'd be happy as a clam. Cosmologists are much more persnickety than artists and sailors.

As we've seen, the big bang also fails to tell us why the universe has lots of radiation, a dollop of matter, and virtually no antimatter. Then there are profound and irresistible questions about the big bang. Why did it happen? What preceded it? Was it a unique event? MIT professor Alan Guth has commented: "the big bang theory says nothing about what banged, why it banged, or what happened before it banged."[4]

SYMMETRY

To make a better big bang model, we have to follow the road back in time. Sturdy tarmac with well-placed signs saying "Galaxy" and "Star" leads to a rutted track marked "Decoupling" and "Nucleosynthesis," which ends in untamed wilderness called "Particle Soup." The edge of cosmology knowledge is the same place as the edge of knowledge in physics. Uncertainty rules the very large and the very small.

We need a compass, a guiding principle to what is simultaneously the birth of space and time and the most fundamental state of matter. The principle is symmetry.[5] Symmetry in art and music means a degree of regularity or harmonious design that's pleasing to the eye or ear. That type of aesthetic sensibility is often rooted in mathematics. The golden ratio, for example, is the relationship of two quantities where the ratio of the sum of the two to the larger quantity is equal to

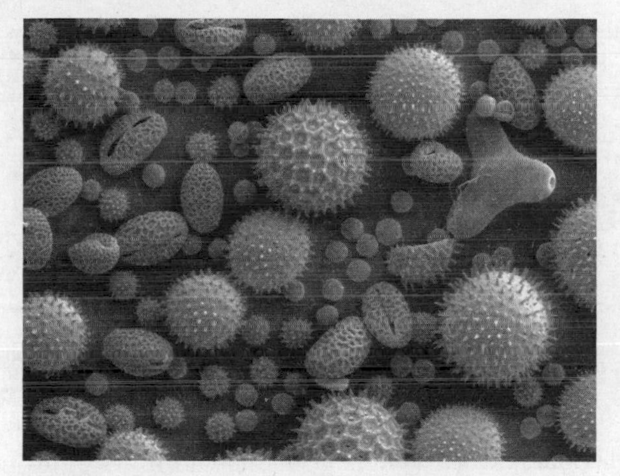

FIGURE 14.1. *Grains of pollen from a variety of plants, seen under high magnification, provide examples of symmetry from nature. The width of the image corresponds to 0.3 millimeters. The overall structures have spherical, bilateral, and trifold symmetry, while the details have fivefold and higher symmetries. The natural world contains many types of symmetry that have a simple mathematical description.*

the ratio of the larger one to the smaller one. Proportions based on a golden ratio are widespread in Western art, music, architecture, and the natural world (Figure 14.1). In geometry, symmetry describes the way shapes can be transformed and still appear to be the same. The Greeks made circles and spheres the basis of their cosmology because they were the most perfect and symmetric shapes. In abstract forms of math, symmetry is related to an economy of formalism, where a general equation accounts for a large number of specific cases.

Symmetry in physical theories combines all of these flavors.[6] Science has been well-served by the premise of an underlying unity in nature that might be obscured or imperfectly realized in the everyday world. How important is symmetry? When the Nobel Prize–winning physicist Richard Feymann was asked to summarize modern science in a sentence he said, "The universe is made of atoms." Allowed a second sentence he added, "Symmetry underlies the laws of nature."[7]

Electromagnetism provides the best historical example. In the 200 years after Newton, scientists experimented with light and discovered invisible forms of shorter and longer wavelength radiation without realizing they were manifestations of the same phenomenon. Electricity and magnetism were thought to be distinct, but Michael Faraday showed that a moving charge could magnetize metal and a moving magnet could generate electric current. Changing electricity and changing magnetism were somehow related. Then came James Clerk Maxwell's extraordinary insight to relate the two phenomena in an elegant set of four equations. Maxwell's equations not only explain the connection between changing electricity and magnetism, they say that changes in either create an electromagnetic wave that travels at 300,000 kilometers per second. Sunlight, the signal used by a radio, and X-rays in a doctor's office are all fundamentally the same.

We've encountered two other forms of symmetry already. Einstein's iconic equation $E = mc^2$ means mass and energy are convertible and are different manifestations of the underlying quantity mass-energy. Dirac found two quantum solutions for the behavior of a particle and his confidence in the symmetry of the descriptions led him to predict antimatter. High-energy physics reactions produce matter and antimatter with equal facility. In yet another example from the early days of quantum theory, Louis de Broglie recognized that electrons can behave like waves and waves can behave like particles—the distinction between them is artificial. This wave-particle "duality" perplexes even physicists, but seems to be the way the natural world works. Particles don't have hard edges and their behavior is ruled by probability not certainty. Symmetry in physics is linked to the idea that disparate phenomena can be linked through a unifying concept.

Before we get carried away (OK, I agree, it's me that's talking up symmetry, you might be sitting there with one eyebrow skeptically raised like Spock), let's approach the idea critically. It's been known for decades that both people and animals have a strong preference for symmetry in their sexual partners.[8] Bilateral symmetry is correlated with strength and health so it's rational to choose a partner with the best genes. One study found that women have more orgasms with highly symmetric men, regardless of their level of romantic attachment or the men's level of sexual experience. Astrophysicist Mario Livio has worried that our biological predisposition for symmetry is biasing our perception and theories of the natural world. He writes: "Because our brains are so fine tuned to detect symmetry, is it possible that both the tools that we use to determine the laws of nature and indeed our theories themselves have symmetry in them partly because our brains like to latch onto the symmetric part of the universe and not because it's the most fundamental thing?"[9] Aware that we might be falling for a pretty face, let's continue the discussion.

What symmetry or hidden unity might we seek in the big bang? Physicists have long cherished the goal of unifying the four forces of nature. In the everyday world, they differ in strength by 40 orders of magnitude. Two have infinite reach and are carried by particles with no mass, while two operate inside the atomic nucleus and are carried by massive particles. The four forces are completely different in their manifestation in experiments. Only an inattentive student of physics would ever confuse them. The protean, but ugly, Standard Model of particle physics has 19 parameters and offers no explanation of the relationship among them. Physicists would love to replace it with something better.

The Standard Model doesn't explain why there are four forces, but it does indicate a fruitful way forward. In the seventies and eighties, experiments at CERN validated a theory that unified the electromagnetic and weak nuclear forces, leading to the award of the Nobel Prize for Physics for that work in 1979. In your living room, electromagnetism is 100 billion times stronger than the weak nuclear force, so it was very bold to speculate that they were the same thing. Electromagnetism is like a ubiquitous, overachieving brother—talented at everything, interacting well with everyone, and having a seemingly infinite range of skills. The weak nuclear force is like a reclusive and slightly weird uncle, who keeps to himself (in this case, the atomic nucleus) and who has only one party trick (in this case, making nuclei decay by radioactivity).

In an accelerator, where the instantaneous temperature can reach 10^{15} K, or 1000 trillion degrees, the weak and electromagnetic forces merge and new particles that carry the combined force can be created from pure energy. In the universe, these two forces parted company 10^{-11} or a ten-trillionth of a second after the big bang. This time corresponds to the limit of tested theories of physics.

Energized by this success, physicists naturally speculated that even higher temperatures might lead to a merger of this "electroweak" force with the strong nuclear force. Since the strong nuclear force is "only" a factor of 100 beefier than the electromagnetic force, you might imagine that it would be fairly simple to merge the three. Unfortunately it's not, because the force that binds quarks inside the atomic nucleus is exceptionally strong and it has a range smaller than a proton. In the domestic analogy we've used so far, the strong force is a sinister cousin who works on the perfect superglue in a lab under his house, and he never emerges into the daylight. Careful calculation shows that for "grand unified theories" to unite electromagnetism, the weak and the strong force must kick in at temperatures trillions of times past the limit of current accelerators, 10^{27} degrees.[10] In the universe, such conditions apply a stunningly brief 10^{-35} seconds after the big bang. For something snappy to impress at a dinner party, that's 100 billion trillion trillionths of a second. It is to a billionth of a second as a billionth of a second is to the age of the universe.

We can see the connection between high energy or temperature and symmetry with a couple of homely examples. As ice melts, the very particular orientations of water molecules in the ice crystals turn into the situation where each water molecule can have any orientation. The result is more symmetry: each point in the liquid is like all other points while each point in the solid isn't like other points. Another example is an iron magnet. At room temperature the atoms in a magnet all have a roughly similar orientation, but when it's heated they orient in every possible direction. Once again, raising the temperature creates a high level of symmetry.

So it is with the early universe. In the cauldron of the universe just after the big bang, it's supposed that there was a "superforce" that combined all four known forces. As the temperature fell, the forces "froze out" one by one to take their familiar and disparate strengths. The unity that was evident for one shining moment was lost forever (Figure 14.2). We'll defer consideration of the superforce to look first at what may have happened when all forces except gravity were united.

Grand unified theories shimmer like the buildings of Oz, just visible on the horizon but requiring an arduous journey to reach. What do they predict and how can they be tested?

A grand unified theory is not like a sturdy car where you can kick the tires, check the paint job, and drive it off the lot. There are a number of variations and it's more like a collection of alluring used cars. Those with lots of chrome are flashy and may be tempting but it's advisable to look under the hood; the last thing you want is a theory that gets you a few blocks and breaks down.

Two generic predictions of most grand unified theories are relics like monopoles and proton decay.[11] Double oops—neither has been seen. Paul Dirac predicted single magnetic charges or monopoles back in 1931 and their nondetection is a problem for the big bang as well as for physics theories. Grand unified theories make connections between heavy particles like quarks and light particles like electrons. That means a proton can decay into lighter products. Normal matter may not be stable! However, no proton has ever been seen to decay, and

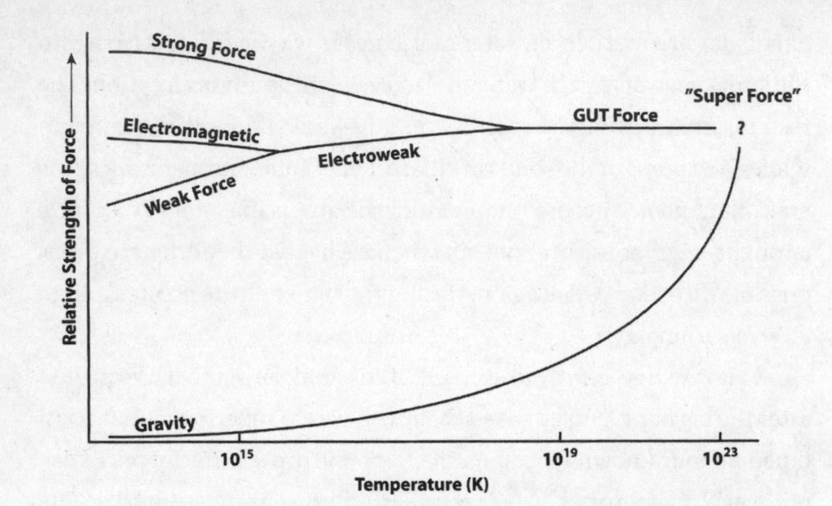

FIGURE 14.2. *The unification of the four distinct forces of nature can be realized at very high temperatures. The electroweak merger has been created momentarily in particle accelerators, but higher unification is only accessible (in principle) through cosmological observations. Grand unified theories are currently untested and theories that unite gravity with the other three forces are even more speculative.*

trying to catch one in the act is much worse than watching paint dry. The current limit to the lifetime is 10^{34} years. Rather than watch one proton for that long, physicists carefully watch a very large number of them for years. The limit already rules out several varieties of grand unified model.

There's good news too. Grand unified theories provide a mechanism for neutrinos to have mass. And the decay of a proton provides a basis for matter disappearing and so the potential to explain the asymmetry between matter and antimatter. Attention has been focused for some years on a class of unified theories called supersymmetry theories. In a supersymmetric theory the distinction between forces and particles is broken down. The theory assigns every particle a supersymmetric partner. For the electron there's a selectron. For each type of neutrino there's a sneutrino, and for each flavor of quark there's a squark. If you could gather enough of these exotic entities you might be able to construct a sperson. The theory also hypothesizes supersymmetric partners for all the particles carrying fundamental forces. For the photon, a photino. For the gluon, a gluino. And for the hypothetical Higgs particle and graviton the even more hypothetical Higgsino and gravitino (Figure 14.3).[12]

And a nice surprise. Supersymmetric partners of known particles are predicted to be very massive, but the lightest of them will be stable, interact weakly with normal matter, and have a mass just beyond the range of current accelerators. That gives it just the right properties to account for dark matter. One huge puzzle in cosmology may be solved in the form of relic particles left over from the unification of three of the fundamental forces 10 ^{35} seconds after the big bang!

If you're feeling a wave of incredulity right now, I don't blame you. But let's look at the strange journey physicists have taken. Faced with the shortcomings in the Standard Model of particles and their interactions, they've conjectured that three of the highly distinct forces of

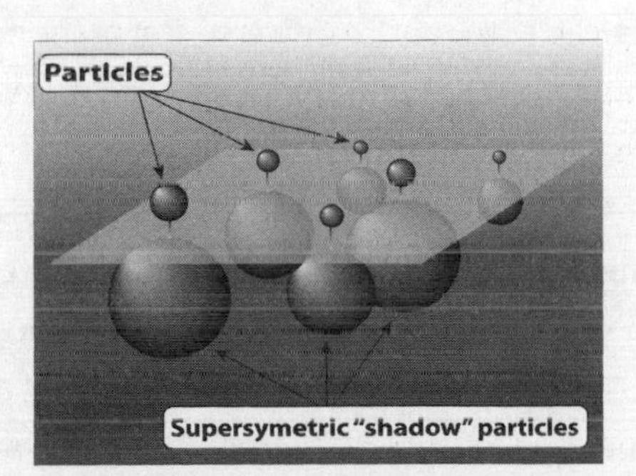

FIGURE 14.3. *Grand unified theories predict a "shadow" force carrier particle for every fundamental matter particle and a "shadow" matter particle for every force carrier. These supersymmetric partners are hypothesized to be very massive and perhaps not accessible with the current generation of accelerators.*

nature are actually tepid manifestations of a grand unified force that's fully realized at energies trillions of times more than any accelerator can reach. So far the relics and proton decay generically predicted by the theory haven't been observed, the number of fundamental particles has doubled, and none of the new ones have been found. Theorists aren't deterred. They're spurred on by principles of symmetry and mathematical elegance, and the goal of explaining the masses of all the families of particles. These new theories *might* also explain why there's much more matter than antimatter in the universe, and why there's more dark matter than normal matter.

I brushed into beauty in its most abstract form when I was a student of physics in London. The year was 1975. I'd fallen in love with physics in college—for its rigor and pristine elegance, and its plausible claim to be the mother of all sciences. But in my second year, the attachment was fraying. I felt ground down by classes and overwhelmed by homework. Nothing was coming easily. One particularly brutal afternoon I spent two hours in a lab trying to derive Gauss's Law from measurement of the electric fields around charged metal plates. My physics textbook told me that Gauss developed a formalism based on symmetry that beautifully described forces on all scales from elementary particles to the entire universe. But in the mundane setting of the lab, nothing worked right and a snotty TA gave me a tongue-lashing for poor experimental technique.

I skulked off to one of my sanctuaries, a lounge upstairs in the Physics building where students in the astrophysics group hung out. Hoping to do a senior project on astronomy, I'd been frequenting the lounge. A few students and a postdoc were chatting amid a chaos of used coffee mugs and journals. The room had a great view north to Albert Hall and Kensington Gardens; my college was in the middle of London.

The conversation was interrupted when Jim Ring, one of the senior professors, walked in and started talking, as much to himself as to anyone in the room, about a graduate student who had gotten stalled halfway to a PhD. Ring was a chain-smoker who didn't even bother removing the cigarette from his mouth when he talked. He's smart, said Ring, but he's not bearing down. He frowned and glanced at us balefully, making sure we got the point. He left, shaking his head. I asked the people in the room who the grad student was. Brian May. The name rang a bell but I wasn't sure why.

I went to my last class of the day, a tutorial on mathematical physics. The professor was Gaspar. We never knew if that was his first or last name—everyone, including students and secretaries, just called him Gaspar. He was of eastern European extraction and very intimidating. Gaspar was a pure mathematician and even though he was supposed to be teaching us how math was used in physical theories, he took us on many detours into the world of pure math. We could tell he thought it was largely wasted on us, that he was casting pearls before swine. There were six of us sitting in the small tutorial room, with Gaspar at the blackboard, covering it with his spidery scrawl. His piercing brown eyes were framed by a shock of hair leaping upward from his forehead and a small goatee thrusting downward.

That day, he was on a roll. Jettisoning any attempt to connect math to the real world, he talked about Euler's identity. This deceptively simple equality, $e^{i\pi} + 1 = 0$, sets the gold standard for mathematical beauty. It combines the three arithmetic operations of addition, multiplication, and exponentiation with the identities 0 and 1, the two most important transcendental numbers π and e, and the symbol that opens the door to the universe of imaginary numbers, i. This equation is high-octane fuel, a turbo-charged distillation of mathematics in pocket-sized form. Luckily, none of us had the temerity to ask what it was useful for.

Then Gaspar surprised us. He reminded us that complex numbers, those composed of a real and an imaginary part, were one of the most fruitful areas of mathematics. In the early days of quantum theory,

Dirac solved the wave equation for an electron and found that there was a second solution involving the square root of a negative number instead of a positive number. Seeing no reason to prefer one over the other, he proposed a mirror form of matter. Gaspar's eyes shone and his guttural voice got raspy as he described how symmetry in the equations led to the prediction that antimatter existed. We stared at him, trying to listen through an unfamiliar stream of squashed vowels and abrasive consonants.

Time was up, but nobody made a move or got restless. Gaspar said, consider this. He wrote on the board: $Z_{n+1} = Z_n^2 + C$. Each number Z is formed by squaring the previous version and adding a constant. The constant C is a complex number, one with a real part and an imaginary part. Compared to what we had to deal with in our physics classes this was a very simple equation. It was recursive, he explained, generating an infinite set of complex numbers. He walked over to a monitor on his desk and typed on the command line for a minute, and then stood aside. A pattern of whorls and loops appeared on the old-style green screen. As he repeatedly hit the space bar it evolved. The view zoomed in on one feature but as it did, new spirals and feathery patterns appeared and grew, and then yet more and different patterns emerged from them. I could do this *forever*, he said. The equation contains an infinitely rich world. It's called a fractal. We listened to his explanation through the impenetrable thicket of his accent, but the endlessly complex pattern told the story. All of us were mesmerized.

That evening, my buddies and I went to a concert at the Hammersmith Odeon. It was an up-and-coming band called Queen. Two years after their debut album, they'd released *A Night at the Opera*, which was making some waves. The lead guitarist was Brian May.[13] The loud, pulsing music washed over me and my mind raced with the myriad possibilities of numbers and the universe.

SIGNS OF INFLATION

Alan Guth was working late when the breakthrough came. He wrote "Spectacular Realization" at the top of his notebook page and he went to bed with his mind still churning.[14] It was 1979, and the young Stanford postdoc had been playing with an idea to get rid of monopoles from the expanding universe, since he knew they were a prediction from the physics of the big bang but had never been observed.

Guth had been bouncing around as a postdoc and was unsure of his future, but he was familiar enough with cosmology to know he was on to something. He knew quantum energy in the vacuum of space might have been able to expand the universe exponentially in the instant of time just after the big bang. That means that in each tiny increment of time, space would double and double again, ballooning the universe in size by many orders of magnitude. If expansion happened very quickly any monopoles (or any glitches in space-time) created earlier would be widely dispersed, making it very unlikely for us to see any in the space visible to us.

The idea, which Guth called "inflation," also solved two very thorny problems with the big bang: flatness and smoothness. Microwaves from the early universe show that the radiation was very uniform and the curvature of space is undetectable. Think of the universe before inflation as the curved surface of a balloon. This is a two-dimensional analogy; our universe would have been curved in three dimensions. A very early view would have detected the curvature of the balloon, but after it has been inflated to enormous size, our patch of it seems to be essentially flat, just as the very large Earth seems flat because we see a small fraction of it. In the standard big bang, the smoothness of the microwave radiation is mysterious because different directions in space were separating at many times the speed of light at the time that the radiation was liberated. In the inflationary model, these regions were close together and at the same temperature before infla-

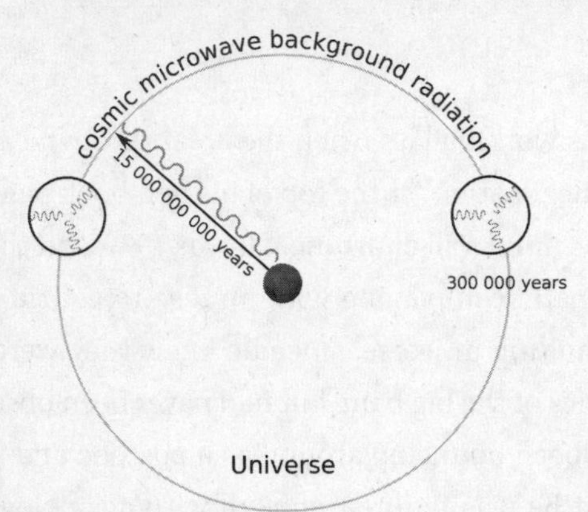

FIGURE 14.4. *Inflation supposedly occurred in the first tiny fraction of a second of the existence of the universe. The exponential expansion of the small curved universe left space that was almost perfectly flat, as in the analogy of a small balloon inflated to enormous size. We see the same microwave signal in all directions in the sky, and very different directions (small circles) represent regions that were very close together before inflation.*

tion, and then were carried apart to enormous distances by inflation (Figure 14.4).

"Important if true." Victorian travel writer Alexander Kingslake said this should be inscribed above the doors of all churches. The same words should be attached to a discussion of inflation. If true, it says that we live in a smooth and flat patch of a much larger space-time that might have a lot of curvature and exotic structure and features. With inflation, the physical universe—all there is—is vastly larger than the observable universe—all that we can see. The totality of space-time is therefore much larger than the 46 billion light-years to the edge of our horizon.

What caused this spectacular event? Guth thought that the expansion was driven by the vacuum of space. To a physicist, a vacuum isn't just boring "nothing" that we might imagine; it seethes with possibility. For a very short time, Heisenberg's uncertainty principle allows energy to be "borrowed" from a vacuum. That energy can briefly make particles and antiparticles and cause other physical effects that are seen in the lab.[15] Quantum theory attributes a tiny amount of energy to absolutely empty space. In his wild theorizing, Guth imagined that if it could be done exceedingly quickly, enough energy could be borrowed from the vacuum to inflate the universe around us. He called this the "ultimate free lunch."

What was the trigger for inflation? The same process that led to the "grand unified" force fragmenting into a strong nuclear force and an electroweak force. Let's use again the homely analogy of ice as it melts. Inflation represents the universe going through a dramatic "phase change" like the abrupt freezing of water supercooled to below 0°C. Heat must be added to the ice to melt it. The converse is that water releases heat when it freezes, and as a solid it has a lower degree of symmetry than when it's a liquid. In the vacuum of the early universe, there was a huge amount of energy released when the temperature dropped enough to break apart the grand unified force. The resulting phase transition left the universe in a state of lower symmetry.

Inflation started 10^{-35} seconds after the big bang, at a temperature of 10^{28} degrees. It expanded the space by 26 orders of magnitude, so that the region we can see now grew from trillions of times smaller than a proton to about the size of an orange. By 10^{-32} seconds after the big bang all the wild action was over. The exponential expansion ceased and the universe began the sedate expansion that continues today (Figure 14.5). In our trip back in time toward the big bang, we have come so far that our nose is pressed hard against the glass of creation, a tiny sliver of time after the origin of everything.

It's amazing, and undeniably clever, but does inflation really work? For a while, it seemed that it wouldn't. Guth wasn't actually the first to contemplate the idea. As in many aspects of cosmology, Russians were pioneers. But it was Guth who connected the physics to the cosmology and set in motion an effort that now involves hundreds of the smart-

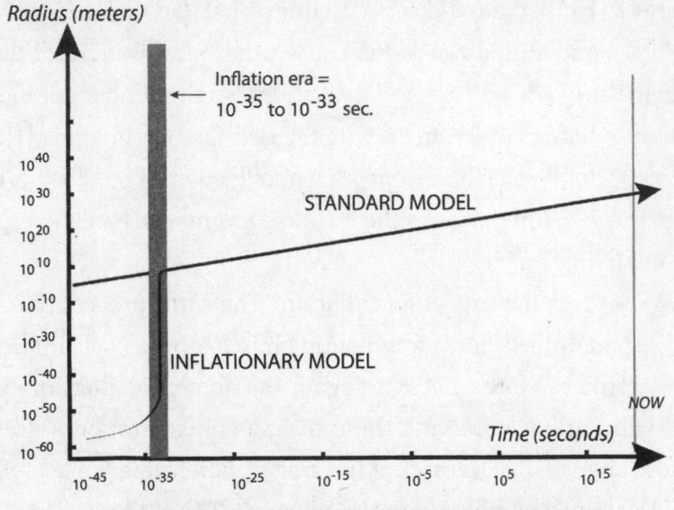

FIGURE 14.5. *Cosmic inflation took today's visible universe from trillions of times smaller than a proton to about the size of a golf ball. After the exponential expansion a slower and nearly linear expansion continued. The graph shows the size versus time on a very compressed scale.*

est people on the planet. As he readily admits, the inflation outlined in his pioneering paper of 1980 didn't work. The initial bubbles of space-time expanded too fast, and inflation ended too quickly, and the result was a tangled and messy universe. Other theorists solved the problem but at the expense of having to "tweak" the vacuum energy that drove the expansion. This was of course the kind of fine-tuning that inflation was designed to avoid! Also, at the end of inflation an enormous amount of energy is dumped into the universe as radiation and a small fraction of it turns into all of the stars and galaxies.

Inflation may or may not be verified by observation, but it vaulted Alan Guth into the elite tier of physicists. After worrying that he would reach the end of the potdoc pipeline and be washed out to sea, he is now the Victor Weisskopf Professor of Physics at his alma mater, MIT.

At major conferences he attracts a crowd of followers that shadow him through the meeting. His office is a chaos of boxes, toys, and papers piled high on desks and undulating down to the floor. He has the mop top, owlish glasses, and smooth face of John Denver circa 1980. Guth virtually defines geek chic. He deserves his fame because big ideas are rare in science and he had one of the biggest.

How long inflation takes and how it ends are still very unclear. Without a tried and tested grand unified theory and a better understanding of vacuum energy it will be hard to make progress. Nevertheless, there are strong signs that something like inflation did happen. The fact that the universe is flat and smooth can't be counted as a success since the idea was designed to solve those problems. But it's telling that better microwave observations since Guth's paper have shown that space is exquisitely flat and smooth, since that didn't have to be the case.

However, the microwave background isn't *perfectly* smooth. There are variations or ripples in the temperature by about 1 part in 100,000. In the standard big bang these ripples are initial conditions and unexplained. A big success of inflation is that it can explain them. Before the epoch of inflation, imperfections in space-time would have been quantum fluctuations like those seen in the subatomic world. As inflation occurred, the quantum fluctuations were stretched to the size that would eventually turn into a galaxy. Quantum fluctuations have a particular property of being scale-free, which means their distribution with strength doesn't depend on scale. A familiar example is a fractal, where the appearance is the same no matter how much you zoom in or out. The fluctuations in the microwaves have the same property.[16] It's amazing to a child that acorns grow into oak trees, but adults and children alike can be amazed that galaxies grew from quantum seeds.

If inflation is correct, maps of the microwave background contain information from when the universe was less than a billionth of a billionth of a billionth of a second old!

Recently, WMAP has been used to test inflation at a very precise level, with impressive results. The microwave fluctuations are not exactly scale-free, and the modest departure is in just the sense predicted by inflation. The successor to WMAP, the Planck satellite, was successfully launched in 2009. *Planck* will aim to detect a signature of inflation 100 times smaller than the temperature fluctuations. The signature is imprinted on radiation by gravity waves and it can discriminate between inflation and alternative theories in a completely new way.[17]

Inflation has passed through its teething phase and is maturing. It makes predictions that can be tested and it has passed most of the tests. The frontier of cosmology has shifted back to the tiny iota of time after the big bang where theoretical progress now depends on deeper understanding of microscopic physics.

THE QUANTUM UNIVERSE

T. S. Eliot might have been talking about the universe when he wrote, "And the end of all our exploring will be to arrive where we started and know the place for the first time." A little later in his *Four Quartets* he talks about "A condition of complete simplicity (Costing not less than everything)."[18]

The ouroboros, a snake eating its own tail, is the visual icon of this closure. It appears throughout Western culture from the time of antiquity to represent self-reference or self-reflexivity. It appears in the Egyptian *Book of the Dead*, and Plato described a circular, self-eating creature as the first living thing in the universe. Physics Nobel Laureate Sheldon Glashow used the image of an ouroboros to represent his hope for the unification of theories of the very small and the very large, something that still eludes us. From the visible universe now to the universe in its first instant spans 60 powers of 10. Humans are poised near the middle of this logarithmic range.

Inflation seems like an adjustment to the big bang model, a clever and esoteric gimmick. But its modern incarnation is much more ambitious. To see why, let's return to the story of how it was invented. Alan Guth had the recipe for a delicious soufflé, but those tiny bubbles of space-time grew too quickly and merged, leaving an unappetizing mess. The idea was further developed by the wryly humorous Russian Andrei Linde, currently a professor at Stanford University, and independently by two other theorists.[19] Linde was a young researcher when he first presented his work at a conference in Moscow in 1981. Ironically, he had just served as the translator of a talk by Stephen Hawking stating that inflation couldn't work! (Hawking later changed his mind.)

Imagine the universe carried an electric charge so that the potential everywhere was 110 volts. We'd never notice it, just as we don't see the state of the vacuum. If the electric field varied over space or time we'd see it in the behavior of charged particles. Inflation involves an analogous situation.[20] It's proposed that the universe emerged from a preexisting space-time vacuum. By conventional physics, the vacuum would be filled with quantum fluctuations. The fluctuations are waves, like undulations in an electric field. The waves have every wavelength possible and move in all directions. They add together and where their strength is sufficient, an exponential expansion is triggered. Inflation. The waves are presumed to arise from the unification of the forces of nature and that physics is so speculative it's untested.

This is an extraordinary example of the harmonics we encountered in patterns of the cosmic microwave radiation. But in the first instant of time, with waves of all sizes and strengths, the result is white noise, the "hiss" from which the music of our universe emerged.

Linde outlined the idea of "eternal" inflation. Some parts of space-time inflate exponentially. Others don't change or are stillborn, col-

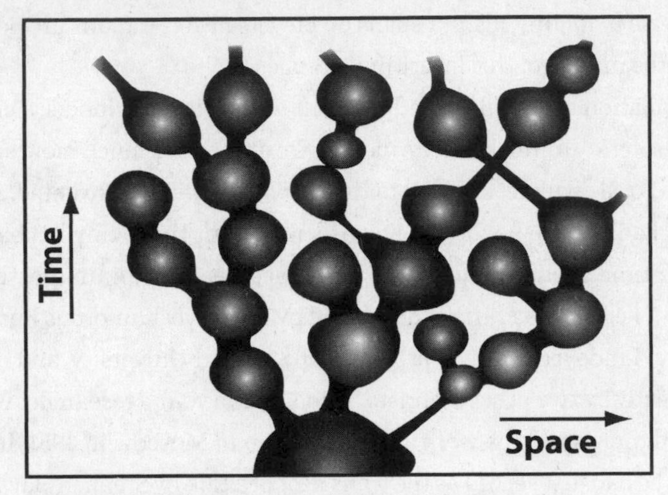

FIGURE 14.6. *On large scales and at low energies space-time appears smooth and uneventful, but on the scale and at an energy where gravity must be reconciled with quantum theory, space-time is in a state of constant flux.*

lapsing like black holes. Even if regions that inflate are very rare, exponential expansion means they are much larger than all other regions.[21] Within any region that inflates, it's the same story. A single sufficiently large region can spawn new inflationary regions. In effect, our universe can spawn new universes, just as our universe may have been spawned by another universe, in a process that had no beginning and that will have no end (Figure 14.6).

This is a profound conceptual shift. In this model, the big bang isn't the beginning of all space-time. Inflation is a mechanism not only to create our universe but also to create other universes like and unlike our own.

The eternal, self-reproducing universe model resembles in some ways the steady state model that was an early rival to the big bang. In the steady state, the universe is eternal and endlessly expanding, with the source of all matter being energy in the expanding space. With eternal inflation, the idea of origins has been finessed. Our beginning isn't the beginning of everything. Fittingly, the ouroboros has a second meaning, that of "eternal return," where a cycle ends and begins anew.

Modern cosmology posits our universe as a quantum event. Vacuum energy can momentarily create particle-antiparticle pairs anywhere, even in front of you (but on an undetectably small scale) in your living room. It can also create a universe opulent enough to contain you and your living room. There's no ducking the weirdness of the microworld so let's remind ourselves how very weird it is.

Complementarity is one key concept. Particles are localized in space and carry energy from one place to another. Waves are extended in space and undergo diffraction and interference. In quantum theory, any entity can behave like a particle or like a wave depending on the situation, and each description is equally valid. In one classic physics experiment, light passes through two narrow slits in a screen, and an interference pattern appears on a second screen, just like ripples that might combine and interfere on a water surface. But when the light source is turned down so that only one photon leaves at a time, the photons still produce the same pattern, as if each passes through both slits and then interferes with itself. Richard Feynman called this the "central mystery" of quantum physics and said that if you understand this, you would understand quantum physics. Then he said, "nobody understands quantum mechanics."[22]

Uncertainty is another key concept. Werner Heisenberg came up with an equation giving a limit to our knowledge of pairs of quantities like momentum and position and time and energy—precision in measuring one comes at the expense of imprecision in measuring the other. The two concepts are linked. A wave has a well-defined direction of motion or momentum, while a particle has a well-defined position. Heisenberg showed that if the position of a quantum entity is accurately measured its waviness is suppressed, but if its direction of motion is accurately measured its waviness dominates.[23] If the measurement involves time and energy, Heisenberg's uncertainty principle means that the law of conservation of energy can be broken momentarily and the swifter the transgression, the larger it can be. That forms the basis for creating everything from particles to universes.

Einstein was exasperated and frustrated by this apparent limitation to our knowledge of the physical world. He was convinced that there was a deeper theory of nature waiting to be discovered or that the limits to measurement implied by Heisenberg's principle would be conquered by better equipment. He engaged in a decade-long intellectual joust with Niels Bohr, trying to circumvent complementarity and uncertainty. He failed. Bohr's view—called the Copenhagen Interpretation—has stood the test of time. Quantum mechanics is nearly a century old and very well tested. Physicists prefer not to dwell on its weirdness but they do have to live with it.

Determinism is dead. Descriptions of nature are probabilistic. There's a deep connection between any observer and the thing being observed. Nothing is real until it's observed. An electron is an extended wave of probability that collapses into finite reality when it's observed. These probabilistic quantum states can extend across space, a phenomenon called entanglement, which implies coupled properties on scales much larger than an atom. If all of this is inescapably true for microscopic entities, and if our universe is itself a quantum event, the philosophical implications are profound.

What does it take to create a universe? Andrei Linde has done the calculation. The mass of the universe now is about 10^{54} kilograms, along with 10^{50} kilograms of radiation (converting from energy into mass by $E = mc^2$). In the very early universe, the equivalent mass in radiation was much larger, about 10^{82} kilograms. A standard big bang must create this out of nothing, whereas inflation "borrows" it from vacuum as a quantum fluctuation. By Heisenberg's uncertainty principle it only takes the mass-energy equivalent of one-hundreth of a gram to almost instantly inflate the universe we know and love—not quite something for nothing, but an impressive return on a modest investment.

Separated at birth. I don't yet exist. I shimmer at the edge of existence, a perturbation on the boundless void. But I have a sense of possibility and I yearn to be more than just potential. The flickering twilight is endless and eternal. Suddenly and with no warning, space and time leap into being. Space balloons in all directions and with it, a part of me is lost, wrenched far beyond the horizon. I'm sad to be incomplete, my virtual doppelganger far from view, though we remain quantum entangled.

I'm distracted from my loss for a long time. There's a universe to fill, galaxies to construct, stars to ignite, life to incubate. Gravity uses its long reach to gather matter against the continuing expansion. The architecture that emerges is diverse and subtle. My other self becomes no more than a vestigial memory.

Eons pass. The visible universe grows a little bigger each day. The expansion slows. To my surprise and delight, my alter ego comes into view, red-tinged by recession. His light has clawed its way back into view from past the limit of vision and he's young, as I remember I was once. Then a new force starts to stretch the cosmic fabric ever faster and my other half slips from view, a second loss more painful than the first. My consolation is the grandeur all around me, forged from a pinch of space-time.

15

MULTIVERSE

THE UNIVERSE NEXT DOOR IS A PUDDLE OF FRIGID SPACE-TIME. *It flickers with particle-antiparticle pairs. The space is quantum-tangled and would fit on the head of a pin; it's barely impressive enough to bear the name* universe.

The universe beyond that is slightly more promising. It oscillates and wobbles like a water-filled balloon. It has tapped a higher space dimension to become infused with matter and radiation. The matter and radiation are interacting to produce ripples and complex harmonies. The sounds of this space are rhythmic and hypnotic. It is music of a kind, and it's sad there's nobody to hear it.

Venturing farther afield, a canyon of featureless space stretches beyond view. This universe is majestic but austere. Gravity has struggled to gather pockets of matter against the galloping void. Here and there, single stars flicker like fireflies in the night. They will die and the canvas will fade to black. I move on.

This is interesting. A universe that's small but complex. Five space dimensions are tessellated in an interlocking pattern. Mat ter forms fleeting arrangements and then quickly disperses. Despite the evanescence of the structure, the organization is rich enough to be on the edge of sentience. I have no frame of reference for understanding it.

I have traveled so far in space and time that I'm beyond space and time. The myriad possibilities of the multiverse enfold me and subsume my sense of self. Is this real? Is anything real?

—

THE LIMITS OF KNOWLEDGE

We've reached the shimmering edge of what we know, and what we may ever know. Science is young. It's been 40,000 years since the last evolutionary advances in human brain function. Hunter-gatherers roaming the African savannah at that time were identical to us. We imagine that the universe was awe-inspiring and inscrutable to them, even though they were as well-equipped to understand it as we are. The tools to empower us to transcend the sheltering sky are very recent. We've had the scientific method for 5 percent of this timespan, telescopes for 1 percent, and a true sense of the size and age of the universe for just 0.1 percent.

Scientists are optimists by nature. They look at the transformation of our knowledge of the universe since Copernicus and don't think the fun will end anytime soon. Telescopes and atom-smashers and computers are bigger and better and faster than ever. More scientists are working now than at any time in the past. At a more superficial level, discoveries in cosmology make the news almost every week. The rate of progress doesn't seem to be slowing down.

The anthem of cosmology is stirring and powerful. In no more than a couple of generations, the hairless apes have done an amazing job of comprehending a universe that dwarfs them in space and time. Only a few discordant notes creep into the music. One is the worry that we've ventured so far from the realm of testable lab physics that we may not have tools sharp enough to discriminate between competing theories. If we crawl too far along the branch of speculation it might break. And a deeper concern intrudes, that our analytic tools and mental faculties might not be good enough to gain a deeper level of understanding.

Then there's apophenia. Apophenia is the perception of meaningful patterns or connections in data when none are present. In statistics, it's called a Type I error or false positive, in which a null hypothesis is not rejected in spite of being false. Swedish playwright August Strindberg exhibited an extreme form of apophenia; he saw the insignia of witches and a goat's horn in a rock, tiny hands praying when he looked at a walnut under a microscope, and a marble head in the style of Michaelangelo when he looked at his crumpled pillow.

Strindberg's apophenia amounted to psychosis, but a milder form is built into all of us by evolution.[1] When our nomadic ancestors thought they saw a dangerous predator in the dappled grass and were wrong, they got a fright; when they failed to see the predator that was actually there, they were lunch! The cost of believing that a false pattern is real is less than the cost of not believing a real pattern. Brains are conditioned for apophenia. That doesn't mean that primal reflexes trump the sophisticated apparatus of hypothesis-testing in science. It just means that we have to beware that we don't overreach with our theories and explanations.

The desire for meaning is most primal of all. The discovery of patterns in nature is a denial of noise. Despite our sophistication, we're fearful of the void and the inevitability of death. Modern cosmology posits the universe as a quantum fluctuation, a random event. That's not—to say the least—very reassuring.

THEORY OF EVERYTHING

How ambitious should scientists be in their hunger for explanation? Is it really possible to explain the universe and everything it contains? In the earliest age of science, Archimedes tried to identify a few axioms or principles from which he could deduce anything, and the Atomists believed that the diversity of all observed phenomena were caused by the collision of atoms. Much later, the power of Newton's universal law of gravitation led the French mathematician Pierre-Simon Laplace to suggest that a sufficiently powerful intellect (we might say, computer) that knew the positions and motions of all particles at one time could calculate their motions and positions at any time in the future.

Many humanists and philosophers abhorred the determinism implied by Newtonian mechanics and gravity because we humans are collections of particles, so perhaps our choices and free will are illusory. Physicists were derided in some quarters for the hubris of their Faustian quest for truth, as if to attain it they were willing to barter their souls with the devil.

The twentieth century reshaped these grand expectations, if not with a dose of humility, then at least with a dose of uncertainty. Determinism is in practice thwarted by the probabilistic nature of quantum theory, by the unpredictable and emergent properties of complex systems, and by sensitivity to initial conditions that leads to mathematical chaos. At a philosophical level, Gödel's incompleteness theorem says that any self-consistent and nontrivial mathematical theory will always be incomplete, or have propositions that can't be decided. Since a "theory of everything" must also be a self-consistent and nontrivial mathematical theory, notable physicists like Freeman Dyson and Stephen Hawking concluded that the search for an ultimate theory with a small number of principles might be fruitless. Even if we find a set of equations that describes everything in the universe, we're still left with unanswered questions about origin and meaning. As Stephen Hawking has asked, "What is it that breathes fire into the equations and makes a universe for them to describe?"[2]

Most scientists have dialed back the grandiosity; they accept that no single theory can be used to understand and predict the behavior of every physical system. Rather, they hope that they can continue the march toward the unification of the four fundamental forces, where those forces that are each distinct in our current low-energy universe are melded into a single superforce at a sufficiently high energy. (We discussed the first steps in the unification in the last chapter.) Grand unified theories try to unite the electromagnetic force with the weak and strong nuclear forces. The most promising theories are based on supersymmetry, where known particles have a shadow partner and there's no distinction between the particles that make up matter and the particles that carry forces.

The final step on this road is the unification of grand unified theories of elementary particles with general relativity, our best theory of gravity. Einstein beat his head against this brick wall for 30 years, until he was on his death bed. The problem's hard because particles are grainy and discrete while gravity is smooth and supple. They're as distinct as wood and marble. Particle theory only works when gravity is so weak we can pretend it doesn't exist and general relativity only works when the graininess and uncertainty of quantum theory are ignored. Quantum gravity has been the holy grail of physics since Einstein's death yet there's no chance of any lab experiment creating the conditions where the four fundamental forces are unified.

Instead, all roads point back to the big bang, and something called the Planck time. Projecting the expanding universe back toward a singularity—a state of infinite temperature and density—the limit of physical understanding is reached at the Planck time. The Planck time is an infinitesimal 10^{-43} seconds after the big bang. It's the smallest interval of time that can be described. At that very early epoch, the universe was a mere 10^{-35} meters across, or 100 billion billionths the

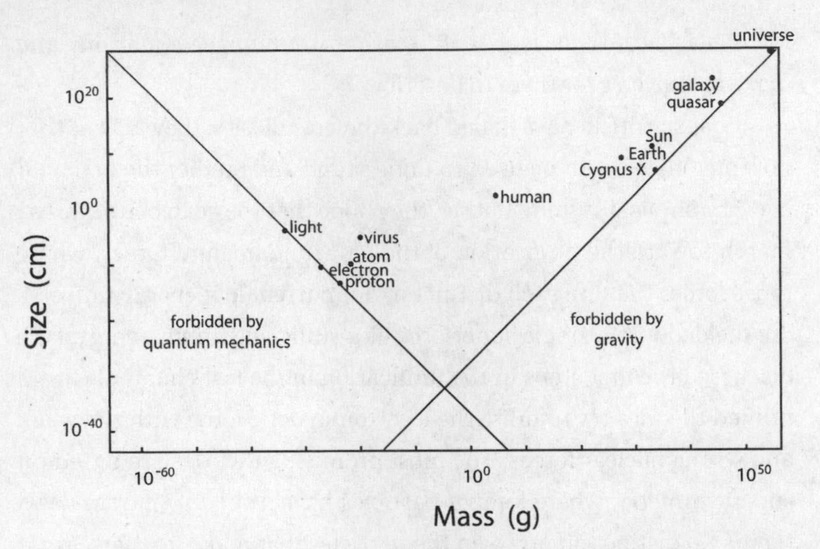

FIGURE 15.1. *The mass-energy and size of objects in the universe. The left-hand wedge is prohibited by quantum theory and the right-hand wedge is prohibited by gravity theory. The point where the diagonals meet is the Planck scale, where the laws of physics break down.*

size of a proton. That distance is called the Planck scale.[3] At that tiny size, space and distance measurement don't have any meaning. Spacetime might be foamy rather than smooth and continuous as general relativity would predict. The temperature back then was a staggering 10^{32} Kelvin, hot enough that the puny force of gravity was equal to the other forces that are far stronger in the current, cold universe. These benchmarks define the limits of both measurement and understanding (Figure 15.1).

Here's another way to think of the Planck time and the Planck scale, independent of whether we are describing the early universe. Heisenberg's uncertainty principle says evanescent or virtual particles come and go all the time, and they can be massive if their lifetime is very short. Einstein's general relativity says enough mass in a small enough space can create a black hole: a region with gravity so strong its escape velocity is the speed of light. Combining these ideas, there is a scale small enough for virtual black holes to exist. It's the Planck scale. In the first instant of time after the big bang, space was curved on the scale of particles, particles had the attributes of black holes, and space-time distortions were governed by quantum uncertainty. Think of a pot of water on a hard boil and that's a very crude approximation for the seething space-time foam of the very early universe.

Einstein's failure to derive a theory of everything was a sobering reminder of how hard the problem is.[4] Physicists have found a way forward by making an audacious leap beyond the Standard Model.

Think of a guitar string. It's under tension, and depending on the amount of tension and how it's plucked, it gives off different sounds. Different harmonics or excitation modes of the string make different musical notes. Now imagine the string is freed from the guitar but it still has tension so it can oscillate and vibrate. Some of these floating guitar strings remain open, with both ends free, and others are closed, forming a loop. Now imagine that these strings are invisibly small and far smaller than any particle—they are one-dimensional objects on the Planck scale, about 10^{-35} meters. That, in a very small nutshell, is what physicists came up with for unifying gravity and quantum mechanics.[5]

In string theory, each different particle is a mode of vibration or "note" of an invisibly small string. The open and closed strings can interact and combine. As a string moves through time, it traces out a sheet or a tube, depending on whether it's open or closed. The vibration modes of the string generate the mass, spin, and charge that a conventional particle would have. By adding supersymmetry to the mix, strings can describe both particles and forces, so an electron is a vibrating string, but so too is a graviton, which carries the gravity force. The theory, renamed superstring theory, naturally includes gravity as well as all the interactions of particles in the Standard Model (Figure 15.2).

The promise of the new theory was so great that many smart young

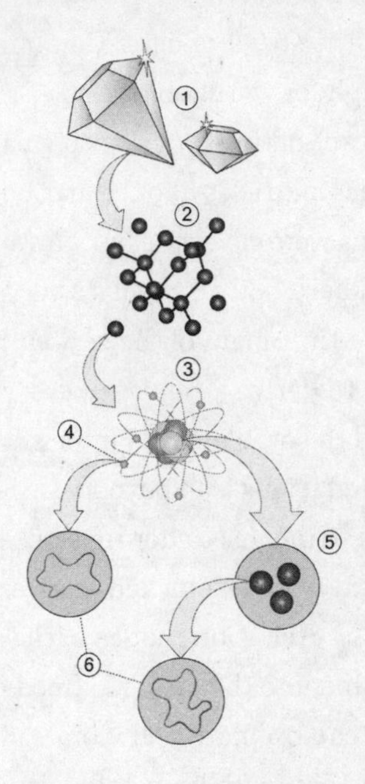

FIGURE 15.2. *The various levels of magnification in the structure of matter. A crystal (1) is made of a lattice of molecules (2), which is made of atoms (3), which all include light electrons (4), and heavier protons and neutrons. Both neutrons and protons are composed of fractionally charged quarks (5). All subatomic particles may be made from tiny one-dimensional entities called strings (6).*

physicists were willing to learn the gruelingly complex and abstract mathematics needed for a quantum theory of interacting strings. But there were two problems and one big surprise. The first problem was that the size and the energy scale of strings are many trillions of times beyond what can be probed by lab experiments or accelerators, so there seemed to be no way to test the theory. Second, detailed work in the 1980s showed there were five different types of string theory, each fiendishly difficult to work on, with no apparent way to decide among them.

And the surprise? All of the supersymmetric string theories involved 10 space-time dimensions!

It doesn't sound much like progress. Having hoped to reach the top of the mountain after a difficult climb, we find ourselves confronted with a set of five vertiginous peaks, with no indication of which is the true summit. Not only that, we're supposed to believe that the familiar three dimensions of space and one of time are fronting for six hidden space dimensions. Yet it's not a ridiculous idea. Successful physical theories are always mathematical in nature, and hypothesizing curved space in higher dimensions dates back to the middle of the eighteenth century.[6] If we assume that six dimensions are curled up like a sock, or compactified, then we might not be aware of their existence. In string theory, those hidden dimensions only manifest on the Planck scale, yet they're built into nature because aspects like the charge of an electron arise from motion in the extra dimensions. More generally, depending on how they vibrate in the hidden dimensions, strings may appear in three space dimensions as matter or light or gravity.[7]

The next breakthrough occurred in the 1990s. Theoretical physicists at several universities, most notably Edward Witten at Princeton, realized that what they thought were five different string theories were actually different ways of looking at the same theory.

Imagine each theory was analogous to a large planet where we only knew of a small island located somewhere on the planet. It's so difficult to explore the theory mathematically that we don't know what else might be found on the planet. As techniques improve, we're able to travel around the seas on each planet and find new islands. Only then is it realized that the five string theories are actually islands on the same planet, not different planets! There's one underlying theory of which all the string theories are different manifestations (Figure 15.3).[8] Witten dubbed the underlying concept "M-theory." He was coy about what the "M" stood for; suggestions have included "mystery," "magic,"

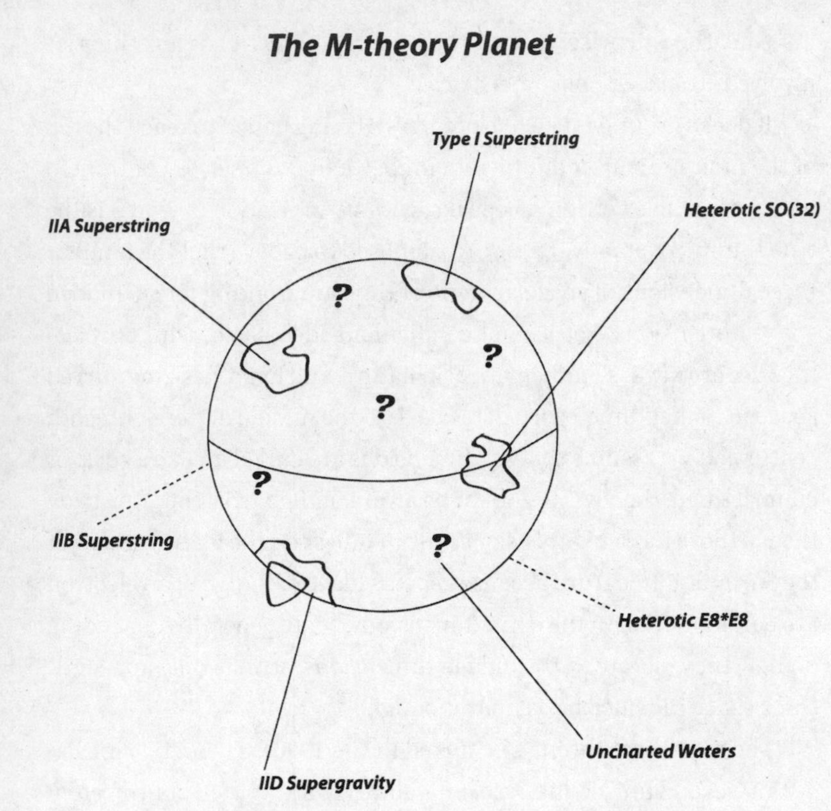

FIGURE 15.3. *The various string theories represent islands in a largely unexplored theoretical and mathematical landscape. It turns out that they are all related through an underlying construct, called M-theory.*

"monster," "matrix," "mother" (as in the "mother of all theories"), and "membrane."

Many theorists think that a formulation using membranes will be the most productive way forward. It includes an extra dimension to make 11 (but really, who's keeping count?). The fundamental object is a membrane or sheet rather than a string. Like a drinking straw seen at a distance, membranes would appear like strings if the eleventh dimension were curled into a small circle. The general object, called a "brane," can range in dimension from zero to nine. A point is a zero-brane, a line is a one-brane, a surface or membrane is a two-brane, and so on up to dimensions that have no name to describe them. M-theory is a beast to work with because the number of different types of membranes in different dimensions increases exponentially. In three dimensions, the theory has to deal with solid objects containing interconnected holes. Over 6 billion three-dimensional knots have been tabulated and knot theory is needed to classify them. You can probably imagine just how gnarly this gets in 11 dimensions.

The number of distinct physical states in a theory with 10 or 11 dimensions is essentially infinite. However, the number of states that correspond to a universe roughly like ours, with just four dimensions of space and time, is a more tractable number, "only" 10^{500}! Each of these states has hidden dimensions on the Planck scale and a unique and different set of forces and particles on the macroscopic scale. This situation is called the string theory "landscape."[9] The question arises: what if our universe represents one of these states, while the others represent other possible universes, radically different from each other and from ours?

If you're dizzy at this point, you have every right to be. Aficionados and proselytizers for string theory like Brian Greene sound euphoric and giddy as they describe its mathematical elegance and beauty.[10] Greene is a New Yorker, the son of a vaudeville performer and high school dropout. He's an expert on the shapes taken by compactified higher dimensions, writing papers with snappy titles like "Duality in Calabi-Yau Moduli Space." He extends himself beyond a mere 10 or 11 dimensions by writing popular books, children's books, and being the founder of the World Science Festival, an annual nexus of art, music, theater, and science that started in New York in 2008.

Lisa Randall, another "rock star" of string theory, was one of his classmates at Stuyvesant School in New York. Randall is supersmart, eloquent, and—unusually in the Y chromosome world of theoretical physics—a woman. She's been the first woman to get tenure in the physics departments of Princeton, MIT, and Harvard. Like Greene, Randall stretches into extra dimensions, having written a libretto for the opera *Hypermusic: A Projective Opera in Seven Planes* with the composer Hector Parra. They have both cleared the high popular culture bar of being interviewed on *The Colbert Report*. Randall and Greene are the intellectual, if not the temporal, progeny of string theory guru Ed Witten.

String theory notched up an important success in 1996 when it was used to explain the surprisingly large entropy of black holes. For the first time, string theory was used to derive a result from "classical" physics, demonstrating the explicit connection between strings and gravity. However, string theory has been subjected to a substantial backlash as its enormous promise seems unfulfilled.[11] In the past decade, several hundred exceptionally talented theoretical physicists have written thousands of papers on string theory, yet it's not been tested so it can't be confirmed or refuted. Some argue that the complexity and nonuniqueness of the theory means that it *can't* be tested and therefore isn't truly science.[12]

Is string theory a theory of everything or a theory of nothing? Is it beauty or the beast? As usual in these heated academic debates, the truth is likely to be somewhere in between. String theory provides real insights into the unification of quantum mechanics and gravity and, although hidden dimensions can't be created in the lab, the hypothesis makes predictions of effects at lower energies. The Large Hadron Collider will test supersymmetry, for example, which is a key component of string theory. We should listen to Edward Witten, who became a full professor at age twenty-nine, won the MacArthur "genius" award two years later, and was the first physicist to win the Fields Medal, considered the "Nobel Prize" of mathematics. "String theory is a part of 21st century physics that fell by chance into the 20th century," he said, and noted that the technical tools required to create the theory were still being invented. We're exploring a vast undiscovered country and are still building the vehicles that we need for the journey. It will take time, and patience is required.

Finally, there's always the lurking doubt, alluded to earlier, that the fundamental truth is beyond our intellectual grasp, just as quantum mechanics is incomprehensible to a dog. Flawed and finite, we may have reached the limit of our understanding.

THE UNIVERSE NEXT DOOR

As E. E. Cummings once said, "listen: there's a hell of a good universe next door; let's go." That sounds like poetic fancy and wishful thinking, but was he on to something? The theory of chaotic inflation proposes that the universe we know and love began as a quantum event. The primordial vacuum that spawned it was a roiling sea of quantum fluctuations, and each of them had (or has, the process may be eternal and ongoing) randomly different physical properties. Most of these fluctuations were stillborn or never evolved into anything more than an iota of space-time. Some of the fluctuations may have inflated to become vast and old expanses of space-time. String theory, which was derived without any reference to cosmology, provides a landscape for the number of possible states of multidimensional space-time (Figure 15.4), and gives an estimate of the number of physically distinct states that are "familiar" in terms of having four space-time dimensions: 10^{500}.

The conjunct of these two largely untested theories has led to an idea called the multiverse. The multiverse hypothesizes a vast number of parallel universes, unobservable by us, each with randomly different properties and laws of physics. As outlandish as it sounds, the concept has the support of a number of eminent cosmologists, including

FIGURE 15.4. *In M-theory, there are six or seven dimensions that are hidden on the Planck scale, in addition to the familiar one dimension of time and three dimensions of expanding space. This is a visualization of a multidimensional surface called a Calabi-Yau manifold that might be relevant to the hidden dimensions.*

Martin Rees, England's Astronomer Royal and member of the House of Lords. He has written: "Our universe may be just one element—one atom, as it were—in an infinite ensemble: a cosmic archipelago. Each universe starts with its own big bang, acquires a distinctive imprint (and its own physical laws), and traces out its own cosmic cycle. The big bang that triggered our universe is, in this grander perspective, an infinitesimal part of an elaborate structure that extends far beyond the range of any telescopes."[13]

Andrei Linde and his colleague Vitaly Vanchurin have recently tried to calculate the number of possible universes and the number dwarfs the number of vacuum states corresponding to familiar universes in string theory. They come up with a frightening number: 10 to the power of 10 to the power of 10,000,000.[14] Even if the number isn't as stupendous as this, quantum fluctuations and inflation can spawn so many universes that every imaginable combination of physical properties must occur somewhere.

With the multiverse we seem to have taken leave of our senses and entered into wild speculation. The last sense available to us is touch. In the everyday world the sense of touch is derived from forces that operate on the scale of an atom. If the universe was once the size of an atom, then a sentient entity in an encompassing space-time might be able to touch our universe, the way we within it would touch a grain of sand. And in the multiverse they might "tweeze" our universe from among many others, because it glitters with potential. All we know for sure is that we can touch our universe with our minds, and that's the thrill of cosmology.

It used to be simple: the universe was everything that there is. Now there seem to be different layers of observation and reality. The first or top level is the observable universe. This is the volume accessible to telescopes, which stretches 46 billion light-years in every direction and contains about 100 billion galaxies. All observations in cosmology take place in this arena; this is very firm ground.

The next level consists of regions of space that we can't observe that are still part of the same big bang "event" as the observable universe. These regions have become visible as the universe decelerated; light from more distant regions reached us as the universe slowed down. But this rosy scenario has been torpedoed more recently by dark energy, which is wrenching galaxies from our grasp until all but the closest objects will vanish from view. How large is the universe at this level? It could be infinite. Space is flat and nothing in our observable patch gives a sign that it doesn't go on without end. (Although we should recall Einstein, who said, "Two things are infinite: the universe and human stupidity, and I'm not sure about the universe.") If space is endless and it teems with galaxies, stars, and planets, Max Tegmark has counted all possible quantum states to deduce the distance to your nearest identical twin.[15] It's 10 to the power of 10 to the power of 28 meters. Ten to the power of 10 to the power of 118 meters away is an identical observable universe. But your doppelganger may be a lot closer given the apparent naturalness of planet formation and biological evolution.

The third level is chaotic or eternal inflation. The parallel or "bubble" universes generated 10^{-35} seconds after the big bang are unobservable because they're expanding so fast that their light could never reach us. This multiverse is very diverse because the bubbles vary in their initial conditions and in physical laws that we consider immutable. Inflation is a theory that makes concrete predictions, so this regime is still within the realm of conventional science, but only barely.

The last level is highly abstract and might not be testable. Several variations have been presented, any or all of which might be valid.

The "many worlds" interpretation of quantum mechanics argues that nature isn't inherently probabilistic. It says that each time there are many viable possibilities, the world splits into many worlds, one world for each possibility.[16] In each world, everything is identical except for one different outcome (Figure 15.5). From then on, they each develop independently and no communication is possible among them, so people living in those worlds are unaware that it's going on. In this way, the "world" branches endlessly. What is "now" to us lies in the pasts of an infinite number of possible futures. Everything that can happen, does, somewhere.

It's easy to map this into multiverses, where each world is a distinct universe with different properties. The idea appeared first in science fiction not in physics, from Olaf Stapledon's 1937 novel *Star Maker*: "Whenever a creature was faced with several possible courses of action, it took them all, creating many distinct histories of the cosmos. Since there were many creatures and each was constantly faced with many possible courses of action, and the combination of their courses was innumerable, an infinity of distinct universes exfoliated from every moment of every temporal sequence." This seems like a crazy violation of Occam's razor, the dictum that physical theories

FIGURE 15.5. *In the Schrödinger's cat experiment, a cat in an enclosed chamber has an equal probability of dying based on the potential radioactive decay of one atom, which will trigger a hammer to smash a flask containing poison. In the "many worlds" interpretation of quantum mechanics, both outcomes happen in different versions of the universe, which are real but cannot communicate with each other. As time passes, the branching and number of universes multiplies in an infinite progression.*

should be simple if possible. But the parallel realities under the many worlds umbrella are part of a single wave function, so the underlying idea is actually very simple.

M-theory, where the fundamental entities have varied numbers of space-time dimensions and manifest physical laws wildly different from our own universe, is another abstraction. The landscape of universes in M-theory is essentially infinite.[17] Tegmark has gone even further and argued that the multiverse doesn't have to be tethered in variations of existing physical laws. Why not throw away the physics rule book and look to mathematics as the basis for physical reality? As I write this I recall Gaspar's piercing eyes and challenging stare 35 years ago.

There's a dichotomy in thinking about the correspondence between physics and math that goes back to the Greek philosopher Plato and his student Aristotle. Aristotle's view, which morphed into the modern scientific method, is that math is no more than a tool, an approximate description of reality described by physics. Plato's view is that math is the true reality and observers perceive it imperfectly. This is counterintuitive since a mathematical structure is an abstract and immutable construct that exists outside of space and time. Tegmark takes an extreme Platonist position that all mathematical structures also exist physically and that each mathematical structure corresponds to a parallel universe.[18] It's audacious, shifting the question, Where do the laws of physics come from? to the new question, Where does mathematics come from? but it contains the unsettling notion that only something that's mathematically describable can be real!

Is the multiverse idea real science? To be good science, it would have to explain what we know and make unique and testable predictions in new areas of explanation.[19] Given its tenuous epistemological status, why has the multiverse idea attracted so much support and attention? Mainly to explain why the universe is the particular way it is.

We saw two chapters ago that the universe is fine-tuned in the sense that if many of its attributes had been different, life as we understand it wouldn't have arisen. Counterfactual universes that have different cosmological parameters or obey different laws of physics are unlikely to host long-lived stars, chemistry, planets, or biology. For example, if the dark energy had been much stronger matter would have dispersed so quickly in the early universe that no structures would have formed. So it goes with many other physical constants—tweak them a bit and atoms don't form or get stable, stars don't shine, and the universe is a junkyard of chaotic matter and radiation. Sterile and dead.

The argument that the universe is "built for life" is called the anthropic principle. It's controversial for many scientists and philosophers since it elevates life to a privileged position in the universe. (In particular it elevates observers, since bacteria see no need to explain fine-tuning.) It's also been critiqued for not making predictions and being post hoc.

Let's scrutinize the argument. At one level it's a tautology. There's no reason to be surprised you exist. I'm not. If the universe didn't have properties consistent with the emergence of carbon-based, intelligent life, neither of us would be here to be surprised. If you were a bridge player who realized the odds of being dealt a particular hand are 1 in 600 billion, it would be absurd to look at your hand and marvel at how unlikely it was.

There's a real argument over how fine-tuned the universe is in terms of permitting atoms and stars to exist. Moreover, without a general theory of biology, there's no agreement on exactly how fined-tuned the universe is for life. It's also likely (or at least hoped for) that the 19 parameters of the Standard Model and the 10 or so primary attributes of the big bang model are related by a deeper underlying theory. Fine-tuning is overestimated if all these parameters aren't independent.

New life was breathed into the anthropic principle by the possibility that our universe is one member of a vast ensemble. In a multiverse, the physical properties vary widely and we happen to live in one of the rare universes with properties hospitable for intelligent life.[20] Our laws of physics are in fact local "bylaws" that favor life. We're winners in a cosmic lottery.[21] As before, the fact that our universe is biofriendly is not surprising, it's just the result of observational selection. Explaining fine-tuning and our existence with the multiverse neatly rebuts theists who argue that fine-tuning is the sign of a Supreme Being or Designer. However, the story won't hang together unless evidence is found for a vast ensemble of universes. As distasteful as it may be to a scientist's unslakable thirst for explanation, the biofriendliness of the universe may be a fluke, with no deeper significance.

Or, there may be nothing to explain because it was never real. The technocrat's version of theism is the idea that we're all living in a computer simulation—a high-end video game. Nick Bostrom proposes this scenario with vigor, and a twinkle in his eye. He's the director of the Future of Humanity Institute in Oxford, and a card-carrying transhumanist. Transhumanism is a broad cultural movement that affirms and explores the idea of enhancing human capabilities and extending life, perhaps infinitely.

A number of futurists have predicted that biological evolution will be superseded by a postbiological phase based on nanotechnology and computation, but Bostrom is a philosopher who has taken such ideas to their logical, and somewhat uncomfortable, conclusion. A modest extrapolation of our current exponential improvement in computer power and speed projects to a time within a century when we could replicate the electrochemical complexity of the brain in silicon.[22] Any alien civilization with this capability could easily replicate the entire history of human thought processes—called an "ancestor" simulation. There are roughly 100 million habitable planets in the Milky Way alone, and there have been up to 12 billion years or more for intelligent creatures on any of these planets to reach or surpass our stage of development. These futurists argue that we're unlikely to be the first or the only civilization with this capability.

Bostrom frames a logical argument based on three propositions, at least one of which *must* be correct. One: Almost all civilizations go extinct or destroy themselves before gaining the capability to create simulated creatures like us. That's a gloomy option because we're approaching that stage. Two: Almost all civilizations choose not to create simulated creatures, even though they could. That's possible, but the $50 billion a year gaming market on this planet indicates a strong desire of humans to create and manipulate artificial entities. Three: Nothing is real, everything is an illusion, and we actually live inside a simulation.[23]

Rebutting the third proposition is surprisingly difficult. Any simulation constructed by a far superior race wouldn't be glitchy, as it was in the movie *The Matrix*. There's no reason we'd know we're simulated unless the creators wanted us to. Your conviction that you're made of flesh and blood and free will is part of the simulation. Since it's easier and cheaper to create computational life-forms than biological organisms, by the Copernican Principle there are many more simulated than real creatures. OK, this argument is more of a provocation than a serious suggestion, but it's no more unfounded or illogical than the multiverse or hidden space-time dimensions.

Perhaps we live in a *Truman Show* type of simulation. The advanced civilization has created an environment for us that maintains a superficial appearance of physical reality. The surface of the Earth is lovingly rendered, plus a few places we've been beyond like the Moon and Mars. Otherwise our planet is hollowed out and stars and galaxies are part of a sophisticated planetarium show. The verisimilitude just needs to be "good enough." Another variation involves a civilization of much greater capability, such that they can create baby universes. Our biggest nuclear bomb packs a punch of 10^{17} joules, or the rest-mass energy of a kilogram of matter. As we saw in the last chapter, a quantum fluctuation equivalent to a 100,000 times smaller mass could generate a universe from the vacuum. We don't have this technology yet but someone or something else might. If we're living in a simulation the idea of cosmic origins is moot.

But for now—for the sake of argument and to conclude our story—let's assume that we and the universe we inhabit are real.

ENDLESS CREATION

At historical sites in Italy you'll often see the outline of a two-headed man above doorways and gates. That's Janus, god of doorways, beginnings, endings, and time. His name gives us the first month of the year. The root of the name suggests an origin in the ancient Middle East, and by the time of the founding of Rome, Janus was the most important god in the archaic pantheon. In civic life, Janus was invoked at marriages, births, plantings, and the beginning of the harvest. He represented the middle ground between civilization and barbarity, the countryside and the city, and youth and adulthood. Janus also symbolized change and transition in more abstract terms, such as the beginning of life, gods, and the universe. With one head looking forward and one head looking backward, Janus is with us now, as we reach the end of our search for the beginning.

The universe around us is middle-aged, active, and resourceful, with tens of billions of years before the lights go out. Looking back, we see someone in their prime, building empires—a bit obsessively—but with speed and efficiency. Earlier still, we see a teenager at the height of their powers, in halcyon days of hangovers and harebrained schemes. Looking farther we see a child, smooth-skinned and tentative. Finally, the universe is just a baby, all promise and potential. With its last (or maybe its first) movement, the infant touches the door, and it opens.

Onto what? Coal-black void? A mirror-image journey through time? An adjacent universe, imperceptibly different? Or a darkened corridor with a single door opposite? Or perhaps a corridor with an endless row of doors on either side? Perhaps the room beyond the door has mirrors on three sides, so the endless doors are an illusion?

In our journey back to the beginning, the Planck era is when theories fail and we enter the realm of speculation. We've seen that "eternal" inflation finesses the idea of a big bang by hypothesizing myriad bubbles of space-time, most of which are stillborn and some of which can grow into universes filled with matter and radiation. Inflation has notched up some successes in explaining our universe, but there's no evidence for eternal inflation, no explanation for the existence and size of dark energy, and no connection between the rapid expansion in the first tiny fraction of a second and the accelerating expansion that kicks in when the universe is 8 billion years old. The door's wide open for competing ideas.

In 1999, Neil Turok and Paul Steinhardt were listening to a lecture on M-theory at a cosmology conference in Cambridge. These sharp young theorists simultaneously had the same thought. Suppose our universe is three-dimensional space embedded on a higher dimensional brane. Branes will generally also move in higher dimensions. What happens when two branes collide? Is it possible for the collision of two branes to be the source of energy for embedded three-dimensional space?

These are difficult calculations. Turok and Steinhardt talked about the idea and did some rough calculations on the train to London (no doubt inducing raised English eyebrows from commuters nearby), and they continued the discussion at a play that evening—appropriately, it was Michael Frayn's *Copenhagen*, about the development of the quantum theory in the 1930s. They had each felt uneasy about the unspecified initial conditions in the inflation idea, and the fact that the big bang is supposed to be the origin of both space and time. To them, that was inelegant and arbitrary. The big bang was a dead end, a door with a sign on it that said, Keep Out. They were unimpressed by anthropic arguments, finding them to be nonpredictive and a cop-out.

So they revived the idea of a cyclic universe, a universe with no end and no beginning.[24] The cyclic universe model was first proposed in the 1930s: a finite universe that expanded, collapsed, bounced, and did that endlessly. Richard Tolman soon pointed out that the model was flawed. Each bounce makes more radiation and so the universe contains more stuff. According to Einstein's relativity, each bounce is bigger and lasts longer than the one before. Tracing into the past, the duration of each bounce is shorter and shorter until it shrinks to zero. So the universe has a finite age and the big bang isn't avoided. In the 1980s astronomers showed there wasn't enough matter to overcome and reverse the expansion, seeming to put a final nail in the coffin of the cyclic universe. But if our three-dimensional world is a surface or brane set into space with an extra dimension, Turok and Steinhardt thought they could explain most features of the universe.

Here's how it might work. String theory explains particle physics pretty well if two branes are separated by a tiny gap (in a higher dimension). The tension between them is a force like dark energy. Quantum fluctuations wrinkle the branes and so provide the seeds for structure formation. The branes collide in a violent process that makes lots of radiation and particles. When the branes are close together, the tension between them, and therefore dark energy, is small, but it grows as they separate until it stretches the universe in an accelerating expansion we see today. As the branes move closer they continue to stretch as they head toward their next collision, explaining why the universe is so flat and smooth. Then they collide again, re-injecting our universe with matter and radiation. A cycle takes about a trillion years. The strength of the dark energy declines in this model; after many cycles it's expected to have the low value we see today.

Is this another clever just-so story? The cyclic universe model explains many features of the universe that motivated inflation. It avoids any singularity and there's no big bang. On the other hand, branes are predicted by string theory, so if string theory unravels, so does the cyclic universe, and it's very difficult to calculate what happens when branes collide in string theory. However, the theory makes predictions, and it can be tested by observations of the gravity waves that ripple through space and time from the early universe. If inflation is right, the Planck microwave satellite and other experiments should detect gravitational waves. If the colliding brane model is correct, these experiments should detect nothing.

If the universe is eternal, we're humbled by our insignificance in both space and time. We are reminded of ancient truths. In the Hindu and Buddhist traditions, the Vedic texts describe a cyclic universe and lay out the measures of time. A kalpa, or "day" of Brahma, is 4.3 billion years, of similar magnitude to the age of the universe. A grand kalpa, or "year" of Brahma, is 3 trillion years, similar to the cycle of time in the colliding brane model. Brahma is reputed to live 100 times longer, an unimaginable 300 trillion years.

I sometimes go to northern India to teach Buddhist monks cosmology. It's a program inspired by the Dalai Lama's desire to have his monks understand the science of the modern world. The monks are gentle and whimsical people, and dedicated learners. I learn as much from them as they do from me. Once, we were talking about the age of the universe and I said that even astronomers have trouble intuiting vast spans of time. A senior monk, or Geshe, told me how the Buddha had described the longest measurable span of time. According to tradition, the Buddha said: imagine a granite mountain 10 miles high. A dove flies past it once a year, brushing it lightly with its wing. The life of Brahma is how long it would take to wear away the mountain.

The Buddha also said that everything is in flux—what you know now will not be true tomorrow. We don't know if scientists will be able to answer our deepest questions about the universe. Albert Einstein said: "All our science, measured against reality, is primitive and childlike—and yet it is the most precious thing we have." We're a young species and our science is also immature. We're too old to be frightened of the darkness, but not clever enough to fully understand it. Our journey is just beginning.

—

The layers of delirium peel back. I lie in bed with a sheen of sweat on my brow. My breathing slows. I can hear birdsong, the water pipes sighing, and the hum of distant traffic. Slowly, I embrace the crispness of here and now.

In my dream, I've been to places more fantastic than Oz. Beyond the sheltering sky, beyond the Moon that tugs the lulling seas, away from the Earth in its curved corkscrew orbit of the Sun and its large meander around the Milky Way. I've seen living and dead worlds, too many galaxies to count, and the death throes of matter at the edge of black holes. I've ventured into the cosmic fireball and witnessed the birth of matter. I've seen structure dissolve into fundamental particles and space and time become gnarled beyond recognition.

Back in my skin, I open my senses and my experience is real enough to get me through today and many days beyond. I'm 70 kilos of blood and gristle, amazed to be meat that thinks. I tightly grasp this universe, this space, this time, this life.

NOTES

1: SEPARATED AT BIRTH

1 Such exquisite accuracy depends not only on superfast electronics but on being able to measure time with even greater accuracy. Since 1967, the second has been defined as the duration of 9,192,631,770 waves of radiation from a hyperfine transition of the ground state of the cesium-133 atom. The best atomic clocks have an accuracy of a few parts in 10^{16}, and they are easily able to calibrate timing experiments with an accuracy of a few parts in 10^{12}.

2 Despite the concerns, the Saturn V rocket worked almost flawlessly. There were 13 successful launches, and only two before the first Apollo crew was sent to the Moon. NASA had to orchestrate 20,000 private firms and 300,000 people in the development effort; the levels of complexity and technological challenges were unprecedented. In 10 years, the agency's launch capacity was boosted by a factor of 10,000. It's as if the Wright brothers had progressed in a decade from their first flyer to a supersonic aircraft.

3 The excitement and broad sweep of the Apollo program is beautifully encapsulated in Andrew Chaikin, *A Man on the Moon: The Voyages of the Apollo Astronauts* (New York: Penguin, 1994). Astronaut perspectives on Apollo include Eugene Cernan and Don Davis, *The Last Man on the Moon: Astronaut Eugene Cernan and America's Race in Space* (New York: St. Martin's Press, 1999) and Michael Collins and Charles Lindbergh, *Carrying the Fire: An Astronaut's Journeys* (New York: Farrar, Strauss, and Giroux, 2009). For a trainspotter's summary, see Richard Orloff and David Harland, *Apollo: The Definitive Sourcebook* (Berlin: Praxis, 2006).

4 J. Bogoshi, K. Naidoo, and J. Webb, "The Oldest Mathematical Artifact," *The Mathematical Gazette* 71, no. 458 (1987), p. 294.

5 M. Rappengluck, "Paleolithic Timekeepers Looking at the Golden Gate of the Ecliptic," *Earth, Moon, and Planets* 85 (1999), p. 391.

6 C. Cheng, "Model of Causality in Chinese Cosmology: A Comparative Study," *Philosophy East and West* 26, no. 1 (1976), p. 3.

7 E. G. Richards, *Mapping Time: The Calendar and its History* (Oxford: Oxford University Press, 2000).

8 The use of the Moon by the world's religions might seem innocuous but it still has the power to inflame. Issues relating to the lunar calendar caused Christianity to veer away from Orthodox Judaism and Eastern Orthodox principles over the centuries and Muslim sensibilities have been inflamed by claims of some Christian scholars that their deity is a pagan "moon" god. Since all the monotheistic religions have been supple enough to co-opt pantheistic and pagan belief systems to gain adherents, the arguments seem doomed to rage without end.

9 Aristarchus calculated the relative sizes of the Earth, Moon, and Sun, which depended on a more difficult observation of the angle between the Sun and the Moon when the Moon is half-lit. He underestimated the distance to the Sun and its size as a result, but he was still able to infer that the Sun was much larger and probably more massive than the Earth. Soon afterward, Eratosthenes used geometry to estimate the size of the Earth. Thus, the early Greeks had assembled all the components of a modern view of the Sun-Earth-Moon system: the fact that they were spheres, their sizes and distances, and the likelihood of the Sun being at the center of the system.

10 E. Chudler, "The Power of the Full Moon. Running on Empty?" in *Tall Tales About the Mind and Brain: Separating Fact from Fiction*, ed. S. Della Sala (Oxford: Oxford University Press, 2007), p. 401.

11 J. Rotton and I. Kelly, "Much Ado About the Full Moon: A Meta-Analysis of Lunar-Lunacy Research," *Psychological Bulletin* 97, issue 2 (1985), p. 286.

12 D. J. Stevenson, "Origin of the Moon—The Collision Hypothesis," *Annual Review of Earth and Planetary Sciences* 15 (1987), p. 271.

13 E. Belbruno and J. R. Gott, "Where Did the Moon Come From?" *The Astronomical Journal* 129, no. 3 (2005) p. 1724.

14 J. Touma and J. Wisdom, "Evolution of the Earth-Moon System," *The Astronomical Journal* 108, no. 5 (1994), p. 1943.

15 Serendipity played a role in deciding who got to make history. The Apollo astronauts' flight chief, fellow astronaut Deke Slayton, had a system of rotating crews. Neil Armstrong was the commander of *Apollo 11* because he'd been backup commander of *Apollo 8*. In fact, Slayton offered the job

to highly experienced Apollo 8 Commander Frank Borman, but Borman was a team player and turned it down. Plus his wife was pressuring him to stop flying and risking his life. Jim Lovell went to the Moon twice without ever setting foot on it. Nine Apollo astronauts never got to the Moon due to the cancellation of the program in 1972. While the Apollo 13 crew had a close call, the only fatalities in the program were the three crew members of *Apollo 1*, who died in a fire on the launch pad.

16 Francis French and Colin Burgess, *In the Shadow of the Moon: A Challenging Journey to Tranquility* (Lincoln: University of Nebraska Press, 2007).

17 It's a losing strategy to debate the Moon deniers since they ignore the vast bulk of the evidence that the Apollo program was real, and home in on the few aspects where the data seem to be interpretable in other ways. Conspiracy theorists have a field day with the fact that NASA accidentally erased much of the TV broadcast tape of the Apollo 11 landing. The most thorough rebuttal of the tropes of lunar deniers is contained in Phil Plait's "Bad Astronomy" blog, now hosted by *Discovery Magazine* online.

18 Charles Murray and Catherine Cox, *Apollo: The Race to the Moon* (New York: Simon and Schuster, 1989).

19 As with any frontier technology, only brave and rich people participate when the activity is expensive and dangerous, as was true in the first decade of commercial aviation. But as economies of scale kick in and the technology matures, it will come down the price curve to be within the reach of a much larger audience. Spaceport Associates did a web survey (the "Adventurer's Survey") in 2006 that showed only 5 percent of Americans are interested in suborbital or orbital flight at current price levels, but almost all would consider it if the costs were reduced to $25,000 for suborbital flight. When I teach freshman and sophomores at the University of Arizona, I administer a survey, and based on the number saying they would do a once-in-a-lifetime suborbital flight for $50,000, the revenue projected for space tourism just from the U.S. college-educated population is $30 billion per year. That comfortably exceeds the annual sales of DVDs and the movie box office, showing that space travel has the potential to be on a par with mass-market entertainment.

2: PLANETARY ZOO

1 Felix Baumgartner plans to free-fall from 120,000 feet in a supersonic space suit. "Fearless Felix" is noted for base-jumping from the world's tallest building and highest bridge, and for crossing the English Channel in free fall using a specially designed fiberglass wing. His main rival is Michel

Fournier, a veteran of 8000 parachute jumps who has already made 100 from over 25,000 feet in preparation for his fall from space. In a heart-breaking reminder of the dangers of near-space, the space shuttle *Columbia* with its crew of seven was lost when it broke up 40 miles above Texas, moving at Mach 20. Yet dozens of pages of the diary of Israeli astronaut Ilan Ramon somehow survived intact and are now on display in a museum in Jerusalem.

2 NASA is a labyrinthine organization made up of nearly a dozen centers, and information of their missions is similarly dispersed and fragmented. The best single source for information on space probes, including all the international players, is Wikipedia's summary article "List of Solar System probes."

3 As in many aspects of space travel and engineering, Russians were pioneers. During his time as a junior army officer in World War I, Yuri Kondratyuk filled four notebooks with his ideas on interplanetary flight. Suggestions included using modular orbiters and landers, as was eventually done with the Apollo program. His 1919 paper "To Whoever Will Read This Paper in Order to Build an Interplanetary Rocket" contained the idea of gravity assist. It was first implemented in 1959 when the Soviet probe *Luna 3* photographed the far side of the Moon.

4 The Voyager care packages each include a "gold record" designed as a message to any alien civilizations that might encounter them. A team led by Carl Sagan made the selection of images, music, and diverse sounds of Earth that are encoded in the grooves of each phonograph record. Some argued that sending a technology that's nearly obsolete on Earth is not very forward-looking, but the team has recently rebutted correctly that the longevity of modern digital technology—CD, DVD, hard drive, and flash drive—is not guaranteed, while a physical gold record should retain its analog information intact for millions of years. See C. Sagan, F. Drake, A. Druyan, T. Ferris, J. Lomberg, and L. Salzman, *Murmurs of Earth: The Voyager Interstellar Record* (New York: Random House, 1978). *Voyager 1* is the probe that took a "family picture" of the planets and a shot of the Earth behind the rings of Saturn that led Carl Sagan to coin the phrase "Pale Blue Dot."

5 Almost every human culture has used an intermediate division of time between the day and the month, which are aligned to astronomical cycles, although it varied around the world between 4 (Central Africa) and 10 days (Egypt). The Western tradition of a 7-day week began with the Jews while they were being held captive in Babylonia in the sixth century BC. Names are assigned to the five planets based on Roman gods since our calendar is a concoction invented under Julius Caesar and modified slightly since then. In Romance languages (and in Hindi, Japanese, and Korean) the

planet names are clear, but in English four Roman names were replaced with Norse gods: Tiw, Woden, Thor, and Freya.

6 C. S. Littleton, *Mythology: The Illustrated Anthology of World Myth and Storytelling* (London: Duncan Baird Publishers, 2002).

7 Galileo's contemporary Simon Marius claimed to have discovered at least one of the moons weeks before Galileo but the claims cannot be substantiated. Intriguingly, a Chinese historian has claimed that Gan De saw one of the moons of Jupiter in 364 BC, 2000 years before Galileo. All four of the Galilean moons are bright enough in principle to see with the naked eye, if clever use were made of a tree branch perpendicular to the orbital plane to occult the planet. Gan De compiled also a star catalog centuries ahead of the Greek astronomer Hipparchus.

8 J. Kelly Beatty, Carolyn Petersen, and Andrew Chaikin, *The New Solar System* (Cambridge: Cambridge University Press, 1999), and also M. Woolfson, "The Origin and Evolution of the Solar System," *Astronomy and Geophysics* 41 (2000), p. 1.

9 As defined by the International Astronomical Union in 2006, a planet is an object that is in orbit around the Sun (ruling out the largest moons), has enough mass to assume hydrostatic equilibrium (meaning it has a nearly spherical shape), and has "cleared the neighborhood" around its orbit. Pluto fails the third criterion. The definition does not have universal acclaim, even among astronomers. The Caltech discoverer of many trans-Neptunian objects, including Eris, wrote a book on the subject: Mike Brown, *How I Killed Pluto and Why It Had It Coming* (New York: Spiegel and Grau, 2010).

10 The use of radiometric dating assumes that neither the parent nuclide nor the daughter product enters or leaves the material after it forms. In geologically active material there can be contamination and alteration over time that compromises the method. To guard against this, multiple samples from the rocky mass are used and ideally multiple decay paths are measured for different isotopes. One gneiss from West Greenland subject to five different methods on 12 different samples gave agreement to within 30 Myr on an age of 3640 Myr. G. B. Dalrymple, *The Age of the Earth* (Stanford: Stanford University Press, 1991).

11 S. A. Bowring, "Priscoan Orthogneisses from Northwestern Canada," *Contributions to Minerology and Petrology* 134 (1999), p. 3.

12 S. A. Wilde, J. W. Valley, W. H. Peck, and C. M. Graham, "Evidence from Detrital Zircons for the Existence of Continental Crust and Oceans on the Earth 4.4 Gyr Ago," *Nature* 409 (2001), p. 175.

13 With reference to the fanciful incident recounted at the end of the last chapter, the "value" of the NASA lunar sample can be estimated by dividing the

current dollar cost of the Apollo missions by the weight of rocks returned. The answer is $1500 per gram, although in a sense they are irreplaceable since we aren't about to go back and get more. So a very heavy premium is paid for the "free" samples that land on Earth. The author may or may not have eaten 2 to 3 grams, worth thousands of dollars. Compare that to the most expensive foodstuffs, such as Almas caviar or saffron, only costing $10 to $20 per gram.

14 M. D. Norman, L. E. Borg, L. E. Nyquist, and D. D. Bogard, "Chronology, Geochemistry, and Petrology of a Ferroan Niritic Anorthosite Clast from Descartes Breccia 67215: Clues to the Age, Origin, Structure, and Impact History of the Lunar Crust," *Meteoritics and Planetary Science* 38 (2003), p. 645.

15 J. Baker, M. Bizzarro, N. Wittig, J. Connelly, and H. Haack, "Early Planetesimal Melting from an Age of 4.5662 Gyrs for Differentiated Meteorites," *Nature* 436 (2005), p. 1127.

16 T. Montmerle, J.-C. Augereau, M. Chaussidon, M. Gounelle, B. Marty, and A. Morbidelli, "Solar System Formation and Early Evolution: The First Hundred Million Years," *Earth, Moon, and Planets* 98 (2006), p. 299.

17 The formation models suggest that the Sun formed as part of a cluster of a few thousand stars, similar to the Orion nebula 1300 light-years away and its embedded star cluster. In a star cluster the most massive stars only live a few million years so their death can trigger the collapse of adjacent regions of gas into the formation of new stars. The composition and structure of the outer Solar System suggest it was influenced by the massiveness of nearby stars soon after formation. Within 100 million years, stars drift from their sites of formation so evidence of the original star cluster is lost; see, for example, S. F. Portegies Zwart, "The Lost Siblings of the Sun," *The Astrophysical Journal* 696 (2009), p. L13. The supernova "trigger" theory has some problems, however, and will be tested by the upcoming Genesis mission.

18 The demarcation between terrestrial planets and gas giants is not arbitrary; it's a division on either side of the "frost line." Inside the frost line volatile materials like carbon dioxide and methane and water are liquid or vapor, and outside it they're solid. In the cool regions beyond the frost line, embryos are larger because they're made of rocks and ices and they can more readily accrete large gassy atmospheres.

19 A resonance occurs any time the orbital periods of two planets or moons are related by the ratio of small integers. Think of it as a version of Pythagoras's "harmony of the spheres," where loud planetary music is a result of harmonics of their motion. Some resonances are unstable and transient while others can persist. In the Solar System, resonance explains the gaps

in the Asteroid Belt and in the rings of Saturn; the simply related orbital periods of Jupiter's moons Io, Europa, and Ganymede; the role of Neptune in stabilizing the orbit of Pluto; and many other subtle features. Resonance also explains why planets clear debris out of their orbits, leading to one part of the definition of a planet.

20 H. Levison, A. Morbidelli, C. Van Laerhoven, et al., "Origin of the Structure of the Kuiper Belt During a Dynamical Instability in the Orbits of Uranus and Neptune," *Icarus* 196 (2007), p. 258.

21 David Portree, "Humans to Mars: Fifty Years of Mission Planning," in *NASA Monographs in Aerospace History Series*, no. 21 (Washington, DC: NASA, 2001).

3: DISTANT WORLDS

1 Greek cosmology was cemented by Aristotle's belief that the Earth was motionless at the center of the universe. However, the Greeks had no problem ascribing rapid motion to the crystalline sphere that carried the stars. At a distance of a million miles and orbiting once a day, the stellar sphere was whirling around at a quarter-million miles per hour.

2 His full name was Abu Abdullah Muhammad ibn Umar ibn al-Husayn al-Taymi al-Bakri al-Tabaristani Fakhr al-Din al-Razi. This Sunni theologian rivaled or eclipsed Aristotle in the breadth of his knowledge, being expert in law, jurisprudence, theology, grammar, history, ethics, metaphysics, logic, mathematics, astronomy, physics, psychology, medicine, and occult arts like astrology and alchemy.

3 English astrophysicist Arthur Eddington, who first developed the theory of stellar structure, called this issue the "whales and the fishes." In an ocean you will more easily see large creatures like whales than smaller fish, and so you will see them to larger distances and overcount them in a census. Similarly, the space around us contains many more feeble dwarf stars than luminous and massive stars, but the latter are so easily visible that we observe them through a much larger volume. So 90 percent of the 50 stars nearest the Sun are cool and red dwarfs at a mean distance of 12 light-years, while 90 percent of the brightest 50 stars are giants, supergiants, or main sequence stars more massive than the Sun, and their average distance is 200 light-years. Thus, there is almost no overlap between the two categories.

4 A hundred years earlier, the hunt for parallax had led to the discovery of another important, but unrelated, effect called aberration. In 1727, James Bradley set up a vertically oriented transit telescope in the garden of his

house in Kew, near London, and observed a seasonal oscillation in the position of a bright star. This was caused by the finite speed of light and the Earth's motion around the Sun. Imagine you're standing in the rain. There's no wind and you hold an umbrella directly overhead. If you start walking forward you have to hold the umbrella slightly ahead to keep the rain off your head. Your forward motion means the rain appears to be coming from a point slightly in front of you rather than directly overhead. A similar effect causes aberration of starlight, but the effect is small since the Earth's motion is small compared to the speed of light. Once aberration was measured and understood, astronomers resumed their search for parallax, although aberration also provides support for the Copernican model by showing that the Earth is in motion around the Sun.

5 Astronomy's most important distance unit is based on parallax. A parsec is defined as the distance of a star whose tiny apex angle is 1 arc second, where the base of the triangle is the Earth's orbit. It's equal to 3.26 light-years, 31 trillion kilometers, or 19 trillion miles. Proxima Centauri has the largest measured parallax, 0.769 arc seconds, corresponding to a distance of 1.3 parsecs or 4.2 light-years. Distance is inversely proportional to parallax angle.

6 The Greeks measured the solar parallax angle, leading to the Earth-Sun distance, at various times, but the first accurate measurement used the transit of Venus across the face of the Sun, as observed from widely separated locations on the Earth. Edmund Halley proposed the method in 1716, but did not live to see its successful use on a pair of Venus transits in 1761 and 1769.

7 The Doppler effect is a shift of the wavelength of light (or any other wave) caused by relative motion. A star moving away from the observer has its light shifted to the red; a star moving toward the observer has its light shifted to the blue. The fractional wavelength shift equals the star's speed as a fraction of the speed of light. The Doppler effect is only observed if there is a component of motion toward or away from the observer, so if a planet is orbiting a star in the plane of the sky, no shift is observed. Assuming planets orbit their stars with random orientations relative to us, some fraction of the Doppler shift will be observed, and on average observation will underestimate the planet velocity (hence the mass) by a factor of 2.

8 Radio astronomers remind optical astronomers, patiently but sometimes with exasperation, that the first exoplanets discovered were found orbiting a millisecond pulsar in 1992. Two are super-Earths and one is lower mass than Earth's Moon. They were not considered "proper" planets because the star they orbit is dead and their formation mechanism is unknown, but

given the increasing diversity of the conventionally discovered exoplanets, it's only fair to consider them as the first discovery. A. Wolszczan and D. Frail, "A Planetary System Around the Millisecond Pulsar PSR 1257+12," *Nature* 355 (1992), p. 145.

9 Absorption lines form a reference grid for measuring the velocity of the star very accurately. Even so, the detection of exoplanets required spectrographs of extraordinary stability, since any tiny wavelength shift or error would bury the signal being sought. In fact, the absorption lines in a star are not narrow enough, they are broadened by gas motions within the star. This is countered by gathering lots of light so the line center can be determined to within a small fraction of the line width. Perhaps the most important innovation was the development of a gas cell in the optical path of the telescope that imprinted absorption lines from iodine or some other element. This provided a steady and stable reference for defining zero velocity.

10 M. Mayor and D. Queloz, "A Jupiter-mass Companion to a Solar-type Star," *Nature* 378 (1995), p. 355.

11 Exoplanet transits have been seen with telescopes as small as 4 inches across and amateur astronomers play an important role in this research. The unexpected existence and abundance of hot Jupiters gave this method a boost because the probability of a transit increases when the exoplanet is larger and orbiting close to its star. D. Charbonneau et al., "Detection of Planetary Transits Across a Sun-like Star," *The Astrophysical Journal Letters* 529 (2000), p. L45.

12 In 2010, researchers announced that many hot Jupiters are orbiting with large inclinations relative to the stellar disk they were born in, and some are rotating in an opposite sense to their parent star. Neither effect is consistent with the migration idea so researchers invoke a mechanism where a distant large body like a planet or a companion star gradually nudges the hot planet until its orbit becomes unstable and it flips over the top of the star like a jump rope.

13 A.-M. Lagrange et al., "A Giant Planet Imaged in the Disk of the Young Star Beta Pictoris," *Science* 329 (2010), p. 57.

14 K. Todorov, K. Luhman, and K. McLeod, "Discovery of a Planetary-mass Companion to a Brown Dwarf in Taurus," *The Astrophysical Journal Letters* 714 (2010), p. L84.

15 Imagine a small object like an asteroid located between two larger planets. The asteroid might go into an orbit of one planet or the other, and it might chaotically shuttle back and forth between the two planets. At any time it would be impossible to predict which planet the asteroid was orbiting.

16 R. Malhotra, M. Holman, and T. Ito, "Chaos and the Stability of the Solar System," *Proceedings of the National Academy of Sciences* 98 (2001), p. 12342.

17 W. Borucki et al., "Characteristics of Kepler Planetary Candidates Based on the First Data Set: The Majority Are Found to be Neptune-sized and Smaller," *The Astrophysical Journal* 728 (2011), p. 117.

18 There is a large intellectual tradition opposed to vitalism, the idea that living organisms are governed by a principal distinct from biochemical reactions, and not reducible to the laws of physics and chemistry. For example, Bechtel and Richardson have stated that vitalism "is often viewed as unfalsifiable, and therefore a pernicious metaphysical doctrine" in the *Routledge Encyclopedia of Philosophy*, ed. E. Craig (London: Routledge, 1998).

19 L. Hood and D. Galas, "The Digital Code of DNA," *Nature* 421 (2003), p. 444.

20 J. Guedes, E. Rivera, E. Davis, G. Laughlin, E. Quintana, and D. Fischer, "Formation and Detectability of Terrestrial Planets Around Alpha Centauri B," *The Astrophysical Journal* 679 (2008), p. 1582.

21 Paul Gilster, *Centauri Dreams: Imagining and Planning Interstellar Exploration* (New York: Copernicus Books, 2004), and also see Gilster's blog at http://www.centauri-dreams.org.

4: STELLAR NURSERY

1 Anaxagoras is probably the most famous scientist and philosopher that few people have heard of. His ideas of the motions in the Solar System show an understanding of centrifugal motion and differentiation of matter, and he understood that the Moon shines by reflected sunlight and what was required for an eclipse to occur. He wrote a treatise on perspective, which must have informed his sense of the three-dimensional nature of the universe and the large size of astronomical objects. Anaxagoras believed that all objects in the sky were stones, and argued a meteorite that was found in Greece must have been a stone that fell from the sky. He apparently also made major progress in mathematics, although few of these writings survived.

2 Scientists get awarded knighthoods on occasion but being named to the House of Lords is rare. The most recent example is the cosmologist Sir Martin Rees, named Baron Rees of Ludlow in 2005.

3 The problem was that geologists and biologists were still far from having reliable methods for measuring the ages of their specimens so physics was the only basis for quantitative calculations. Kelvin's preeminent reputation meant that his erroneous calculations held sway throughout the second

half of the nineteenth century. Darwin wrote to his colleague and rival Alfred Russel Wallace that "Thompson's views on the recent age of the world have been for some time one of my sorest troubles." Kelvin assumed the Sun worked using the physics known at the time, and to be fair to him, he added a caveat when he said that "inhabitants of the Earth cannot continue to enjoy the light and heat essential to their life for many million years longer, unless sources now unknown to us are prepared in the great storehouse of creation."

4 Arthur Eddington, 1920 presidential address to the British Association for the Advancement of Science, London, England.

5 The three-stage nuclear reaction that causes sunlight is called the proton-proton chain. We can't see into the center of the Sun, so direct verification of the nuclear reactions presumed to be taking place there didn't come until the late 1980s, when an underground detector in Japan was able to detect neutrinos, the weakly interacting subatomic particles that are predicted as a by-product of the fusion of hydrogen into helium. The detection led to a new mystery, since neutrinos were detected at half of the rate predicted by solar models. That discrepancy resulted in a modification of the Standard Model of particle physics.

6 The visible edge of the Sun is called the photosphere. It seems sharp and well-defined, but the Sun is a smooth ball of hot gas whose temperature, density, and pressure all increase smoothly throughout. There's no physical discontinuity at the photosphere; a space probe falling in would feel no "bump." However, the decreasing density moving out means the radiation interacts less and less. Deep in the Sun, high-energy photons can't move very far before encountering and bouncing off an electron. The surface marks the place where on average a photon will not interact anymore and will travel freely through the vacuum of space to Earth. So we see an edge. The edge of a cloud works in a similar way, except in this case the light is hitting water droplets. It marks transition between light not traveling freely (the dark and murk inside the cloud) and light traveling freely (the clearly defined surface of the cloud).

7 There is a common confusion between "heat" and "infrared radiation," but heating can be caused by all kinds of electromagnetic radiation, not just the wavelengths somewhat longer than the eye can see.

8 Infrared wavelengths are preferred for the direct detection of exoplanets, an experiment discussed in the last chapter. A star is hot and its radiation declines toward longer wavelengths, while a planet is cool so it has intrinsic radiation that peaks in the infrared (as opposed to the tiny portion of the

star's visible light it reflects). By observing in the infrared, the "contrast," or the visibility of a planet with respect to its parent star, can be improved by a factor of hundreds.

9 B. Zuckerman et al., "Detection of Interstellar Trans-Ethyl Alcohol," *The Astrophysical Journal Letters* 196 (1975), p. L99.

10 A gas cloud will not change its shape as long as the potential energy of the internal gravitational force is twice the kinetic energy of motion of the gas particles; this equilibrium is described mathematically as the virial theorem. If the mass is too high, or the temperature is too low, for this condition to be satisfied, the cloud will collapse by gravity. The mass at which this happens in a nebula was derived by Sir James Jeans in 1902, and it's the physical basis for star formation. J. Jeans, "The Stability of a Spherical Nebula," *Philosophical Transactions of the Royal Society of London* 199 (1902), p. 1.

11 When two quantities are related by $y = e^x$, that's an exponential relationship. Logarithmically, $\log y = x \log e$, so when quantity x varies linearly, quantity y varies logarithmically. Natural or base e logarithms are normally used, and this relationship is typical of biological phenomena such as population and cell growth rates. A power law relationship has the form $y = x^n$, where n is a pure number, called the power law index. Logarithmically, $\log y = n \log x$. This very general formalism describes a very wide range of phenomena, since x and y can represent any physical quantity and n can be any number, positive or negative. The power law can also describe the distribution of quantity x, so $n(x) = x^n$. In this case, with $n = 1$, if there is one object with quantity x, there are 10 with quantity $x/10$, and 100 with quantity $x/100$, and so on. With $n = 2$, if there is one object with quantity x, there are 100 with quantity $x/10$, and 10,000 with quantity $x/100$, and so on. So the larger the power law index (positive or negative), the more dramatically the quantity x varies across the sample.

12 The ubiquity of power laws in physical process is intriguing because it has no grounding in fundamental theory. In fact, power laws are often seen in dynamical phenomena that are difficult to model and predict, like turbulence and phase transitions. The prevalence of power laws in biological systems only deepens the mystery. For some insights, see Per Bak, *How Nature Works: The Science of Self-Organized Criticality* (New York: Springer-Verlag, 1999).

13 To put this in perspective, imagine Sirius, a star not much more massive than the Sun 9 light-years away, to be like the bulb of a small flashlight at a distance of 50 feet. Eta Carinae would be a light source almost as bright 6 miles away.

14 Brown dwarfs have central temperatures of 2 to 3 million Kelvin and densities 10 to 100 times that of water. Their mass range goes up to 75 to 80 Jupiter masses; above 65 Jupiter masses they fuse some lithium, and above 13 Jupiter masses they fuse some deuterium. The formal demarcation between a brown dwarf and a planet is 13 Jupiter masses, since below that no fusion of any kind is possible. It turns out there are more than 20 exoplanets above this limit, so the definition of a planet includes the fact that it must be orbiting a star. Most brown dwarfs are about the same size as Jupiter.

15 Published in 1997, the first of J. K. Rowling's seven novels for children was called *Harry Potter and the Philosopher's Stone* in the United Kingdom, but the title was changed to *Harry Potter and the Sorcerer's Stone* in the United States, presumably because American children were less likely to be familiar with the real-world mythology surrounding the Philosopher's Stone.

16 Classical alchemy centers around the Great Work (or *magnum opus* in Latin), which refers both to the transmutation of base matter into gold and spiritual transformation in the Hermetic tradition. In Newton's time it had three stages: (1) nigredo, or blackening by putrefaction or dissolution, (2) albedo, or whitening by purification and the burning out of impurities, and (3) rubedo, or reddening, representing the unification of man with God or the limited with the unlimited. Alchemy may be misguided early theorizing but it's not completely unreasonable. In their natural states, lead and gold are both heavy, dull, malleable metals, and it seems plausible that it might be possible to convert one to the other.

17 In fact, there are two ways to make helium from hydrogen by fusion. Stars like the Sun and up to 50 percent more massive than the Sun use the three-stage proton-proton chain described earlier. More massive stars can make helium using carbon, nitrogen, and oxygen isotopes as catalysts. This process dominates above 16 million degrees and forms only a few percent of the energy production in the Sun.

18 Most of the helium in the universe was created in the first few minutes after the big bang, 13.7 billion years ago, when the infant universe was at a temperature of 10 million degrees. The deep trough for very light elements is a consequence of the rapid cooling of the expanding universe, and its diminishing ability to continue nuclear fusion. Also, beryllium is radioactive and lithium is consumed in stars so the result is greatly reduced abundance of elements between helium and carbon. Carbon acts like a bottleneck so is a peak in abundance, followed by another decline.

19 E. M. Burbidge, G. R. Burbidge, W. A. Fowler, and F. Hoyle, "Synthesis of the Elements in Stars," *Reviews of Modern Physics* 29 (1957), p. 547.

20 The theory of solar energy production was put in place by Hans Bethe in

1938, but for many scientists the decisive evidence that the proton-proton chain was operating was the detection of neutrinos from the Sun in 1964 by Ray Davis and John Bahcall, and more refined measurements by a Japanese group in 1986. For a modern summary, see J. N. Bahcall, M. C. Gonzales-Garcia, and C. Pena-Garay, "Does the Sun Shine by pp or CNO Fusion Reactions," *Physical Review Letters* 90 (2003), p. 131301. Proof that stars create heavy elements preceded the full development of the theory by Hoyle and Fowler. Technetium was detected in the atmosphere of a red giant in 1952. Since it's radioactive, with a half-life much less than the age of the star, it must be created by the star in its lifetime. S. P. Merrill, "Spectroscopic Observations of Stars of Class S," *The Astrophysical Journal* 116 (1952), p. 21.

21 Stellar nucleosynthesis includes a lot of subtleties and complexities. The main processes have now been simulated in computers. Iron is a bottleneck because for lower elements nuclear binding energy is released when they are combined (fusion) while for higher-mass elements nuclear binding energy is released when they decay (fission). Silicon fusion actually produces radioactive nickel, which decays into cobalt and then iron. When the star has no more pressure support from energy release in nuclear reactions, it collapses, leading to the dramatic end phase of its life: supernova, and neutron star or black hole.

5: THE EDGE OF DARKNESS

1 The human eye is a remarkable device; it can accommodate brightness ranges of a factor of a billion and is 10,000 to 1 million times more sensitive in total darkness than it is in bright sunlight. Among the photoreceptor cells, the 6 million cones primarily detect color at high light levels and the 120 million rods don't detect color well but are adapted for low light levels. Some nocturnal species like owls have many more rods than humans, and their low light level sensitivity and visual acuity is 10 times better than ours.

2 In the long "dark" days for Western astronomy between Archimedes and Copernicus, Arab scientists kept the flame alive, making many important contributions to mathematics, astronomy, and optics. In the early eleventh century, Alhazen attempted to measure the parallax of the Milky Way and failed, showing that it could not be a nearby phenomenon of the atmosphere, as Aristotle had claimed. Around the same time, Persian astronomer Abu Rayhan al-Biruni proposed that the galaxy was a collection of countless nebulous stars.

3 The difficulty of measuring distances is a recurring theme in the history of astronomy. Trigonometry and the use of parallax provide the most direct

method, but the best satellite can only measure parallax out to 1500 light-years, just a percent of the diameter of the Milky Way. Shapley attempted to use a particular type of variable star in globular clusters called an RR Lyrae star, where the period of variation relates in a known way to luminosity. But he was actually observing a different type of variable star called a Cepheid, so he grossly overestimated the size of the galaxy as a result.

4 It was thought that pure neutron material would be unobservable, so it was a surprise when Jocelyn Bell and Anthony Hewish discovered a rapidly pulsing radio source, whose period was more precise than the best atomic clock. Theorists realized that a magnetized neutron star would emit radio waves that could be seen each rotation; only an extremely compact star can rotate in a second or less, as observed in most pulsars. Because the radio beam is narrow, we detect only a small fraction of all pulsars.

5 General relativity is almost mythically difficult, the theory people invoke when they want to allude to something so incomprehensible that almost nobody understands it. This perception was cemented by a story told about Sir Arthur Eddington, one of Einstein's champions and one of the scientists who organized an expedition to test the theory during a solar eclipse in 1919, as recounted in John Waller, *Einstein's Luck* (Oxford: Oxford University Press, 2002). After one of Eddington's lectures, a physicist in the audience asked, "Professor Eddington, you must be one of only three people in the world who understands general relativity." Eddington paused, apparently unable or unwilling to answer. The questioner continued, "Don't be modest, Eddington!" Finally, he replied, "On the contrary, I'm trying to think who the third person is." In fact, thousands of physicists have taken a graduate-level course on general relativity, and hundreds of people do research that involves solving Einstein's equations. The theory is challenging, but certainly not incomprehensible and, to practitioners, its beauty repays the effort in coming to grips with it.

6 Wheeler is widely reported to have coined the evocative term "black hole" but he actually never claimed this. He said he was referring to a "gravitationally completely collapsed object" at a meeting in 1967 when a person in the audience suggested saying "black hole" instead. Wheeler thereafter popularized the term. But there's a report by Ann Ewing in the *Science News Letter* of January 18, 1964, on a general relativity talk at an AAS meeting that used the phrase, so its true origin may never be known.

7 These are idealizations on thought experiments. In practice, someone in a free-falling elevator could do experiments to determine their situation (assuming they had the presence of mind to do experiments while plunging to their death). The Earth would create a tidal force on a person, or slightly

more gravity on their feet than their head. In addition, objects floating in the elevator would slightly converge as they fell, since they would be heading toward the center of the Earth on radial paths. Tidal effects are difficult to avoid in most normal gravitational situations, so in practice the equivalence principle is true only in the limit of arbitrarily small volume.

8 Intuition is subverted by the paradigm shift of general relativity, because many of its physical effects are so remote from everyday experience. Some commentators have claimed that Einstein's theory "does away with gravity" or renders it an illusion. It's certainly true that Einstein's gravity is very different from that of Newton's, but Newton himself had no explanation for a force that could operate instantly across an absolute vacuum, saying famously, "I frame no hypotheses." In general relativity, light follows geodesics, paths that represent the shortest distance between two points in space-time of arbitrary curvature. If the geodesic is curved, light follows that curved path. This is familiar to anyone who has taken a nonstop flight from the western U.S. to Europe. The flight goes well above the Arctic Circle; if you drew the path on a flat map it would not be a straight line! However, on the curved two-dimensional space of a globe, that path is really the shortest one.

9 J. Michell, "On the Means of Discovering the Distance, Magnitude, etc. of the Fixed Stars, in Consequence of the Diminution of the Velocity of Their Light, in Case Such a Diminution Should be Found to Take Place in Any of Them, and Such Other Data Should be Procured from Observations, as Would Be Farther Necessary for That Purpose," *Philosophical Transactions of the Royal Society of London* 74 (1784), p. 35.

10 R. Ruffini and J. Wheeler, "Introducing the Black Hole," *Physics Today* (April 1971), p. 31.

11 G. Brown and H. Bethe, "A Scenario for a Large Number of Low Mass Black Holes in the Galaxy," *The Astrophysical Journal* 423 (1994), p. 659.

12 A. Celotti, J. Miller, and D. Sciama, "Astrophysical Evidence for the Existence of Black Holes," *Classical and Quantum Gravity* 16 (1999), p. 301.

13 J. Casares, "Observational Evidence for Stellar Mass Black Holes," *Black Holes: From Stars to Galaxies Across the Range of Masses*, IAU Symposium 238, Prague, Czechoslovakia (Cambridge: Cambridge University Press, 2007), p. 3.

14 R. Penrose, "Gravitational Collapse and Space-Time Singularities," *Physical Review Letters* 14 (1965), p. 57.

15 Kip Thorne, *Black Holes and Time Warps* (New York: Norton, 1994).

16 Sean Carroll, *Space-Time and Geometry* (Reading, MA: Addison-Wesley, 2004).

17 Hawking's playful sense of humor has spawned several bets. In 1975, he bet the Caltech physicist Kip Thorne a one-year magazine subscription that black holes do not exist. Hawking called this a kind of insurance policy, since much of his life's work would be diminished if Thorne were right. That bet was recently settled; both parties considered the evidence for the existence of black holes compelling. Thorne received a year's subscription to *Penthouse*, his original choice, much to the chagrin of his wife. The 1997 bet had Thorne and Hawking betting that the impermeability of the event horizon was absolute and information was lost, and Thorne's Caltech colleague John Preskill bet against them. Hawking conceded the bet in 2004, and gave Preskill a baseball encyclopedia from which "information could be extracted at will." Thorne hasn't yet conceded the bet. Hawking also commented: "I gave John an encyclopedia of baseball, but maybe I should have just given him the ashes." In 2008, Hawking made a modest $100 bet that the Large Hadron Collider *would not* find the Higgs particle since "It will be much more exciting if we don't find the Higgs. That will show something is wrong and we need to think again."

18 S. Hawking, "Information Loss in Black Holes," *Physical Review Letters* D72 (2005), p. 4013.

19 A. Ghez et al., "Measuring Distance and Properties of the Milky Way's Supermassive Black Hole with Stellar Orbits," *The Astrophysical Journal* 689 (2008), p. 1044, and M. Reid, "Is There a Supermassive Black Hole at the Center of the Milky Way Galaxy?" *International Journal of Modern Physics D* 18 (2009), p. 889.

20 S. Doeleman et al., "Event-Horizon-Scale Structure in the Supermassive Black Hole Candidate at the Galactic Centre," *Nature* 455 (2008), p. 78.

21 The scale of the problem is established by realizing that it takes 5×10^{20} joules, an amount equal to the world's annual energy consumption, to accelerate 1000 tons to one-tenth the speed of light. That mass is barely adequate to hold the fuel and life support for a small manned mission to a nearby star. Solar sails don't work far from a star so the fuel has to be carried for the journey. At a Joint Propulsion Conference in 2008, the participants were pessimistic about interstellar travel, arguing that it will remain science fiction for a long time, perhaps centuries.

6: ISLAND UNIVERSE

1 For more than a century, astronomers used photographic emulsion for recording images of the sky, and plates coated with emulsion were so stable that they were used for some purposes even after CCDs became available

in the 1980s. Silver halide in an emulsion interacts with incoming light, and areas receiving the most light have the largest chemical reactions and so turn the darkest. The recorded image is therefore a "negative" and for an astronomical plate like those taken by Edwin Hubble, stars are smudges of dark not smudges of light.

2 Galileo's best and largest telescope was a refractor with a diameter of 1.5 inches, designed in 1620. Then Christian Huygens held the record for nearly 50 years, from 1686 to 1734, with his ingenious "aerial" telescope with an aperture of 8.5 inches. The optical elements were mounted in short tubes and connected by a 200-foot-long string that was kept taut. The objective lens swiveled from the top of a tall pole to allow the contraption to be pointed most places in the sky. From the beginning of the eighteenth century all the largest telescopes in the world were reflectors rather than refractors. William Herschel next held the record, for 50 years, with his 40-foot telescope, completed in 1789. Then came the Leviathan of Parsons and the Hooker 100-inch. George Ellery Hale trumped himself with a 200-inch reflector on Mount Palomar in 1948, which was only eclipsed in size by a Russian 240-inch telescope in 1976. In the modern age, telescopes of 8 meters (320 inches) and larger aperture are commonplace.

3 Thomas Wright, *An Original Theory or New Hypothesis of the Universe* (London: MacDonald, 1750). Wright made no observations and his ideas were influenced by theological speculation. He supposed that the appearance of the Milky Way indicated a spherical shell of stars, where the Sun was midway between the inner and outer edges of the shell.

4 Adriaan van Maanen was a colleague of Shapley's at Mount Wilson Observatory, and a close friend. He compared plates of spiral nebulae taken 10 to 20 years apart and he claimed to see evidence of rotation in many of them. If true, and if the nebulae were external systems of stars, the implied rotation velocities would have been faster than light, which was impossible. It was more likely that he had systematic errors he didn't recognize, or simply that he saw what he expected and wanted to believe. In fact, galaxies rotate every few hundred million years so he could not possibly have detected rotation within a human lifetime. This false observation muddied the debate over the nature of the nebulae, and Shapley was exasperated when he learned of the error, saying, "I believed in van Maanen's results . . . after all, he was my friend!"

5 Hubble's initial measurement was in error because he hadn't realized there were two types of Cepheid variables. Using a calibration appropriate to the type of stars he was measuring, his distance estimate goes from 1 million to 2.5 million light-years.

6 By a strange coincidence, Cepheid variables were discovered in 1784 by John Goodricke, the English astronomer who was also deaf, having contracted scarlet fever as a young boy. Goodricke tragically died from pneumonia at the age of twenty-one.

7 K. Haramundanis, ed. *Celia Payne-Gaposchkin* (Cambridge: Cambridge University Press, 1996), p. 209.

8 The word *galaxy* derives from the Greek words for our own system of stars, "milky circle," in an allusion to its appearance in the sky. In astronomical usage, *galaxy* in lower-case is any remote system of stars and *Galaxy* in upper-case refers to our system of stars, the Milky Way.

9 Edwin Hubble, *Realm of the Nebulae* (New Haven: Yale University Press, 1936).

10 Charles Messier didn't know the true nature of the 103 objects he cataloged in 1771, but his list contains many spectacular nearby galaxies. Messier made his observations in the Northern Hemisphere, so did not include nebulae from the far southern sky, including the Magellanic Clouds, two dwarf companion galaxies to the Milky Way. His list was eventually expanded to 110 based on his notes, with the last addition made in 1966. There are 39 galaxies in the final catalog.

11 The different classification elements are combined, which leads to designations complicated enough that only an expert would be able to figure them out. For example, an SAB(r)c galaxy is a spiral galaxy with a weak bar, loosely wound arms, and a ring.

12 The story is more complicated than this simple description and is still the subject of active research. The density wave idea probably can explain "grand design" spirals but it has trouble with the more chaotic structures. A second idea is that star formation triggers star formation in adjacent regions, independent of the actual density. The triggering occurs with a certain probability, so it's not deterministic. Rotation of the galaxies winds this pattern of star formation into a spiral. The physics of star formation is nonlinear and very difficult to understand through simple equations that can be solved analytically, which is why this is a challenging problem.

13 In 1959, Louise Volders showed that the nearby spiral M33 doesn't rotate as expected from Kepler's laws applied to the visible light, but this single observation was viewed, like Zwicky's, as an uncomfortable "anomaly" and didn't cause astronomers to take up the issue.

14 V. Rubin, N. Thonnard, and W. Ford, "Rotational Properties of 21 Sc Galaxies with a Large Range of Luminosities and Radii," *The Astrophysical Journal* 238 (1980), p. 471.

15 The number of rotation curves for spiral galaxies has grown from a few

dozen when Vera Rubin and Ken Freeman started their research to tens of thousands today. All of them show evidence for dark matter but the proportions of visible and dark matter vary in systematic and interesting ways with galaxy properties. Dwarf galaxies have a larger proportion of dark matter than giant galaxies. Spiral galaxies usually have small satellite galaxies—the Milky Way has a dozen—and they can act as "test particles" to measure the dark matter out to very large distances. A spiral galaxy like the Milky Way has dark matter that extends to about 200 kiloparsecs or about 650,000 light-years, 20 times larger than the radius of the visible matter. Elliptical galaxies also have dark matter; rather than rotation in a disk, it's measured by the mean velocity of stars in their elliptical orbits of the galaxy nucleus.

16 Steve Soter and Neil Tyson, eds., *Cosmic Horizons: Astronomy at the Cutting Edge* (New York: New Press, 2000).

17 O. Eggen, D. Lynden-Bell, and A. Sandage, "Evidence from the Motions of Old Stars that the Galaxy Collapsed," *The Astrophysical Journal* 136 (1962), p. 748.

18 L. Searle and R. Zinn, "Compositions of Halo Clusters and the Formation of the Galactic Halo," *The Astrophysical Journal* 225 (1978), p. 357.

19 A delightful example of how you can make a spiral galaxy in the workshop, if not the kitchen, appeared in the July 1936 edition of *Popular Science* magazine. All you need is a hand drill, a thumbtack, and some oil! First, inject a globule of machine oil just under the surface of a large beaker of wood alcohol, using an eyedropper. Impart a whirling motion with a thumbtack attached to the end of a hand drill. Twin spiral arms form, and they will often even break up into tiny blobs of star formation. It beats any supercomputer, hands-down.

20 M. Steinmetz and J. Navarro, "The Hierarchical Origin of Galaxy Morphologies," *New Astronomy* 7 (2002), p. 155; and F. Hammer, H. Flores, M. Puech, Y. Yang, E. Athanassoula, M. Rodrigues, and R. Delgado, "The Hubble Sequence: Just a Vestige of Merger Events?" *Astronomy and Astrophysics* 507 (2009), p. 1313; and M. Martig and F. Bournaud, "Formation of Late-Type Spiral Galaxies: Gas Return from Stellar Populations Regulate Disk Destruction and Bulge Growth," *The Astrophysical Journal Letters* 714 (2010), p. L275.

21 The gas reservoir that makes galaxies and fuels their subsequent star formation was difficult to detect because diffuse gas in intergalactic space is kept ionized (extremely hot) from the ultraviolet radiation of massive stars in galaxies. Gas at a temperature of 100,000 to 1 million Kelvin emits

strongly at far ultraviolet and soft X-ray wavelengths, which is just where the Earth's atmosphere is opaque. This gas was finally detected with sensitive X-ray telescopes, and by seeing its "shadowing" of the light of distant active galaxies.

22 J. Binney, "Fitting Orbits to Tidal Streams," *Monthly Notices of the Royal Astronomical Society* 386 (2008), p. L47.

23 Only a very advanced civilization would be noticeable in another galaxy. We can generate enough radio or optical energy to momentarily outshine our star, but that's a negligible "blip" of a signal in the ocean of stars in a galaxy. Hypothetically, an advanced civilization could orchestrate stellar cataclysms and generate artificial signals that would stand out even with so many competing natural signals. Regardless of the actual value of *N*, for a full census of technological life-forms it must be multiplied by 100 billion, the number of galaxies in the visible universe.

7: COSMIC ARCHITECTURE

1 It would be inappropriate to take a revisionist stance and downplay Hubble's importance in the history of modern cosmology. However, his discoveries rested firmly on the work of Leavitt and Slipher, who are far less well known, even among many astronomers! To the outsider or nonscientist, it's natural that a subject gets cast in terms of heroes or "giants," but it can do a disservice to reputations or to the understanding of how science actually works. Think, for example, of how less feted Alfred Russel Wallace is than Charles Darwin. In practice, Hubble hit a single and cleverly stole second and third, but was given credit for a triple in the record books.

2 V. Slipher, "Nebulae," *Proceedings of the American Philosophical Society* 56 (1917), p. 409.

3 Astronomers had also cataloged smooth elliptical "nebulae" that were assumed to be distinct stellar systems, but they weren't easily resolvable into stars so it wasn't possible to locate Cepheid variables in them and derive reliable distance. The pivotal papers are: V. Slipher, "Nebulae," *Proceedings of the American Philosophical Society* 56 (1917), p. 403; K. Lundmark, "The Determination of the Curvature of Space-Time in de-Sitter's World," *Monthly Notices of the Royal Astronomical Society* 84 (1924), p. 747; and E. Hubble, "A Relation Between Distance and Radial Velocity Among Extra-Galactic Nebulae," *Proceedings of the National Academy of Sciences* 15 (1929), p. 168.

4 This phrase is so famous it has become folklore in physics and it might be

suspected of being an urban legend. In fact, Einstein never wrote or was heard to say this phrase; evidence for it rests solely on George Gamow's autobiography *My World Line* (1970), where he writes, "Much later, when I was discussing cosmological problems with Einstein, he remarked that the introduction of the cosmological terms was the biggest blunder of his life."

5 E. Hubble, "The Effects of Redshift on the Distribution of Nebulae," *The Astrophysical Journal* 84 (1936), p. 517.

6 Doppler derived the effect of relative motion on light waves to try to explain the colors of binary stars. Only the radial component of the three-dimensional-space motion of a star manifests as a Doppler shift, so it will not in general give the full velocity. Astronomers therefore interpret a Doppler shift as a radial velocity. Mathematically, if the motion is slow compared to the speed of light, the fractional frequency shift equals the speed as a fraction of the speed of light. The Doppler shift was measured for sound waves for the first time by Dutch chemist Buys Ballot in 1845, using a group of musicians playing a calibrated note on the Utrecht-Amsterdam train line.

7 This guide to intuition is only valid in the comfy, traditional world of classical physics. In the early twentieth century, Einstein deduced a relativistic form of the Doppler effect, which acknowledged the speed of light as a fundamental limit and was valid for any speed up to 300,000 kilometers per second. Even with the classical Doppler effect, intuition sometimes has to be tossed out the window. Lord Rayleigh wrote a textbook on sound where he noted that by moving appropriately it would be possible to hear a symphony played backward.

8 As the poet Robert Frost once said, "All metaphors are imperfect, and that's the beauty of them." While the material of a balloon can serve as a reasonable facsimile of expanding space, the universe we live in is flat, not detectably curved (if you like, just imagine the balloon as vast and we make our measurements on a small patch of it). Also, a wave drawn on an expanding balloon will grow in every direction, including its amplitude, while a wave traveling through space has amplitude that reduces in accordance with the inverse square law. However, the analogy is useful because it overcomes the "ballistic" way of thinking about the motion of galaxies and replaces it with a sense that they are "carried apart" by the expansion of space.

9 Many of the world's cultures, from the ancient Greeks to the Vedic tradition in India to the Mayans and the Australian aborigines, believed that time and the universe were cyclic. Even if it's infinite, cyclic time doesn't have to deny the possibility of progress, but the common thread in these traditions

is that physical existence has no limit going back in time. The expanding universe model, on the other hand, postulates a true origin to both space and time.

10 We further must assume that the laws of physics are the same everywhere. This is difficult to test but it's crucial to try, and a subfield of astrophysics involves clever measurements designed to test whether the law of gravity or the strengths of the fundamental forces and constants are the same in remote galaxies. Despite some provocative measurements, there is no clear evidence that our physics is special to us or to our corner of the universe.

11 The exact solutions of general relativity for an expanding (or contracting) universe were only possible because of the simplifying geometric effects of homogeneity and isotropy. With this in place, Einstein's field equations are only needed to calculate the size of the universe as a function of time. Alexander Friedmann, Georges Lemaître, Howard Robinson, and Arthur Walker did the hard work and so the metric that describes expanding space-time is named the FLRW metric after them. It was derived in the 1920s, and the formalism is so successful it's still called the "standard model" of cosmology. Of course, the universe is lumpy so not homogeneous, but the calculations still work. Cosmologists like to say the universe is "nearly FLRW."

12 The Palomar Observatory Sky Survey took a decade to complete, and it was the core source material of research and discovery in astronomy for nearly three decades, until digital CCD detectors matured in the 1980s. Much more recently, a Caltech astronomer used the 48-inch telescope to discover the dwarf planet Eris, which led to the demotion of Pluto. A second epoch of the survey was done in the 1980s and 1990s using an upgraded telescope and more sensitive photographic emulsions. The southern counterpart of the Palomar Sky Survey was carried out in the 1980s by a twin of the 48-inch telescope operating at Siding Spring Observatory in New South Wales, Australia. Each of these major photographic surveys has been digitized, giving them a longer "shelf life" and allowing them to make scientific contributions well into the age of the CCD.

13 Gerard de Vaucouleurs, a Frenchman transplanted to Texas, was a pioneer in the study of the large-scale structure of the universe. He had made great efforts to find distance indicators to complement and use with Cepheid variables and he steadily accumulated data for bright galaxies out to a redshift of about 3000 kilometers per second, corresponding to a distance of 40 megaparsecs or 130 million light-years. In addition to the clusters that had been identified previously on the basis of counting galaxies, he showed evidence that the Milky Way is part of a huge flattened "supercluster"

containing many thousands of galaxies and extending over hundreds of millions of light-years. For a modern summary, see R. B. Tully, "The Local Supercluster," *The Astrophysical Journal* 257 (1982), p. 389.

14 K. Abazajian et al., "The Seventh Data Release of the Sloan Digital Sky Survey," *The Astrophysical Journal Supplement* 182 (2009), p. 543.

15 Colorful descriptions of large-scale structure shouldn't obscure the fact that galaxies are rarely in close proximity. Even in the densest clusters they're not cheek by jowl but are separated by 3 to 5 times their diameters. In lower-density regions like that occupied by the Milky Way, the separations are 10 to 20 times the typical galaxy diameters. Since gravity is a long-range force with infinite reach, it sculpts structures on enormous scale without the individual galaxies being close together.

16 M. A. Aragon-Calvo, R. van de Weygaert, and A. Szalay, "Multiscale Phenomenology of the Cosmic Web," *The Astrophysical Journal* 723 (2010), p. 364.

17 M. Fleenor, J. Rose, W. Christiansen, R. Hunstead, M. Johnson-Hollitt, M. Drinkwater, and W. Saunders, "Large-Scale Velocities Structures in the Horologium-Reticulum Supercluster," *The Astronomical Journal* 190 (2005), p. 957; and A. Kopylov and F. Kopylova, "Search for Streaming Motions of Galaxy Clusters Around the Giant Void," *Astronomy and Astrophysics* 382 (2002), p. 389; and J. Gott, M. Juric, D. Schlegel, F. Hoyle, M. Vogeley, M. Tegmark, N. Bahcall, and J. Brinkmann, "A Map of the Universe," *The Astrophysical Journal* 624 (2005), p. 463.

18 This barb is a cutting riff on an inside physics joke. Among theoretical physicists, there's a tendency to simplify any complex situation to make calculations easier. But sometimes the loss of detail strays into caricature and oversimplification. They joke about this by saying, "Imagine a spherical cow..." Despite his aggressive demeanor toward some of his senior colleagues, Zwicky had a lot of patience for students and young researchers and he also made some significant humanitarian gestures.

19 D. Walsh, R. Carswell, and R. Weymann, "0957+561 A,B: Twin Quasi-Stellar Objects or a Gravitational Lens," *Nature* 279 (1979), p. 381.

20 The nature of lensing is that odd numbers of images are always produced. Mathematically, the surface of equal light travel times gets "folded" and each fold adds two images, so one image becomes three and then five and then seven and so on, as the mass distribution causing the lensing gets more complex. In practice, the central or undeviated image in a lensing geometry is always demagnified and is difficult to detect so astronomers tend to detect even numbers of images. The maximum number of images seen for one galaxy is 13! Each image represents a different path through

the cluster of dark matter and suffers a different angle of deflection, so the total number of probes and constraints on the mass distribution is large. Lensing gives excellent information on how dark matter is distributed in the universe.

21 D. Clowe, M. Bradac, A. Gonzales, M. Markevitch, S. Randall, C. Jones, and D. Zaritsky, "A Direct Empirical Proof of the Existence of Dark Matter," *The Astrophysical Journal Letters* 648 (2006), p. L109.

22 R. Genzel, L. Tacconi, D. Rigopoulou, D. Lutz, and M. Tecza, "Ultraluminous Infrared Mergers: Elliptical Galaxies in Formation?" *The Astrophysical Journal* 563 (2001), p. 527.

8: NUCLEAR POWER

1 The mechanism is analogous to fluorescent light tubes that contain a low-pressure mercury or sodium gas. When the tube is illuminated, it never becomes hot, because the gas in it is such a good vacuum, but the gas is excited by electricity and produces most of its emission in narrow spectral lines. The color purity of neon lights is caused by almost all the emission coming out in the few red lines. In a nebula, ultraviolet radiation from a young star plays the role of electricity in a lamp.

2 C. Seyfert, "Nuclear Emission in Spiral Nebulae," *The Astrophysical Journal* 97 (1943), p. 28.

3 Spectral lines have a "natural width" set by quantum uncertainty in the emission process. However, this is much smaller than the Doppler width caused by the random motions of atoms and molecules in any gas. In a gas, as the temperature goes up, the velocity of the particles goes up, as does the spectral line width, with a well-understood scaling. In astronomy, line width is often used to infer temperature. In Seyfert galaxies, the line width is so large that it corresponds to a temperature higher than that of any star, where the highest is about 100,000 Kelvin. Stars don't emit enough ultraviolet radiation to excite a gas with lines this broad.

4 The transition from military to civilian applications was seamless. In Australia and England, radar operators had detected interference fringes between radar coming directly from a plane and radar that arrived after reflecting off the sea. In 1946, an Australian group used a similar "sea cliff interferometer" to observe radio waves from the Sun at sunrise traveling directly and after being reflected off the sea. Modern radio interferometers create interference fringes by gathering radiation from widely separated dishes and bringing the signals together with coaxial cables, waveguides, or other transmission lines. The interference depends on the time delay of

signals reaching the different dishes and that information can be used to reconstruct the position of the source with a precision set by the spacing between dishes rather than the size of any dish. Angular resolution is proportional to the wavelength of radiation being collected. Radio waves are so much longer than optical waves that, per meter of aperture, radio telescopes have far worse angular resolution than optical telescopes. However, modern radio interferometers have dishes with such wide separations (often the baselines are intercontinental) that their angular resolution exceeds that of the best optical telescopes.

5 F. Smith and B. Lovell, "On the Discovery of Extragalactic Radio Sources," *Journal for the History of Astronomy* 14 (1983), p. 155.

6 The process of science has many twists and turns. Schmidt's discovery depended on the clever use of lunar occultation by Cyril Hazard to get an extremely accurate radio position for 3C 273. Without that, he wouldn't have known where to point the telescope. Meanwhile, down the hall at Caltech, Jesse Greenstein already had a spectrum of a similar radio source, 3C 48, but it was so puzzling that he simply stuffed it in his desk drawer while he worked on other things. 3C 48 had a recession velocity 37 percent of the speed of light and an inferred distance of over 5 billion light-years. Once Schmidt had figured out that the lines could be identified by looking for a large redshift, Greenstein and Tom Matthews published their data alongside Schmidt's, but the Dutchman got the credit for "discovering" quasars.

7 M. Schmidt, "3C 273: A Star-like Object with a Large Redshift" *Nature* 197 (1963), p. 1040; C. Hazard, M. Mackey, and A. Shimmins, "Investigation of the Radio Source 3C 273 by the Method of Lunar Occultation," *Nature* 197 (1963), p. 1037; and J. Greenstein and T. Matthews, "Redshift of the Unusual Radio Source 3C 48," *Nature* 197 (1963), p. 1041.

8 The limited resolution of optical telescopes and the great distances of quasars mean that imaging doesn't give a true sense of the compactness of a quasar. However, a light travel time argument says that a single object can't vary on a timescale faster than the time it takes for light to traverse the object (otherwise the variations would be smoothed or averaged out). Quasars have light that varies irregularly on timescales of several weeks, so the light must be coming from a region no larger than a few light-weeks.

9 D. Lynden Bell, "Galactic Nuclei as Collapsed Old Quasars," *Nature* 223 (1969), p. 690. Lynden Bell had the most fully developed treatment, but papers by Ed Salpeter and Yakov Zel'dovich in 1964 were the first to present the supermassive black hole hypothesis.

10 M. Rees, "Black Hole Models for Active Galactic Nuclei," *Annual Reviews of Astronomy and Astrophysics* 22 (1984), p. 471.

11 This technique is called reverberation mapping. It's a very direct method for measuring mass within a small region because the gas clouds trace the gravity close to the black hole and the physical scale is set by the time delay of the gas responding to changes in emission near the black hole. This type of data can only be gathered for a few dozen nearby active galaxies. For more distant objects, including almost all quasars, the mass estimate is based on the width of the emission lines, an indirect method tethered in the reverberation of nearby objects. For inactive black holes, stellar velocities must be used to diagnose the black hole, as we saw for the center of the Milky Way. Kepler's laws and Newtonian gravity adequately describe motions of gas and stars light-weeks to light-months away from the black hole; general relativity only needs to be used close to the event horizon.

12 The role of technology in our understanding of the universe has been explored in detail by Martin Harwit in *Cosmic Discovery: The Search, Scope, and Heritage of Astronomy* (New York: Basic Books, 1981), and several articles, including W. van Breugel and J. Bland-Hawthorn, eds., "Instrumentation and Astrophysics: How Did We Get to Be So Lucky?" *Imaging the Universe in Three Dimensions: Astrophysics with Advanced Multi-Wavelength Imaging Devices* (San Francisco: Astronomical Society of the Pacific, 2000), p. 3. Harwit was a professor and head of the Astronomy Department at Cornell University and director of the National Air and Space Museum in Washington, DC. He argues that it is technological innovation rather than theoretical insight that has led to progress in astronomy. Certainly objects like quasars were not predicted, and observations beyond the visible spectrum were critical in understanding and even defining the phenomenon.

13 In a magnetized plasma, a high-temperature gas containing a magnetic field, electrons spiral in the magnetic field and emit synchrotron radiation. The particular attributes of nonthermal radiation are a smooth or power law spectrum and linear polarization. Compact objects like neutron stars and black holes have strong magnetic fields and are the most likely sources of synchrotron radiation. Synchrotron radiation was first created artificially in the lab in 1946.

14 H. Alfven and N. Herlofson, "Cosmic Radiation and Radio Stars," *Physical Review* 78 (1950), p. 616, and G. Burbidge, "On Synchrotron Radiation from Messier 87," *The Astrophysical Journal* 124 (1956), p. 426.

15 The imaging method looked for ultraviolet "excess" relative to the hottest stars in the Milky Way galaxy. The spectroscopic method was sensitive to

both ultraviolet excess and emission lines superimposed on the spectrum. Quasars are rare, so large areas of sky had to be surveyed, and in the 1970s and 1980s this was done with photographic plates. The Palomar and U.K. Schmidt telescopes followed their pioneering imaging surveys with surveys for emission line objects, such as quasars, but also included planetary nebulae and peculiar stars. Currently, the same methods are used with CCDs, reaching even fainter levels.

16 G. Richards et al., "Efficient Photometric Selection of Quasars from the Sloan Digital Sky Survey. II. A Million Quasars from Data Release 6," *The Astrophysical Journal Supplement* 180 (2009), p. 67.

17 M. Rowan-Robinson, "On the Unity of Activity in Galaxies," *The Astrophysical Journal* 213 (1977), p. 635.

18 The "bestiary" of active galaxies contains 20 or more categories, and while unification schemes plausibly reduce this multiplicity, they can't reduce all active galaxies to one fundamental set of components. Active galaxies are governed by different scales of their environment: the Hubble type of the underlying galaxy, the amount of star formation within a thousand light-years of the nucleus, the availability of gas and dust in the central couple of light-years, the prominence of the accretion disk and jets, and the spin and mass of the black hole. Orientation alone can't account for all, or even most, of the differences among the categories. The most important gap in our understanding is the fact that no single wavelength of selection recovers the "true" population of active galaxies.

19 More precisely, a luminous quasar in the galactic center would be 2 to 3 percent of the Sun's brightness. That's quite enough to be prominent in the daytime sky, and millions of times brighter than any planet or star. Of course, the speculation is fantasy because dust between us and the center would dim it back down to the level of a nighttime star; its true brightness would only be manifest if we could move out of the disk.

20 P. Hopkins, L. Hernquist, T. Cox, T. Di Matteo, P. Martini, B. Robertson, and V. Springel, "Black Holes in Galaxy Mergers: Evolution of Quasars," *The Astrophysical Journal* 620 (2005), p. 705.

9: THE GROWTH OF GALAXIES

1 To play out this imaginary scenario, we've had to leave Einstein bound and gagged in a corner and play fast and loose with his theory. Relativity is based on the speed of light as an absolute limit to motion. In the real world, it's not possible to catch up with a beam of light and light always has the same measured speed regardless of relative motion. In the fantasy world

of slow light, relativistic effects would kick in at very low speeds, and the distortions of time, space, and mass would blow away the mild dissonance of the scenario described in the text.

2 The measurement accuracy improved to 1 percent in 1728 when James Bradley realized that starlight would arrive at the Earth with a slight deflection depending on which way the Earth was moving. To see this, imagine the thankless task of walking in circles on a windless day in a steady downpour of rain. The rain will not arrive vertically—more will hit your front than your back, and the sense of the deflection depends where in your circular path you are. The effect is called the aberration of starlight. The next improvement was more in the spirit of Galileo. In 1850, French rivals Hippolyte Fizeau and Leon Foucault used timing measurements of rapid flashes of light to improve the accuracy to 0.5 percent.

3 In the 1970s, having reached a precision of a few parts per 100 million, measurement of the speed of light began to be limited by the precision of the meter. So, in a judo move, the international body governing weights and measures redefined the meter as the distance traveled by light in 1/299,792,458 of a second. The speed of light is now a defined constant in SI units, and improvement in experimental technique simply helps to better define the meter.

4 Albert Michelson and Edward Morley's experiment is rightly considered one of the most famous experiments in the history of science. It ruled out the idea of an "ether," an invisible substance through which light propagates, which had been hypothesized since the time of the Greeks. It was actually the lack of evidence for the ether that guided Einstein rather than the Michelson-Morley experiment, plus the central role of the speed of light in Maxwell's theory for the propagation of electromagnetic waves.

5 All of the effects of special relativity are seen thousands of times daily in physics labs around the world. Macroscopic objects can't be made to move at significant fractions of the speed of light, so their relativistic effects are mild (but measurable). Atoms and subatomic particles can be accelerated to within a fraction of a percent of the speed of light, where the effects of relativity are dramatic. Despite many attempts to detect them, no tachyons—particles traveling faster than light—have ever been observed.

6 G. Benford, D. Book, and W. Newcomb, "The Tachyonic Antitelephone," *Physical Review D* 2 (1970), p. 263.

7 In fact, light traveling at infinite speed has as many conceptual and logical problems as superslow light. If light arrived instantly from all locations in the universe, causality would be affected because all time frames would be compressed to an instant and all events everywhere would be seen simul-

taneously. More fundamentally, light is an electromagnetic wave, where accelerating charges create a disturbance consisting of oscillating and coupled electric and magnetic fields that propagate through space. Light as a wave and light as a signal would be meaningless if the speed were infinite; Maxwell's equations would have to be tossed out and replaced with something different.

8 In the late 1940s some theorists were dissatisfied with the implications of the cosmic expansion and an origin to the universe. Fred Hoyle, Herman Bondi, and Thomas Gold proposed the "steady state" theory. This theory did not deny the expansion, but proposed that matter was being created at a modest rate in the growing spaces between galaxies, leading to new galaxies and a universe that always looked the same. The theory extended the cosmological principle into the "perfect" cosmological principle, where the universe was the same in all directions and at all places and at all times.

9 M. Blanton et al, "The Galaxy Luminosity Function and Luminosity Density at Redshift z = 0.1," *The Astrophysical Journal* 592 (2003), p. 819.

10 It was realized in the 1930s that these "extra" effects could provide a very important test for the reality of the expansion (since there are other hypothetical mechanisms involving "tired light" that could generate the observed redshifts). Light from galaxies is extended and is characterized by the surface brightness, or brightness within a particular area. In a static universe the light received from a galaxy drops as the inverse square of the distance, but the apparent area also drops as the inverse square of the distance, so the surface brightness is constant and independent of distance. In an expanding universe, the two extra effects reduce the light received from a distant galaxy. By comparing galaxies of similar sizes at different redshifts, the signature of the expansion should be detectable. This is called the Tolman test, after the physicist Richard Tolman. A recent and successful application of the test is A. Sandage and L. Lubin, "The Tolman Surface Brightness Test for the Reality of the Expansion. IV. A Measurement of the Tolman Signal and the Evolution of Early-Type Galaxies," *The Astronomical Journal* 122 (2010), p. 1084.

11 In 2010 Nelson and Angel shared the million-dollar Kavli Prize in Astrophysics in recognition of their work as telescope pioneers, with Raymond Wilson, who developed the technique of adaptive optics used on all large telescopes to compensate for the blurring effects of the Earth's atmosphere.

12 Near-Earth orbit is above the Earth's atmosphere but it's not perfect for optical imaging. There is Earth glow that must be avoided, and machinery in space can "out-gas" and contaminate optical surfaces, plus the risk of damage from space debris is not insignificant. The Hubble Space Telescope

is in a low Earth orbit since it has to be serviced by space shuttle astronauts, but NASA is putting other telescopes at a Lagrange point (where gravity from the Sun and Moon balance) a million miles from Earth. Probably the best place for optical astronomy would be the far side of the Moon—there's no atmosphere or geological activity and it's exceptionally dark.

13 When moving beyond the "local" universe, the conceptual distinction between a cosmological redshift and a Doppler shift becomes very important. The only pure observed quantity for a galaxy or a quasar is redshift; distance and age are dependent on the physical mechanism for the redshift. If the redshift is small the Doppler formula is a good approximation, where redshift equals the recession velocity as a fraction of the speed of light. For large velocities in an inertial frame, a relativistic version of the Doppler formula is needed, one that takes into account time dilation. But special relativity is inappropriate for describing light propagation where recession velocity is caused by expanding space; general relativity is required.

14 Although the existence of some form of matter that doesn't interact with radiation and dominate normal "baryonic" matter—protons and neutrons—is not doubted by most cosmologists, its physical nature is still completely unknown. With black holes, brown dwarfs, planet, rocks, and dust all ruled out, the best bet is a ubiquitous subatomic particle not yet detected in a physics lab. Theoretically, dark matter particles can be characterized as hot, where they move relativistically or close to the speed of light, and cold, where they move at a fraction of the speed of light. Hot dark matter travels so fast that in the early universe it erases or smoothes out structure, meaning that the first things to form are supercluster-sized. Galaxies then form by fragmentation from the larger object. This is called "top-down" structure formation. If the dark matter is by contrast cold, then smaller objects like dwarf galaxies can form quickly and they subsequently merge to form normal-sized galaxies, with clusters and superclusters forming late, at the present epoch. These two prescriptions for dark matter make quite different predictions for the appearance of structure in the mature universe. The universe has the hallmarks of bottom-up or hierarchical structure formation so the cold dark matter option has been the standard paradigm since the 1980s.

15 We've seen that dark matter outweighs normal matter by six to one. But three-quarters of normal matter is in the form of diffuse gas in the space between galaxies, much of which is so hot that it emits at short ultraviolet wavelengths that are difficult to observe. Even when gas falls into a galaxy, its efficiency for forming stars is low. When all this is taken into account, the shining lights that optical astronomers observe represent just 1 percent

of the cosmic mass. It's difficult to be confident of telling the full story with such limited information. In a sense, it's amazing that we can tell a plausible and coherent story at all.

16 Foundational papers on structure formation with cold dark matter are G. Blumenthal, S. Faber, J. Primack, and M. Rees, "Formation of Galaxies and Large Scale Structure with Cold Dark Matter," *Nature* 311 (1984), p. 517, and M. Davis, G. Efstathiou, C. Frenk, and S. White, "The Evolution of Large-Scale Structure in a Universe Dominated by Cold Dark Matter," *The Astrophysical Journal* 292 (1985), p. 371.

17 L. Ferrarese and D. Merritt, "Supermassive Black Holes," *Physics World* 15 (2003), p. 41.

18 The nature of galaxy evolution is the subject of intense (and occasionally argumentative) debate and the active research by hundreds of astronomers. The fact that lower-mass galaxies have their star formation peaking at later times mirrors the fact that less luminous active galaxies (in other words, those with lower-mass supermassive black holes) have their activity peaking at later times. The shift of star formation and nuclear activity to lower-mass objects as time goes by has been called "downsizing" and it runs counter to the simple expectations of the growth of structure in a cold dark matter cosmology. The extra ingredient that may resolve this conundrum is "feedback" in an active galaxy, which couples the growth and activity of the supermassive black hole to the star formation rate in the surrounding galaxy. An example of a paper that makes this connection is E. Scannapieco, J. Silk, and R. Bouwens, "AGN Feedback Causes Downsizing," *The Astrophysical Journal Letters* 635 (2005), p. L13.

19 Chen Guying, ed., *Zhuangzhi* (Beijing: Chinese Press, 1983).

20 A. Riess et al., "Observational Evidence from Supernovae for an Accelerating Universe and a Cosmological Constant," *The Astronomical Journal* 116 (1998), p. 1009, and S. Perlmutter et al., "Measurements of Omega and Lambda from 42 High Redshift Supernovae," *The Astrophysical Journal* 517 (1999), p. 565.

21 M. Kowalski et al., "Improved Cosmological Constraints from New, Old, and Combined Supernova Datasets," *The Astrophysical Journal* 686 (2008), p. 749.

22 Physics has a concept of vacuum energy as part of quantum theory, where the pure vacuum of space has a tiny component of quantum energy. But if standard physics is used to predict the dynamic effect on space, the value is 120 orders of magnitude larger than the tiny value observed to give cosmic acceleration. This huge discrepancy is an embarrassment of course, but just a reflection of the fact that physics isn't complete. Einstein added a term to the solution of his equations of general relativity to suppress the expansion

that is a natural solution (because at the time he was told that the universe was static). The dark energy as observed by cosmologists seems to have the character of the cosmological constant invoked by Einstein, which is to say it has negative pressure equal to its energy density and it doesn't vary with time or across space. Dark energy has distressingly few observable properties, making research progress very difficult.

23 T. Davis and C. Lineweaver, "Expanding Confusion: Common Misconceptions of Cosmological Horizons and the Superluminal Expansion of the Universe," *Publications of the Astronomical Society of Australia* 21 (2004), p. 97.

10: LIGHT AND LIFE

1 The Hubble Space Telescope provides a great example of the long and arduous road space projects often have to travel, and it had its own story of adversity and triumph. Yale professor Lyman Spitzer first floated the idea of a space telescope in 1946 and he barely lived to see it launched. After being endorsed by august scientific bodies, design work started in 1969, but Congress cut funding in 1975. After shrinking in size and budget, launch was set for 1983. Delays followed and then the tragic loss of the shuttle *Challenger* in 1986 threw the whole project into limbo. It finally was launched by the shuttle *Discovery* in 1990, when NASA found to its dismay that the primary mirror had been machined to the wrong spec. Hubble's blurry images were an embarrassment and a public relations disaster for the space agency. Astronauts provided "glasses" for the telescope in 1993, restoring it to perfect vision, and a series of five servicing missions have shown why it's important to have astronauts in space as well as rejuvenating the telescope every few years with new instruments and gyros and solar panels and computers. Hubble is well into its third decade as a frontier research facility. Public affection for the telescope became clear when NASA administrator Sean O'Keefe decided to let Hubble die a natural death rather than service it for a fifth time; another shuttle had been lost and O'Keefe thought the risk to the astronauts from a servicing mission was too high. An outpouring of public support for the facility (plus lobbying by key legislators who had NASA facilities in their states) caused him to backtrack and the servicing mission in 2009 went off without a hitch.

2 R. Williams et al., "The Hubble Deep Field: Observations, Data Reduction, and Galaxy Photometry," *The Astronomical Journal* 112 (1996), p. 1335. This was followed up by a counterpart Deep Field in the southern sky, R. Williams et al., "The Hubble Deep Field South: Formulation of the Observing

Campaign," *The Astronomical Journal* 120 (2000), p. 2735. Normal successful proposals for the telescope get a propriety period of one year for the data, when nobody other that the proposers can work with it. The Hubble Deep Fields were made available immediately.

3 S. Beckwith et al., "The Hubble Ultra Deep Field," *The Astronomical Journal* 132 (2006), p. 1729. The proliferation of deep fields would seem to mitigate the advantage of putting a lot of resources in one patch of sky, but in practice astronomers need surveys of varying depth, with a trade-off between depth and area of sky covered. The conceptual result is a "wedding cake" with a shallow, wide-field tier at the base and the Ultra Deep Field as the tiny pinnacle layer of greatest depth.

4 Spectroscopy is much more demanding of light than simple imaging, so many of the faintest and most interesting galaxies in the Ultra Deep Field are too faint for spectroscopy, even with ground-based 10-meter telescopes. The "poor man's" method for measuring a redshift from colors is called a photometric redshift. The accuracy of the redshift is much lower than a spectroscopic redshift, but that's not a problem if the goal is to measure evolution across a wide redshift range. A bigger problem is reliability: some galaxies have unusual energy distributions so the standard templates of spiral and elliptical galaxies that are used to convert colors into redshift can fail badly. Astronomers consider spectroscopy essential for a truly reliable redshift, but for statistical work photometric redshifts are fine as long as the reliability is above 90 percent.

5 R. Bouwens et al., "Discovery of $z \sim 8$ Galaxies in the Hubble Ultra Deep Field from Ultra-Deep WFC3/IR Observations," *The Astrophysical Journal Letters* 709 (2010), p. L133.

6 Over the years, a number of groups have been burned by claiming very high redshifts for galaxies that weren't confirmed by spectroscopy. In such a competitive field, the urge to be first to publish a record-breaking redshift is very strong. The highest galaxy redshifts with rock-solid confirmation by spectroscopy are 6 to 7. Redshifts of 8 and above are based on photometry so are necessarily more uncertain. Since such objects emit only at near-infrared wavelengths, confirming spectroscopy isn't even possible technically. The other issue that affects this field is dust obscuration. In the earliest wave of galaxy formation, some activity may be shrouded in dust, which would remove it from an optical survey. This could lead to underestimates of the star formation rate or the number of very high redshift galaxies. Current indications are that obscured galaxies exist but we are not being badly misled by dust.

7 Newton should have been worried, but apparently wasn't. He formalized gravity as an inverse square law force in an infinite universe. Since both light and gravity diminish with the inverse square of distance, the Olbers' paradox exists for gravity as well as light: an infinite universe contains both an infinite amount of gravity and an infinite amount of light.

8 From Edgar Allen Poe, *Eureka: A Prose Poem* (1848); for a definitive account of the history and resolution of Olbers' paradox, see Edward Harrison, *Darkness at Night: The Riddle of the Universe* (Cambridge, MA: Harvard University Press, 1987).

9 The distance to the edge of the Hubble sphere has to take into account the expansion history, since we are looking back in time to that point. Relative to the expansion rate measured by the local Hubble relation, the expansion rate was slower (recent acceleration) and before that faster (deceleration for the first two-thirds of the universe's age). The correct calculation gives about 14 billion light-years. Hubble spheres are relative, so if a galaxy is just beyond our Hubble sphere we are just beyond their Hubble sphere too.

10 C. Lineweaver and T. Davis, "Misconceptions About the Big Bang," *Scientific American* (March 2005), p. 36.

11 S. Sigurdsson, H. Richer, B. Hansen, I. Stairs, and S. Thorsett, "A Young White Dwarf Companion to the Pulsar B1620-26: Evidence for Early Planet Formation," *Science* 301 (2003), p. 193.

12 The fact that astronomers can reconstruct the history is almost as amazing as the history itself. The giant planet was only found because of a decade of careful monitoring of the timing of the pulsar. Data from the Hubble Space Telescope of the white dwarf's color and temperature established its age and mass. This was combined with wobble of the neutron star in its orbit to give the mass of the neutron star. Tiny irregularities in the pulsar timing revealed the third body, a Jovian planet. All this information together generated the tilt of the orbits of both dead stars and the planet. The planet's wide orbit means that mass transfer from the Sun-like star, now a white dwarf, onto the neutron star, must have taken place after the planet was already orbiting the pair. The planet's wide orbit means it is fragile to interactions with nearby stars, so its current position gives an indication of its orbit in the globular cluster, to not have passed through the core in the few billion years since the system formed. Although there is a passing resemblance to the first exoplanets ever detected, Earth-like planets in the PSR 1257+12 system, those original pulsar planets formed after the death of a massive star, so are not habitable and probably never have been.

13 Alan Boss has argued that the Methuselah planet supports his idea of grav-

itational instability for forming planets, since it can happen much more quickly than core accretion and doesn't require preexisting heavy elements and a rocky core.

14 R. Salvaterra et al., "GRB090423 at a Redshift of z = 8.1," *Nature* 461 (2009), p. 1258, and N. Tanvir et al., "A Gamma-Ray Burst at a Redshift of z = 8.2," *Nature* 461 (2009), p. 1254.

15 Gamma ray bursts represent the death throes of a massive star that will leave behind a black hole. Models of the first stars, which are made of only hydrogen and helium, imply that they were very massive, 100 to 200 times the mass of the Sun. Such stars would live only a few million years and leave behind sizable black holes. This production of black holes 10 to 50 times the mass of the Sun helps with another issue in astrophysics: the fact that luminous quasars are seen past redshift 6. This leaves not much more than half a billion years to grow a black hole a billion times the mass of the Sun. With beefy "seed" black holes from the first generation of stars, mergers, and copious fuel in the young, dense universe, it's not considered a problem to grow such beasts that quickly, although it would be nice to have evidence of how it happens.

16 D. Fischer and J. Valenti, "The Planet-Metallicity Correlation," *The Astrophysical Journal* 622 (2005), p. 1102.

17 C. Lineweaver, Y. Feener, and B. Gibson, "The Galactic Habitable Zone and the Age Distribution of Complex Life in the Milky Way," *Science* 303 (2004), p. 59; C. Lineweaver, "An Estimate of the Age Distribution of Terrestrial Planets in the Universe: Quantifying Metallicity as a Selection Effect," *Icarus* 151 (2001), p. 307; and for a dissenting view, N. Prantzos, "On the Galactic Habitable Zone," *Space Science Reviews* 135 (2008), p. 313.

18 C. Mordasini, Y. Alibert, W. Benz, and D. Naef, "Extrasolar Planet Population Synthesis. II. Statistical Comparison with Observations," *Astronomy and Astrophysics* 501 (2009), p. 1161.

19 Bertrand Russell, *Why I Am Not a Christian and Other Essays on Religion and Related Subjects* (New York: Simon and Schuster, 1957).

11: BIG BANG

1 Jennifer Isaacs, ed., *Australian Dreaming: 40,000 Years of Aboriginal History* (Sydney: Lansdowne Press, 1980).

2 Bruce Chatwin, *The Songlines* (London: Franklin Press, 1986).

3 Hubert Vecchierello, *Einstein and Relativity: Lemaître and the Expanding Universe* (Paterson: St. Anthony Guild Press, 1934).

4 Quoted by A. Deprit in "Monsignor Georges Lemaître," *The Big Bang and Georges Lemaître*, ed. A. Barger (London: Reidel, 1984).

5 In fact, Russian mathematician and physicist Alexander Friedmann derived the first "expanding universe" solutions to the equations of general relativity in 1922, and he also explored the full range of models with positive, negative, and zero space curvature in a paper in 1924. Lemaître's special contribution was to connect the expansion to the observed redshift of galaxies, and derive the first estimates of the age and expansion rate of the universe. Like Lemaître, Friedmann served in the military in World War I. His contributions are less well known because of the isolation of Russian scientists in the early part of the twentieth century and the heavily mathematical nature of his work. As director of the Geophysical Observatory in Leningrad, Friedmann set a high altitude record for balloon flight in 1925, and he died from typhoid fever that same year at the age of thirty-seven.

6 The current uneasy relationship between science and religion, epitomized by the "new atheists" like Richard Dawkins and Christopher Hitchens and by fundamentalist Christians who adhere to a 6000-year-old Earth, is challenged by the life of a man who was a senior religious figure and a front-line scientist. Dan Brown, author of the best-selling *Angels and Demons*, claimed that Lemaître was a monk who proposed the big bang theory to reconcile science and religion. Brown was wrong on every count. Lemaître thought that the Bible had nothing to say about scientific matters and might indeed be full of errors, but he held that it was the source of wisdom on the correct path to salvation and immortality. Forty years after his death, a good biography finally appeared: John Farrell, *The Day Without Yesterday: George Lemaître, Einstein, and the Birth of Modern Cosmology* (Emeryville, CA: Thunder's Mouth Press, 2005).

7 Quote on the American Institute of Physics website, in the History section called "Ideas of Cosmology," at http://www.aip.org/history/cosmology/ideas/bigbang.htm.

8 It turns out Hubble overestimated the expansion rate and underestimated the distances to galaxies and the age of the universe by not realizing that there are two types of Cepheid variable star. The Cepheids in the Milky Way that Shapley used to calibrate the period-luminosity relation in the Milky Way are different from the Cepheids Hubble found in distant galaxies.

9 Oscar Godart and Martin Heller, *Cosmology of Lemaître* (Tucson: Pachart, 1985).

10 A cloud is distinct enough from the Sun that the analogy shouldn't be pushed too far. In a cloud, the light is scattered off microscopic water

droplets, while in the Sun the gas is a high-temperature plasma where the electrons have been stripped from the atomic nuclei and it's the electrons that interfere with light traveling freely. The transition from no scattering or absorption—which is a transparent medium—to total scattering or absorption—which is an opaque medium—occurs gradually. By convention, the boundary is defined as the place where the probability of light interacting on its way out is 50 percent.

11 Cosmology uses the evolution of the scale factor R to describe the history of the universe. The scale factor is the distance between any two points in space; the isotropy and homogeneity embodied in the cosmological principle means it doesn't matter which two points are used for the measurement. Redshift, which is the observable, denoted by z, is related simply to the scale factor by $R = 1/(1+z)$. So redshift is the fractional change in the size of the universe, and at early times when the redshift is much larger than 1, it roughly equals the factor by which the universe was smaller at the time the light was emitted.

12 Examples of nonlinear physics that must be included in a realistic model of star formation are the role of shocks, cooling of gas by emission from spectral lines of heavy elements, and magnetic fields that can act to support gas clouds against gravitational collapse. On the scale of galaxy formation, all of these essential ingredients are either ignored or included with very simplistic physical prescriptions. Another consequence of the nonlinear physics is sensitivity to initial conditions, sometimes known as the "butterfly effect." As the starting conditions in a computer simulation are varied, the results can vary strongly and in a way that's impossible to predict. As a result, astronomers are limited to broad brushstroke descriptions of how the structure in the universe emerged.

13 R. Alpher, H. Bethe, and G. Gamow, "On the Origin of the Chemical Elements," *Physical Review* 73 (1948), p. 804, and also R. Alpher and R. Hermann, "Remarks on the Evolution of the Expanding Universe," *Physical Review* 75 (1949), p. 1089. Gamow was not an author of the second paper that first quoted a temperature for the relic radiation from the big bang, but his fingerprints are all over it. Two versions of the big bang model were quoted, one yielding a predicted temperature of 1 Kelvin and the other yielding 5 Kelvin. In fact, the theory was imprecise enough that a range of values was quoted in these early years. In 1950, Alpher and Hermann reestimated a temperature of 28 Kelvin, and in the mid-1950s, Gamow published values as different as 5 Kelvin and 50 Kelvin. But the general idea of a low-temperature radiation "bath" permeating space was a common feature of all this work.

14 While the early universe does share some of the characteristics of the photosphere of a very cool star, the analogy is not perfect. In the early universe, there are two related but distinct processes: recombination and decoupling. Recombination occurs when photons are reduced enough in energy so that they can't liberate the electron from a hydrogen atom (this ignores helium, which has to be included in a detailed calculation). The hydrogen goes from being a plasma to being a neutral gas. Associated with this transition is the process of decoupling, where the probability of a photon interacting with a hydrogen atom drops to zero, making the universe transparent. Photons in the early universe have a wide range of energies, and the high energy tail of the energy distribution kept the universe ionized after the temperature had dropped below the level that would generally allow neutral hydrogen atoms to form. Also, the transition from opaque to transparent didn't happen instantly, but took about 20,000 years.

15 The first observation of the cosmic microwave background actually preceded Penzias and Wilson by 25 years. In 1940, Andrew McKellar saw an interstellar absorption line from the carbon-nitrogen (CN) molecule in the spectrum of the bright star Zeta Ophiucus and realized that the feature could only be created by exposure to radiation with a temperature of 2.3 Kelvin. But he made no comment on what might have produced this radiation, so his observation wasn't reinterpreted in terms of cosmology until 1966.

16 In physics, this type of spectrum is called a blackbody spectrum, and it results from a nonreflecting and opaque object that is in perfect equilibrium with its surroundings. A blackbody spectrum has a steep fall to short wavelengths and a more gradual fall to long wavelengths. The wavelength of peak emission and the shape of the spectrum are characterized by just one physical quantity: the temperature. It was a consideration of the properties of blackbody radiation in 1900 that led Planck to the idea of quanta.

17 C. Lineweaver, L. Tenorio, G. Smoot, P. Keegstra, A. Banday, and P. Lubin, "The Dipole Observed in the COBE DMR 4-Year Data," *The Astrophysical Journal* 470 (1996), p. 38.

18 A large amount of theoretical effort has been expended to understand how structure in the universe evolved. One question is the origin of the very low level ripples in temperature (and density); that will be covered later in the book. The initial density variations are a very small fraction of the mean density and are called perturbations. Early on, they grow slowly and linearly, with Newtonian gravity providing a simple framework. Gravity begins to form structures, although it has to act against the rapid expansion that's thinning out the gas. Without any dark matter, galaxies wouldn't

form at all! When the variation or contrast in density is a substantial fraction of the mean density, the growth of structure accelerates and becomes nonlinear. The gravitational formalism is still Newtonian, but the mathematical description is more complex and shocks and complex gas dynamics must be taken into account when stars and galaxies form.

19 George Smoot and Keay Davidson, *Wrinkles in Time* (New York: Morrow, 1993), and John Mather and John Boslough, *The Very First Light: The True Inside Story of the Scientific Journey Back to the Dawn of the Universe* (New York: Basic Books, 1997).

12: WHITE HEAT

1 Nothing written by Pythagoras survived and most of the information about him was written centuries later. Unfortunately, a book about him by Aristotle was also lost. It's difficult to translate the world of Pythagoras and his followers to the present day, but his operation resembled a secret society or a cult, where mathematical knowledge had a mystical power. Pythagoras was influential and fell out of favor with the ruler of Samos, the tyrant Polycrates, so was forced to flee to Croton in Italy. There he continued his work with a new set of followers and adherents. Again he became involved in political intrigue; his temple was sacked and his followers banished. There's no agreement on how he met his end. Pythagoras greatly influenced Plato, and that's how Pythagoras came to influence the history of Western philosophy.

2 Jamie James, *The Music of the Spheres: Music, Science, and the Natural Order of the Universe* (New York: Springer-Verlag, 1993).

3 Both in his use of three-dimensional shapes to fit planet distances and his use of harmonics to fit planet orbital speeds, Kepler was indulging in numerology that did not have a basis in physical theory. And in fact both patterns fail to hold when the full set of planets is considered. But the method of looking for patterns based on geometry and harmonics is well-motivated, and in modern planetary science phenomena such as the orbits of planet moons and the patterns in their ring systems are accurately described by harmonics or ratios of whole numbers through the process of gravitational resonance.

4 Walter Isaacson, *Einstein: His Life and Universe* (New York: Simon and Schuster, 2007).

5 W. Hu and M. White, "The Cosmic Symphony," *Scientific American* (February 2004), p. 44.

6 G. Hinshaw et al., "Five Year Wilkinson Microwave Anisotropy Probe Observations: Data Processing, Sky Maps, and Basic Results," *The Astrophysical Journal Supplement* 180 (2009), p. 225.

7 The microwave spectrum and its evolution have been subject to "sonification" by Mark Whittle, where he converts the harmonics and the waveform into an audible signal by shifting it higher in frequency by 50 octaves; see http://www.astro.virginia.edu/~dmw8f/BBA_web/index_frames.html.

8 The early universe had both sound and light. The sound was low-frequency waves released 380,000 years after the big bang. We can think of them leaving the cosmic vocal chords at that time. The universe at that time had a temperature of 3000 Kelvin and glowed dark red. Earlier it was hotter, and its radiation shifted through the visible spectrum from 10,000 years after the big bang to 380,000 years. Therefore, the universe would have shown a slow "flash" as its radiation passed from ultraviolet radiation to visible light and on down to infrared waves. Son et lumière.

9 D. Eisenstein et al., "Detection of the Baryon Acoustic Peak in the Large Scale Correlation Function of SDSS Luminous Red Galaxies," *The Astrophysical Journal* 633 (2005), p. 560.

10 Like Mary Queen of Scots in the famous Monty Python sketch, the steady state theory didn't die very gracefully. Hoyle's sharp critique that the steady state only requires a slow and steady creation of matter, while the big bang requires the extraordinary instantaneous creation of all matter in the universe, was well-taken. However, there is no natural way to explain why we're immersed in a uniform 3 Kelvin radiation bath in the steady state theory without introducing a new concept, while the big bang explains this naturally. More recent versions of the theory are called "quasi-steady state cosmology" and they try to explain the cosmic microwave background radiation temperature and power spectrum. However, the adjustments to the theory are ad hoc and don't fit the best data very well; Ned Wright of UCLA provides a critique on his web pages. The big bang has also seen off theories that hypothesize mechanisms for the redshift other than expansion of the universe, such as the chronometric cosmology and the "tired light" theory.

11 John Mather and John Boslough, *The Very First Light: The True Inside Story of the Scientific Journey Back to the Dawn of the Universe* (New York: Basic Books, 1996).

12 A star is a stable fusion reactor, while the universe is expanding and has a rapidly changing temperature, so there are many subtleties with what is called "big bang nucleosynthesis." One major one is the fact that free neu-

trons decay with a half-life of 17 minutes, so the ratio of neutrons to protons is changing during the expansion and that affects how fusion progresses.

13 The lithium abundance is a slight fly in the ointment since the observed lithium in stars is lower than predicted by the big bang. However, stars destroy lithium as well as create it, and testing the big bang with lithium data has spurred better calculations of fundamental atomic physics. In practice, early universe observations and theory have now advanced to the level where they can help improve lab physics; see, for example, A. Coc, E. Vagnioni-Flam, P. Descouvement, A. Adahchour, and C. Angulo, "Updated Big Bang Nucleosynthesis Confronted to WMAP Observations and the Abundance of Light Elements," *The Astrophysical Journal* 600 (2003), p. 544.

14 Simon Singh, *Big Bang: The Origin of the Universe* (New York: HarperCollins, 2004).

15 M. White and W. Hu, "The Sachs-Wolfe Effect," *Astronomy and Astrophysics* 321 (1997), p. 89.

16 Carl Sagan, *Cosmos* (New York: Random House, 1980).

17 In our sensory exploration of the universe, there's sound and light, but smell can't operate until there are molecules, so it's only possible after first light and a few generations of star birth and death. A hundred million years or so after the big bang, we get the first whiff of soot and brimstone as massive stars forge and eject heavy elements.

18 Jerry Ostriker of Princeton University is another top-tier theorist on topics of large-scale structure and galaxy formation who began to use computers in his research group when they were powerful enough. In the mid-1980s he became the provost of the university, and even while he carried out his duties as the chief academic officer, he kept teraflops of computing power humming in the basement of the Provost's Office.

19 Sverre Arseth, *Gravitational N-Body Simulations: Tools and Algorithms* (Cambridge: Cambridge University Press, 2003).

20 One example takes advantage of the inverse square law of gravity. Although each particle in a simulation exerts a force on every other particle, the force is much weaker for widely separated particles. It turns out that the accuracy of the simulation isn't reduced noticeably by ignoring particles far from the one you are considering. That reduces the number of calculations enormously. Another shortcut involves the fact that when particles merge by gravity they can be considered as one entity in terms of their force on all the other particles. Analogously, when particles are close to each other their motions must be tracked very precisely while particles far from the one under consideration can be viewed crudely and lumped together, with

little effect on the accuracy of the simulation. In general these strategies don't reduce the computational scaling from N^2 to N, where N is the number of particles, but they do reduce it to $N \times \log N$.

21 V. Springel et al., "Simulations of the Formation, Clustering and Evolution of Galaxies and Quasars," *Nature* 435 (2005), p. 629.

13: SOMETHING RATHER THAN NOTHING

1 Philosophers have worried about why there's something rather than nothing for a very long time. In philosophy it's called the Primordial Existential Question or PEQ. The hypothesis "There's nothing" couldn't be supported by any experiment because any observation implies the existence of an observer. The Greek Parmenides thought it was even meaningless to talk about nothing, since the referent implied existence. If we say "Atlantis doesn't exist" it's still a statement about something, and a statement can only be about something if that thing exists! Lined up against Parmenides were the Atomists, who posited invisible and indivisible things moving around in empty space as the source of matter and motion. Even mathematicians have trouble postulating nothing. Set theory can describe almost all mathematics, and we could describe a set containing you, then a set that contains that set, and so on. But an assumption still has to be made that something exists. This is what the metaphysician and philosopher of science Welsey Salmon said about the empty set: "The fool says in his heart there is no empty set. But if it were so, the set of all such sets would be empty, hence *it* would be the empty set."

2 The classic "Fermi problem," one actually used by Fermi with undergraduate students, is to estimate the number of piano tuners in Chicago. At first sight, this seems impossible. But with a series of reasonable assumptions, it's possible to come up with a ballpark number. There are 3 million people in the greater Chicago area and roughly 3 people per household. Assuming one piano per 20 households and pianos tuned about once per year, and guessing that a piano tuner can tune a piano in two hours (including travel time) and that they'll work a nearly full day each day of the year, gives a "market" of 50,000 tunings per year and a "capacity" of 1000 tunings per year per tuner. Dividing the two numbers gives about 50 tuners. The online Chicago Yellow Pages returns 34 tuners, and this number is low if not all advertise there, so the method works quite well. Another Fermi classic is his 1950 speculation that the abundance of worlds with the potential for biology and the large amount of time for evolution to take place made it

unlikely that we would be alone in the universe. So he posed the question: "Where are they?"

3 Graham Farmelo, *The Strangest Man: The Life of Paul Dirac* (London: Faber and Faber, 2009).

4 The observation led to Anderson's getting the Nobel Prize in Physics at the age of thirty-one, the youngest person ever to win the prize. Several others just missed out on the discovery. In 1923, Dmitri Skobeltsyn saw the same effect and was puzzled enough by it to share the result with other scientists, but he didn't pursue it. He missed the Nobel Prize, but won six Orders of Lenin in his illustrious career. Chung-Yao Chao, a Caltech grad student, saw an anomaly in the lab that was also due to a positron, but didn't follow it up.

5 The Sun is overwhelmingly made of matter, but intriguingly, it's one of the most prodigious antimatter factories we know of. The Sun and other stars emit high-energy particles called cosmic rays, and when they hit other particles they spawn new by-products, including antiparticles. These antiparticles don't live long before they annihilate with normal matter. A single solar flare in 2002 created a pound of antimatter, or enough to power the United States for two days and billions of times more than can be produced in particle accelerators each year.

6 A. Cohen, A. De Rujula, and S. Glashow, "A Matter-Antimatter Universe?" *The Astrophysical Journal* 495 (1998), p. 539.

7 Particle accelerators showed that while electrons were pointlike and fundamental, neutrons and protons were not. The proton is composed of two "up" quarks, each with electric charge +2/3, and one "down" quark of electric charge –1/3, while the neutron is composed of one up quark and two down quarks. The quarks are strongly confined to composite particles and are never observed moving freely. The term was proposed by Murray Gell-Mann, one of the physicists who came up with a model for the interactions of massive particles. It was intended to represent the sound made by ducks and it came from a nonsensical phrase in James Joyce's *Finnegan's Wake* (which many people consider to be full of nonsense): "Three quarks for Muster Mark."

8 For a while the third generation or level of quarks were named "truth" and "beauty" but even physicists saturated on cute and whimsical names for particles, so they switched to the more prosaic labels of "top" and "bottom." And even physicists have trouble keeping track of the names. Nobel Prize–winner Enrico Fermi once said to his student Leon Lederman (who would also one day win a Nobel Prize), "Young man, if I could remember the names of all these particles, I would have been a botanist."

9 Protons and neutrons are made of triads of quarks, but there are also par-

ticles made of quark-antiquark pairs. They're called mesons and they're all unstable, decaying in tiny fractions of a second. The addition of mesons to the particle "zoo" explains why even physicists have trouble keeping it all straight.

10 Robert Oerter, *The Theory of Almost Everything: The Standard Model, the Unsung Triumph of Modern Physics* (New York: Pearson Education, 2006).

11 The Higgs boson is the most sought-after particle in physics, named after Peter Higgs, an Emeritus Professor at the University of Edinburgh who proposed its existence as a by-product of his work on the mechanism by which fundamental particles get mass. Although intensive efforts are being made at both CERN and Fermilab, the Higgs had not been detected by mid-2011, despite several false alarms.

12 Neutrinos move at close to the speed of light, are electrically neutral, and interact so weakly with matter that they could pass through miles of lead without flinching or interacting. Their feeble interactions involve the weak nuclear force and they are produced in beta decay. In the Standard Model, neutrinos are massless but they've been observed "oscillating" among the flavors of the three families—electron, muon, and tau—which means they must have mass. Experiments so far measure mass differences among the flavors, not absolute masses. Neutrinos are abundant enough, and may be massive enough, to account for the dark matter, but they have the wrong behavior. Neutrinos are "hot" and move relativistically and so would wash out the large-scale structure of galaxies, while the hypothetical dark matter particle is "cold" and leaves the structure intact.

13 This is true of all fermions. When the particle is charged, the antiparticle has the opposite charge. When the particle is neutral, like a neutron or neutrino, the antineutron and antineutrino are also neutral. Bosons, including the photons that carry the electromagnetic force, act as their own antiparticles.

14 An even bigger tour de force was the creation of an isotope of antihelium in 2003 at CERN. But there's no realistic way to ever gather enough to make it levitate, put it in an antiballoon, or use it at a children's birthday antiparty or antibirthday party.

15 In fact, the experiment showed a violation of the combination of mirroring in space and replacing matter by antimatter. If the triad of space, time, and charge is to remain as a universal symmetry, then the reversal of time must be violated as well, so the K meson is a particle that can distinguish the arrow of time. Some of the most sensitive tests for these effects now involve experiments carried out on antiatoms.

16 Also around this time, the remaining quarks link up in pairs to form mesons

and triplets to form baryons and as the universe cools they remain tightly bound in these configurations by the strong nuclear force. In the old and cold universe free quarks are never observed.

17 Nick Bostrom, *Anthropic Bias: Observation Selection Effects in Science and Philosophy*, (New York: Routledge, 2002).

18 Another thread in the history of fine-tuning arguments is the idea that some pure numbers are anthropic coincidences. In 1919, Hermann Weyl wondered why the ratio of the electromagnetic to the gravitational forces between two electrons was the large number 10^{39}. A little later, the astrophysicist Arthur Eddington noted that the square of that number is the number of particles in the universe, and Paul Dirac noted that 10^{39} is also equal to a stellar lifetime divided by the time it takes light to cross a proton. In 1962, Robert Dicke also elucidated the large-number coincidences. These men were all great scientists, but in the absence of an underlying theory to explain the pure numbers, anthropic explanations carry a whiff of both teleology and numerology and have fallen into disfavor.

19 Roger Penrose, *The Emperor's New Mind* (New York: Oxford University Press, 1989).

20 Stephen Hawking, *A Brief History of Time* (New York: Bantam, 1988).

21 John Leslie, *Universes* (New York: Routledge, 1989).

22 V. Stenger, "Natural Explanations for the Anthropic Coincidences," *Philosophy* 3 (2000), p. 50. Stenger has a program called "MonkeyGod" with which he varied the physical parameters of hypothetical universes to see how often they might produce heavy elements, long-lived stars, and biology. He asserts that for variation over plausible ranges of physical parameters, life-bearing universes are not that rare. The first scholarly book to detail the litany of instances of fine-tuning and corresponding anthropic arguments was J. Barrow and F. Tipler, *The Anthropic Cosmological Principle* (New York: Oxford University Press, 1986).

14: UNIFICATION AND INFLATION

1 Steven Weinberg, *The First Three Minutes* (New York: Basic Books, 1977).

2 A gas, liquid, or solid at the same temperature is said to be in thermal equilibrium. Heat is carried by collisions of atoms and molecules in a normal gas, liquid, or solid, and this lets materials get to thermal equilibrium over time. The early universe was a high-temperature gas, but it was expanding so fast that initial variations in temperature could not have been smoothed out by collisions. Radiation travels at 300,000 kilometers per second and is the fastest way energy can be transported to bring a material into thermal

equilibrium. But the early universe was expanding much faster than light, so any variations in temperature from the initial state should have been "frozen in."

3 Other relics include one- and two-dimensional ruptures in space-time called strings and domain walls and complex space-time glitches called textures. Magnetic monopoles are the consequence of so-called grand unified theories, extensions of the Standard Model designed to unite the electromagnetic, weak, and strong nuclear forces. Introduced in the 1970s, these theories predicted exotic relics to be ubiquitous and to outnumber normal particles. Cosmologist Martin Rees takes a slightly jaundiced view of these predictions: "Skeptics about exotic physics might not be hugely impressed by a theoretical argument to explain the absence of particles that are themselves only hypothetical. Preventative medicine can readily seem 100% effective against a disease that doesn't exist!" See Rees's popular book *Before the Beginning* (New York: Basic Books, 1998).

4 A. Guth, "Was Cosmic Inflation the 'Bang' of the Big Bang?" *The Beamline* 27 (1997), p. 14.

5 Leon Lederman and Chris Hill, *Symmetry and the Beautiful Universe* (New York: Prometheus, 2004).

6 In physics, Emmy Noether showed in 1918 that symmetries were related to underlying conservation laws, such as the conservation of energy or momentum. With special relativity, Einstein elevated the symmetry of space-time to a fundamental premise, whose consequence was the strange distortion of mass, length, and time seen at speeds close to that of light, and he continued this thinking with general relativity. Symmetry principles were also central to the development of quantum mechanics and are foundational in the current search for unified laws of physics that can explain the microscopic and cosmological realms.

7 Quoted in A.V. Voloshinov, "Symmetry as a Superprinciple in Science and Art," *Leonardo* 29 (1996), p. 109.

8 There is a large literature, in particular, on facial symmetry and beauty; see, for example, D. Perrett, "Symmetry and Human Facial Attractiveness," *Evolution and Human Behavior* 20 (1999), p. 295.

9 Mario Livio, *The Equation That Couldn't Be Solved* (New York: Simon and Schuster, 2005).

10 There may be hints of the unification well below the temperature required for the forces to merge. But all reasonable speculation awaits discovery of the Higgs particle, since it represents a minimal extension to the Standard Model that sets the mass scale of fundamental particles, including the quarks that participate in the strong interaction. Apart from the central

role of the Higgs particle for prescribing masses, there are so many varieties of theory for unification, and so many potential "low-energy" signatures, that it will be very difficult for current accelerators to provide a strong constraint. In fact, there is not a single temperature or energy for grand unification; it depends on the particulars of the theory.

11 Grand unified theories embody symmetry in complex and bewilderingly abstract mathematics. They use a form of algebra designed to capture any sort of transformation that embodies symmetry; rotation about an axis would be one example. The algebra has a corresponding geometric description in terms of manifolds, or curved multidimensional spaces in one, two, three, or more dimensions. The physical theory is fronting an underlying symmetry that is purely mathematical, and in fact the same formalism is used in math that has nothing to do with the "real world."

12 Supersymmetry hypothesizes interactions between fermions or particles and bosons or carriers of force. Bosons have integer spins and fermions have half-integer spins, and in quantum theory they do not share any interactions. Naturally the supersymmetric partners of fermions have spin zero and the supersymmetric partners of bosons have half-integer spin.

13 Brian May and Queen went on to sell 300 million albums and May was one of the top 10 guitarists of all time in a 2005 *Planet Rock* poll. He is a commander of the British empire and in 2007 he finally received a PhD in astrophysics from my alma mater Imperial College. Fractals based on the Julia set were explored by several mathematicians in the mid-1970s and were first visualized and popularized by the French mathematician Benoit Mandelbrot in the late 1970s.

14 Alan Guth, *The Inflationary Universe: The Quest for a New Theory of Cosmic Origins* (New York: Perseus, 1997).

15 Vacuum energy has negative pressure if it's like Einstein's cosmological constant. As we saw earlier when we encountered the "dark energy" that's driving the current acceleration of the universe, the energy of the vacuum is far too large to explain the current acceleration. It can more plausibly explain the early inflation, but currently there is no quantized theory of gravity to make a convincing explanation.

16 More precisely, the microwave fluctuations are governed by random or Gaussian statistics and they have the character of a power law, where logarithmically there are smaller contributions from yet smaller scales. Power laws are described by only two parameters: strength and slope. A power law slope of –1 corresponds to scale-free fluctuations.

17 The new measurement depends on detecting polarization of the microwave background radiation. Fluctuations in the radiation, observed at the

time the universe became neutral and matter and radiation interacted for the last time, have three different behaviors. The ripples at 1 part in 100,000 that lead to galaxy formations are caused by compression and rarefaction, where a good analogy is sound waves. A second mode is like a vortex or an eddy. The third mode is caused by gravity waves that distort space by different amounts in different directions, and this imprint makes the radiation slightly polarized. These modes might be 1000 times smaller than the standard temperature variations, in which case *Planck* won't detect them. If they are only 100 times smaller, as many inflation models predict, there is the chance of a major new test of the physics of the very early universe.

18 T. S. Elliot, *Four Quartets* (Orlando: Harcourt, 1943).

19 Well before Guth's insight, theorists had speculated about the role of vacuum energy in cosmology. Ed Tryon pointed out that a vacuum fluctuation could have generated the universe for zero overall energy cost. In 1979, before Guth's pivotal paper, Alexei Starobinsky came up with a viable mechanism for inflation, but he failed to recognize how it could solve several major cosmological problems. Guth's "graceful exit" problem was solved independently by Paul Steinhardt and Andreas Albrecht. In the early 1980s some two dozen theorists made major contributions to the field. A majority of them were Russians, which was a sign of the vitality of the theoretical tradition in physics in the former Soviet Union, a tradition that has largely, and sadly, dissipated over the last few decades. Guth and Linde went on to attain academic superstardom and iconic status for their work, a typical situation in science where there are sung and unsung heroes, new ideas are rarely unique, and the distinction between major and minor contributors can be subtle. They have each written popular books and give frequent public lectures to bring these esoteric ideas to a wider public audience.

20 Electrical potential is called a scalar field. In the formalism of particle physics, there are scalar fields associated with symmetry-breaking. So space is permeated with a scalar field created at the time of the electroweak unification, and as a result of their interactions with the field, W and Z bosons carrying the weak interaction are heavy while the photons carrying the electromagnetic interaction are light. Similarly, the hypothesized Higgs particle is associated with a scalar field that gives particles their masses in the current low-temperature universe. Variations in another type of scalar field are considered the cause of inflation, and particle physicists often ascribe it to symmetry-breaking in grand unified theories.

21 Space-times bounded by horizons are similar to black holes. A variant of the eternal inflation scheme has the budding process taking place through black holes. Every time a black hole collapses into a singularity it bounces

out into another space-time, and a new inflationary universe is created. This is the baby universe scenario.

22 Richard Feynman, *The Character of Physical Law* (London: BBC Publications, 1965).

23 Jim Al-Khalili, *Quantum: A Guide to the Perplexed* (London: Weidenfield and Nicholson, 2005).

15: MULTIVERSE

1 Michael Shermer and Stephen Jay Gould, *Why People Believe Weird Things: Pseudoscience, Superstition, and Other Confusions of Our Time* (New York: Henry Holt and Company, 2002).

2 Stephen Hawking, *A Brief History of Time* (New York: Bantam, 1998).

3 Although these scales of time and space are incredibly small, and the corresponding scales of energy and temperature are incredibly large, the Planck scale is in some sense a "natural" scale of physics. Each of the quantities is a different algebraic combination of three fundamental constants: the speed of light, Planck's constant, and the gravitation constant. There is only one way to combine these three constants to derive a length, a time, and an energy; that's why the Planck scale is considered so important for physical theories. But since it corresponds to a situation where quantum mechanics and relativity meet, and there's no tested theory that combines them, it's also the edge of our physical understanding. Planck units are often referred to mock-ironically as "God's units" since they are derived without reference to any human objects (unlike the metric SI system, for example). Some scientists even speculate that intelligent aliens would use such a system to communicate with other intelligent aliens. Planck units allow physicists to reframe questions about nature. For example, Frank Wikczek has written, "The question is not 'Why is gravity so feeble?' but rather 'Why is the proton's mass so small?' For in Natural or Planck units, the strength of gravity is simply what it is, a primary quantity, while the proton's mass is the tiny number" (*Physics Today* [June 2001], p. 12).

4 Reconciling quantum mechanics with general relativity runs aground in a sea of mathematical problems, most notably a set of "infinities" that arise in the calculations. In pure mathematics, infinities cause little concern and in fact they define a rich subject for investigation. In physics, infinities are a problem because physical quantities should always be bounded or determined. For example, in the Standard Model, electrons have no size and so must have infinite mass density and infinite charge density. That's one simple example, but many infinities arose in the otherwise highly success-

ful theory of quantum interactions. To allow them to do useful calculations, physicists developed a strategy called renormalization, which was a clever way to get the infinities to cancel out. Paul Dirac was particularly unhappy with this situation, saying that sensible mathematics involves neglecting a quantity when it's very small, not neglecting it when it's infinitely larger and you don't want it! Richard Feymann was also unnerved, calling renormalization a "dippy process" and "hocus-pocus." Despite their reservations, renormalisation is part of the standard toolkit of physics, but it completely fails on the Planck scale.

5 George Musser, *The Complete Idiot's Guide to String Theory* (Indianapolis: Alpha, 2008).

6 In the 1820s, Carl Friedrich Gauss worked out the general geometry of curved three-dimensional space and this work was pivotal in helping Einstein develop the general theory of relativity 100 years later. In the 1850s, Bernhard Riemann extended Gauss's theory to higher-dimensional spaces called manifolds. There's no limit to the number of dimensions possible in the field of differential geometry. The work was considered to be esoteric mathematics, without any application to the real world, until the 1920s, when Theodor Kaluza and Oscar Klein attempted to unify gravity and electromagnetism with five-dimensional space-time. Kaluza and Klein's research echoes in current work on superstrings and 10-dimensional space-time.

7 Lisa Randall, *Warped Passages: Unraveling the Mysteries of the Universe's Hidden Dimensions* (New York: Ecco Press, 2005).

8 Different string theories are related by a series of transformations called dualities. Dualities not only map one string theory onto another one, but they link quantities that were always thought to be distinct: large and small distance scales, and strong and weak coupling constants. The dualities are fascinating because they imply, for example, that the distinction between very small and large scales in physics isn't fixed but is fluid and depends on how we choose to measure distance!

9 Ten to the power of 500 is the number one with five hundred zeros after it. So far, the highest concrete number we've encountered is 10^{80}—the number of particles in the observable universe. The number of string theory vacuum states corresponds to each particle in the universe having 10^{80} particles associated with it, which in turn had 10^{80} particles associated with it, which in turn had 10^{80} particles associated with it, which in turn had 10^{80} particles associated with it, which in turn had 10^{80} particles associated with it, which in turn had 10^{100} particles associated with it. Quite a hefty number!

10 Brian Greene, *The Elegant Universe: Superstrings, Hidden Dimensions, and*

the Quest for the Ultimate Theory (New York: Norton, 2003), and Brian Greene, *The Fabric of the Cosmos: Space, Time and the Texture of Reality* (New York: Alfred A. Knopf, 2004).

11 Lee Smolin, *The Trouble with Physics: The Rise of String Theory, the Fall of a Science, and What Comes Next* (New York: Houghton Mifflin, 2006), and Peter Woit, *Not Even Wrong: the Failure of String Theory and the Search for Unity in Physical Law* (New York: Basic Books, 2006).

12 The acrimony of the debate over string theory is painful to many physicists and it's surprising to most outsiders, who imagine that theoretical physics is an arena of pristine intellectual debate. Two decades ago Richard Feynmann criticized string theory as "crazy" and "the wrong direction" for physics, and Sheldon Glashow, who won a Nobel Prize for one of the last great advances in particle physics in the prestring era, tried to keep string theorists out of his department at Harvard (and failed). More jaundiced suggestions for the "M" in M-theory include "murky" and "masturbation." As string theory has matured, it has lost some of the simplicity (and youthful innocence) that made it appealing on aesthetic grounds, and its detractors have accused the physics community of "group think" and creating a situation where young theorists who aren't pursuing string theory have little chance of getting jobs. Some rhetoric against string theory is overdone but if the theory remains unverified and if sociology sculpts the problems that physicists work on, it will be an unhealthy situation.

13 Martin Rees, *Before the Beginning* (Reading, MA: Helix Books, 1997).

14 A. Linde and V. Vanchurin, "How Many Universes Are in the Multiverse?" *Physical Review D* 81 (2010), p. 83525. The number of universes allowed in their calculation is 10 to the power of 1 with 10 million zeros after it, as opposed to merely 10 to the power of 500. The largest number in common use is a googol or 10 to the power of 100, which is the number one followed by a hundred zeros (amusingly, Larry Page and Sergey Brin misspelled the word when they used it for the name of their new startup, search leviathan Google). A googolplex is 10 to the power of a googol, which astronomer Carl Sagan estimated would be impossible to write out since it would take more space than is available in the known universe (presumably the multiverse would have enough space). Even a googolplex is vastly smaller than the number of universes allowed in inflation. However, late in their paper, Linde and Vanchurin hedge their number down to a more modest 10 to the power of 10 to the power of 16, or 10 to the power of 10,000,000,000,000,000, based on the number of configurations the human brain can distinguish.

15 M. Tegmark, "Parallel Universes," *Scientific American* (May 2003), p. 41. Tegmark calculates the nearest doppelgänger under the assumption that all

possible universes are equally probable, but as the text argues, naturalness of planet formation and biological evolution may well render some universes more probable than others, and so the distance to your alter ego diminishes.

16 The "many worlds" interpretation of quantum mechanics was formulated by Hugh Everett in his PhD thesis in 1957, and popularized by Bryce DeWiit in the 1960s and 1970s. It aimed to reconcile the fact that individual quantum events like radioactive decay are inherently probabilistic, while the equations that govern quantum mechanics are deterministic. Physicists have a schizophrenic reaction to the idea. Most of them accept it as just as valid a description of reality as the more conventional Copenhagen interpretation, but it also makes them uncomfortable and they'd usually rather not talk about it.

17 Leonard Susskind, *The Cosmic Landscape: String Theory and the Illusion of Intelligent Design* (New York: Little Brown, 2005).

18 The examples of pure mathematics that subsequently are found to describe aspects of nature are legion. Mathematicians tend to take Plato's view. Roger Penrose was amazed by the infinite and infinitely complex "world" revealed when a simple recursive relation of complex numbers, the Mandelbrot set, is mapped into two dimensions. He voted for discovery rather than invention: "The Mandelbrot was waiting to be discovered; like Mount Everest, it's just there!" (*The Emperor's New Mind* [Oxford: Oxford University Press, 1989], p. 95). Over the past century, physicists have been struck by the elegance, beauty, and simplicity of theories that describe complex aspects of nature. Eugene Wigner said in a 1959 lecture, "the enormous usefulness of mathematics in the natural sciences is something bordering on the mysterious" (*Communications in Pure and Applied Mathematics* 13, no. 1 [1960], p. 13).

19 The Planck scale is many orders of magnitude beyond the reach of the Large Hadron Collider or any other proposed physics experiment. However, the situation is not totally bleak because glimmers of high-energy behavior appear imprinted on lower-energy phenomena. Beyond supersymmetry, physicists hope to be able to indirectly detect the hidden dimensions that are central to string theory and its expression in the multiverse. Another way to diagnose the same scales would be to create miniature black holes either in accelerators or using cosmic rays. Recently, Craig Hogan has proposed a lab experiment to detect noise magnified from space-time "pixels" on the Planck scale. His idea utilizes the "holographic principle"—the idea that any volume of space can be completely described by what happens on its boundary. Hogan hopes to use a high-precision optical interferometer to measure the "jitter" of pixelated space-time on the Planck scale, although his prediction is only valid if special relativity is violated on the quantum gravity scale.

20 M. Livio and M. Rees, "Anthropic Reasoning," *Science* 309 (2005), p. 1022.

21 A small but growing research literature tries to estimate the probability distribution of the fundamental physical parameters in a multiverse where the underpinning is string theory or M-theory. Since there's no tested "theory of everything" this is needless to say difficult to do, but it's required if the fine-tuning of our universe as a multiverse is to be understood.

22 The premise of this argument is "substrate independence," the idea that the brain is nothing more than the sum of all its neuronal connections and activity. Many humanists and some philosophers find a materialist premise like this abhorrent, insisting that consciousness, sentience, and the ineffable quality of being human cannot be reduced to a neural net and a computational abstraction. Nevertheless, scientific experiments do not rule out the possibility of substrate independence, where "minds" could be stored in silicon, modified, and then uploaded and downloaded at will.

23 N. Bostrom, "Are You Living in a Computer Simulation?" *Philosophical Quarterly* 53, no. 211 (2003), p. 243.

24 Paul Steinhardt and Neil Turok, *Endless Universe: Beyond the Big Bang* (New York: Doubleday, 2007).

ILLUSTRATION CREDITS

Figure 1.1: T. A. Rector, I. P. Dell'Antonio / NOAO / AURA / NSF.
Figure 1.3: Greg L. Ruppel.
Figure 1.4: NASA / MSFC / Renee Weber.
Figure 1.5: NASA / Apollo 11.
Figure 2.1: NASA / Jet Propulsion Laboratory.
Figure 2.2: NASA / Jet Propulsion Laboratory.
Figure 2.3: Larry McNish / Royal Astronomical Society of Canada.
Figure 2.4: B. Dalrymple, V. Murthy, C. Patterson, D. York, and R. Farquhar.
Figure 2.5 Subaru Telescope / National Observatory of Japan.
Figure 3.2: Paul Butler et al. / Department of Terrestrial Magnetism.
Figure 3.3: NASA, ESA, and D. Lafrieniere / University of Toronto.
Figure 3.4: Greg Laughlin / California and Carnegie Planet Search.
Figure 3.5: Greg Laughlin / California and Carnegie Planet Search.
Figure 3.6: NASA / Kepler Mission.
Figure 4.1: Wikimedia Commons / Borb.
Figure 4.2: NASA / JPL-Caltech / J. Pyle / Spitzer Science Center.
Figure 4.3: NASA / Goddard Space Flight Center.
Figure 4.4: Orionis / Wikimedia Commons.
Figure 5.1: National Park Service / United States Government.
Figure 5.2: NASA / STScI / David Kaplan / University of California Santa Barbara.
Figure 5.3: Johnson / Wikimedia Commons.
Figure 5.4: European Space Agency.
Figure 5.5: Andrea Ghez / UCLA Galactic Center Research Group.
Figure 6.1: NOAO / AURA / NSF / Bill Shoening, Vanessa Harvey.
Figure 6.2: NASA / Space Telescope Science Institute.
Figure 6.3: European Southern Observatory.
Figure 6.5: NASA / JPL Caltech / R. Hurt / Spitzer Science Center.

Figure 7.1: *Proceedings of the National Academy of Science, 2004*, vol. 101, p. 8.
Figure 7.2: Brews ohare / Wikimedia Commons.
Figure 7.4: Schaap / Wikimedia Commons.
Figure 7.5: NASA / JPL-Caltech / GSFC / SDSS.
Figure 7.6: NASA / ESA and the Hubble Heritage Team.
Figure 8.1: NRAO/AUI.
Figure 8.2: NASA / CXC / M. Weiss.
Figure 8.3 Marc Turler / Geneva Observatory.
Figure 8.4: NASA / ESA / John Hutchings.
Figure 9.1: LSST Corporation / Howard Lester.
Figure 9.2: NASA / Space Telescope Science Institute.
Figure 9.4: NASA / ESA / E. Hallman.
Figure 9.5: Saul Perlmutter / Physics Today.
Figure 10.1: NASA / ESA / Space Telescope Science Institute.
Figure 10.2: NASA / ESA / Space Telescope Science Institute.
Figure 10.3: Wikimedia Commons / Htkym.
Figure 10.4: NASA / Spitzer Space Telescope.
Figure 10.5: Exoplanet Encyclopedia / J. Schneider / CNRS.
Figure 11.1: Physics of the Universe / Luke Mastin.
Figure 11.2: NASA / Goddard Space Flight Center / WMAP.
Figure 11.3: NASA / United States National Parks Service / Bell Labs.
Figure 11.4: NASA / Goddard Space Flight Center / COBE.
Figure 11.5: NASA / Goddard Space Flight Center / COBE.
Figure 11.6: NASA / Goddard Space Flight Center / WMAP.
Figure 12.1: NASA / Goddard Space Flight Center / WMAP.
Figure 12.2: NASA / Goddard Space Flight Center / WMAP.
Figure 12.3: NASA / Goddard Space Flight Center / WMAP.
Figure 12.4: Volker Springel / Max-Planck Institute for Astrophysics.
Figure 13.1: NSF / Kirk Woellert.
Figure 13.2: Fermi National Accelerator Laboratory.
Figure 13.3: Fermi National Accelerator Laboratory.
Figure 13.4: Maximillien Brice / CERN.
Figure 14.1: Dartmouth Electron Microscope Facility / Wikipedia.
Figure 14.3: Particle Adventure / NSF / DOE.
Figure 14.4: Wikimedia Commons / Chrkl.
Figure 14.5: Wikipedia Foundation.
Figure 15.2: Wikimedia Commons / MissMJ.
Figure 15.3: Carlos Herdeiro.
Figure 15.4: Wikimedia Commons / Jbourjai.
Figure 15.5: Wikimedia Commons / Dc978.

INDEX

Page numbers in *italics* refer to figures.
Page numbers beginning with 363 refer to notes.